Western E[...]

A Systematic Human Geography

Brian W. Ilbery

Western Europe

A Systematic Human Geography

Second Edition

Oxford University Press

Oxford University Press, Walton Street, Oxford OX2 6DP
Oxford New York Toronto
Delhi Bombay Calcutta Madras Karachi
Petaling Jaya Singapore Hong Kong Tokyo
Nairobi Dar es Salaam Cape Town
Melbourne Auckland
and associated companies in
Berlin Ibadan

Oxford is a trade mark of Oxford University Press

Published in the United States
by Oxford University Press, New York

First published 1986
Paperback reprinted 1990

British Library Cataloguing in Publication Data
Ilbery, Brian W.
Western Europe: a systematic human geography.—2nd ed.
1. Anthropo-geography—Europe
I. Title
304.2'094 GF540
ISBN 0–19–823278–0
ISBN 0–19–823277–2 (Pbk)

Library of Congress Cataloging in Publication Data
Ilbery, Brian W.
Western Europe: a systematic human geography.
Bibliography: p.
Includes index.
1. Anthropo-geography.—Europe. 2. Europe—Economic conditions. 3. Europe—Social conditions.
I. Title
GF540.143 1986 306'.094 86–738
ISBN 0–19–823278–0
ISBN 0–19–823277–2 (Pbk.)

Printed in Hong Kong

To Lynne, Gareth, and James

Preface

Whilst the first edition of this book examined the forces creating and remaking the human landscape of western Europe during the post-war era of almost continuous economic growth, the second edition has been set within a more pessimistic environment of stagnation and decline. The oil crisis of 1973/4 was an important turning point in the human geography of western Europe and the late 1970s and 1980s have been characterized by a reversal of many of the processes which were so important in the 1960s. Therefore, as well as emphasizing the complex and diverse nature of western Europe, with its regional disparities, core–periphery relationships, temporal changes, and varied government policies, this book describes such processes as counterurbanization, return migration, deindustrialization, the urban–rural manufacturing shift, increasing unemployment and regional differences, and planning orthodoxies for decline rather than growth. The definition of western Europe remains unchanged and for the purposes of this study includes all those countries which are not members or associates of the east European socialist bloc.*

Traditional texts often adopted a regional approach to the geographical subject matter of western Europe, which tended to be both descriptive and repetitive. However, this second revised edition continues a trend which the first edition helped to initiate by producing a systematic coverage of selected topics, thus facilitating a greater understanding of the processes which have created spatial variations in the human activity of western Europe. The broad topics selected for study remain unaltered and reflect the author's personal interests and need to create a measure of balance within the book. Nevertheless, changes within these subject areas are examined, with emphasis placed upon developments which have occurred as a consequence of the economic recession.

In more detail, the first chapter sets the scene by placing the study area into its wider European context and examining both its distinctiveness and core–periphery relationships. National and regional inequalities are shown to exist in a range of phenomena, from economic and social welfare to relief and climate. Chapter 2 deals with population geography and analyses the demographic background in terms of the density, distribution, and structure of population, population growth, and the components of natural change. Migration and urbanization are examined as factors affecting the redistribution of population, as are the more recent trends of return migration and counterurbanization. The chapter on agriculture stresses both structural change and the diversity of types which exist in western Europe. Post-war developments are discussed in terms of the approaches adopted by different governments toward agricultural problems and productivity, and the evolving nature of the Common Agricultural Policy is explored. Chapter 4 examines the changing patterns of energy supply and demand and discusses how the different types of power have affected the spatial nature of economic development in western Europe. The relative importance of the various modes of transport and attempts to develop an integrated transport network are the main themes of Chapter 5, and in Chapter 6 factors affecting the distribution of industry are

*This book was written before Spain and Portugal entered the European Community in 1986 and consequently discussion often refers to the 'Ten' rather than the 'Twelve'. It will take one or two years for the main data sources to be adjusted to accommodate these changes.

examined and their importance stressed in a study of the major regions of relative growth and decline, as well as in case-studies of the iron and steel, chemical, and electrical and engineering industries. Deindustrialization and a shift in manufacturing employment to rural areas are important modern trends and increasing regional disparities, between centre and periphery, are in evidence when a model of economic potential is applied to western Europe. The various trading organizations and arrangements affecting the study area are analysed in Chapter 7 and trends in tourism, both international and in Austria and Switzerland, are discussed in Chapter 8. Variations in social well-being are the main focus of attention in Chapter 9 and emphasis is placed on inequalities at the national, regional, and urban levels. Chapter 10 is concerned with regional policy and different approaches adopted in western Europe are examined, before attention is devoted first to the Regional Development Fund of the European Community and secondly to the Mezzogiorno problem region. Finally, Chapter 11 uses the systematic topics of the previous chapters to emphasize the characteristics which distinguish western from eastern Europe.

In preparing this second edition, reference has been made to numerous statistical sources, as well as to many recent books and papers. However, the book is still intended as an introductory volume, of use to undergraduates, sixth-form students and teachers, and all those with an interest in the European landscape. I alone am responsible for the book, but acknowledge the help of the Geography Department at Coventry Polytechnic and Andrew Schuller of Oxford University Press. Last but not least, I would like to pay particular tribute to Mrs Shirley Addleton, for her cartographic skills and continued enthusiasm through the pressure of tight schedules, and to various people who typed parts of the manuscript, including the Faculty of Business, Typing Bureau, and Geography Department.

August 1985 Brian W. Ilbery

Contents

List of Plates

List of Figures

Acknowledgements

The author and publishers gratefully acknowledge permission to reproduce copyright material.

Fig. 1.1: from N. J. G. Pounds: *Europe and the Soviet Union*, 2/e., Copyright © 1966 by McGraw-Hill Inc. Reprinted with permission.
Fig. 1.4: from R. L. King, 'Southern Europe: dependency or development?', *Geography* 67, 221–34 (1982). Reprinted by permission of The Geographical Association.
Fig. 2.3: from P. L. Knox: *The Geography of Western Europe: A Socio-Economic Survey* (1984). Reproduced by permission of Croom Helm Limited.
Fig. 2.7: from H. D. Clout *et al.*: *Western Europe: Geographical Perspectives* (1985). Reprinted by permission of Longman Group Limited.
Fig. 2.8: after A. J. Fielding: 'International migration in Western Europe' in L. Kosinski, and R. M. Prothero (eds), *People on The Move* (1975). Reprinted by permission of Methuen and Company.
Fig. 2.9: from P. G. Hall and D. Hay: *Growth Centres in the European Urban System* (1980). Reprinted by permission of Gower Publishing Company Limited.
Fig. 2.10: after A. J. Fielding: 'Counterurbanisation in Western Europe', *Progress in Planning 17*, pp. 1–52 (Pergamon Press Inc.).
Fig. 2.11(a): From P. Hall: *The World Cities* (2nd ed. 1977). Reprinted by permission of Weidenfeld and Nicolson Limited.
Fig. 2.11(b): From D. Burtenshaw: *The City in West Europe* (1981). Reprinted by permission of John Wiley & Sons Limited.
Fig. 3.3: From M. Chisholm: *Rural Settlement and Land-use* (Hutchinson Publishing Group Limited).
Fig. 3.4(a): From H. D. Clout: *Geography of Post-War France: A Social and Economic Approach* (1972). Reprinted by permission of Pergamon Press Limited.
Fig. 3.5: From D. Burtenshaw: *Economic Geography of West Germany* (1974). Reprinted by permission of Macmillan, London and Basingstoke.
Fig. 3.6: after I. R. Bowler, 'Recent developments in the agricultural policy of the EEC', *Geography* 61, 28–31 (1976). Reprinted by permission of The Geographical Association.
Fig. 3.7: after I. R. Bowler, 'Some consequences of the Industrialisation of agriculture in the European Community', from P. Henry (1981), *Study of the regional impact of the Common Agricultural Policy* (Regional Policy Series 21, Commission of the European Communities, Brussels).
Fig. 4.2: after P. R. O'Dell: 'The EEC energy market: structure and integration' in R. Lee and P. E. Ogden (eds): *Economy and Society in the EEC* (Saxon House, Teakfield Limited 1976). Reprinted by permission.
Fig. 4.3: after P. R. O'Dell, 'The energy economy of Western Europe: a return to the use of indigenous resources', *Geography* 66, 1–14 (1981). Reprinted by permission of The Geographical Association.
Fig. 5.2: From J. Tuppen: 'Canals and Waterways in the EEC' in *European Studies* vol. 21 1975 pp. 1–4. By permission of *Exploring Europe*.
Fig. 5.6: From G. Parker: *The Logic of Unity* (1981). Reprinted by permission of Longman Group Limited.
Fig. 6.2(a): From P. Hall: *The World Cities* (2/e 1977), p. 136. Reprinted by permission of Weidenfeld and Nicolson Limited.
Fig. 6.2(b): From D. Burtenshaw: *Economic Geography of West Germany* (1974). p. 214. Reprinted by permission of Macmillan, London and Basingstoke.
Fig. 6.5: From G. N. Minshull: *The New Europe: An Economic Geography of the EEC* (2nd ed. 1980) p. 72. Reprinted by permission of Hodder and Stoughton.
Fig. 6.6: From K. Chapman: 'Corporate Systems in the U.K. Petrochemical Industry' p. 129, Fig. 1 in *Annals of the Association of American Geographers*. Vol. 64, 1974. Reproduced by permission.
Fig. 6.7: after D. Keeble *et al.* 'Regional accessibility and Economic Potential in the European Community', *Regional Studies* 16, pp. 419–31 (1982). Reprinted by permission of Cambridge University Press.
Fig. 8.6: From H. G. Kariel, and P. E. Kariel: 'Socio-cultural impacts of tourism: an example from the Austrian Alps', *Geografiska Annaler* 64B, 1–16 (1982). Reprinted by permission of the editor.
Figs. 9.1, 9.2 and 9.3: from *Geography* 69, pp. 298–302 (1984). Reprinted by permission of The Geographical Association.
Fig. 9.4: From *Le Point*, 68 (7 January 1980). Copyright © 1980 Le Point. Reprinted with permission.
Fig. 9.5: after P. L. Knox and A. Scarth: 'The Quality of Life in France', *Geography* 62, 9–16 (1977) p. 12. Reprinted by permission of The Geographical Association.
Fig. 9.6: From A. C. MacLaren: *Spatial Aspects of Relative*

Deprivation, mimeographed paper presented to the Urban Geography Study Group of the Institute of British Geographers 1975. Reproduced by Permission of the Secretary, IBG Urban Geography Study Group.
Fig. 10.2: From H. D. Clout: *The Regional Problem in Western Europe* (1976). p. 11. Reproduced by permission of Cambridge University Press.
Fig. 10.3: From Peter Hall: *Urban and Regional Planning* (Allen & Unwin 1985). Reprinted by permission of the author.
Fig. 10.9: after M. Pacione: 'Development Policy in Southern Italy: Panacea or Polemic?', *Tijdschrift Voor Economische en Sociale Geografie* 67, 38–47, p. 45. Used by permission.
Table 1.1: From T. G. Jordan: *The European Culture Area.* (Harper and Row Inc. 1973).
Table 3.2: after I. R. Bowler, 'Some Consequences of the Industrialisation of agriculture in the European Community' in M. J. Healey and B. W. Ilbery (eds): *The Industrialisation of the Countryside* (1985). Reprinted by permission of Geo Books.
Table 4.4: From *Nuclear News*, February 1979. Used with permission.
Table 5.2: after J. Tuppen, 'Canals and Waterways in the EEC', *European Studies*, 21, 1–4 (1975) and United Nations Statistical Year Book.
Table 6.1: after D. Keeble *et al*: 'The Urban-rural manufacturing shift in the European Community' *Urban Studies*, 20, 405–18 (1983), pp. 410–11. Reprinted by permission of Longman Group Limited.
Table 6.3: From D. Burtenshaw: *Economic Geography of West Germany* (1974) p. 122. Reprinted by permission of Macmillan, London and Basingstoke.
Table 9.1: From P. L. Knox: *Social Well-Being A Spatial Perspective* (1975), p. 26. Reprinted by permission of Oxford University Press.
Tables 9.3, 9.4, 9.5, and 9.6: From B. W. Ilbery: 'Core–periphery Contrasts in European Social Well-being', *Geography* 69, 289–302 (pp. 292, 294, & 297). Reprinted by permission of The Geographical Association.
Table 9.7: after B. W. Ilbery: 'Core–periphery Contrasts in European Social Well-being', *Geography* 69, 289–302 (1984), pp. 298–9. Reprinted by permission of The Geographical Association.
Table 9.8: adapted from P. L. Knox and A. Scarth: 'The Quality of Life in France', *Geography* 62, 9–16 (1977). Reprinted by permission of The Geographical Association.
Table 9.9: From B. E. Coates *et al*: *Geography and Inequality* (1977) p. 76. Reprinted by permission of Oxford University Press.
Table 11.1: From *Atom* 1984, p. 25. Reproduced by permission of UKAEA, Information Services Branch.
Plates 3.1, 3.2, 3.3, 4.1, 5.1, 5.3, 5.4, 5.5, 6.1, 7.1, 8.1. By permission of the European Communities Commission.
Plate 4.2. By permission of Der Oberstadtdirektor, Stadt Gelsenkirchen.
Plates 10.1, 10.2. By permission of Alan B. Mountjoy.

Although every effort has been made to trace and contact copyright holders this has not always been possible. We apologize for any apparent negligence thus caused.

1

The European Background

This text is essentially concerned with the human landscape of western Europe. Before analysing and attempting to explain the various patterns of human activity, it is necessary to place the study area into a wider European context. Consequently, this introductory chapter provides some insights into the broad human and physical contrasts that exist within Europe.

1.1 Europe as a spatial entity

'The whole of Europe appears to be a big peninsula attached to Asia by a wide base in the east, jutting westward toward the Atlantic Ocean, and steadily narrowing down until it looks like a chain of isthmuses between the Mediterranean Sea and the ocean' (Gottman 1969, p. 7). It would thus appear that Europe is part of the continent of Eurasia, occupying approximately one-fifth of its total land surface. Despite this, Europe is a separate entity, in both physical and human senses.

In a physical sense, the distinctiveness of Europe can be assessed with reference to plate tectonic theory. Geologically, Europe and Asia were separate continents from between 600 (late Precambrian) and 300 million years before present (Carboniferous), and it was not until the Uralian orogeny occurred around 300 million years before present (Permo-Carboniferous) that the two continents came together (Ager 1975; Anderson 1978). Further evidence was afforded by Hamilton (1970), who deduced a subduction zone operating along the eastern frontier of Europe.

In a human sense, Europe is occupied by peoples with particular traits and cultural characteristics (Table 1.1). The distinctive features of the continent 'belong to the pattern of culture rather than to the natural environment' (Gottman 1969, p. 3), and Europe is a clearly definable cultural area where the inhabitants have a religious tradition of Christianity, speak one of the related Indo-European languages, and are of caucasian race. Europe is further distinguished by its urban–industrial economy and its healthy, well-fed, and well-educated population.

A recent essay by Mead (1982) suggested that the distinctiveness of Europe lies in a combination

Table 1.1 European traits

1. Majority of population speaks an Indo-European language
2. Majority of population is Caucasian
3. Majority of population has Christian heritage
4. 90 per cent or more of population is literate
5. Infant mortality rate is less than 25 per 1000 live births
6. Annual rate of population increase is 10 or less per 1000 population
7. Per capital income is $1000 or more
8. 60 per cent or more of population lives in towns and cities of 2000 plus inhabitants
9. 35 per cent or more of workforce is employed in manufacturing, mining, and construction industries
10. Density of railway network is 6 or more km of rail per 100 square km
11. No violent or illegal overthrow of government since 1950
12. At least 100 kg of fertilizer applied to each hectare of arable land per year

Source: Jordan (1973)

of both physical and human characteristics. Five features in particular were emphasized:

1. *Physical geography*, with a varied resource base and fewer natural disasters and less insecurity than most of Eurasia.
2. *Ethnography*, with a variety of cultures in close juxtaposition.
3. *Technology*, with Europe for a long time being the centre of technological innovations which were diffused throughout the world. Now second only to North America, technology has helped to reduce physical impediments and environmental risks.
4. *Values*, with a high value placed upon the rational manipulation of the environment and the ability to build or rebuild speedily from scratch.
5. *Territorial organization*, with the greatest concentration of independent nation-states in the world. The nation-state is one of the most positive and negative institutions that Europe has given to the world. In Europe, over thirty independent political units foster their own concepts and formulate their own categories.

The core–periphery relationship (Friedmann 1973) is central to the idea of Europe existing as a spatial entity. This concept dates back to 1919, when Mackinder published his heartland theory in order to account for the dominance of Eurasia as a base from which a campaign of world conquest could be launched. Mackinder divided Eurasia into an interior heartland and a marginal or peripheral crescent. The heartland, centred on the east European plain and immune to sea power, was dominant and possessed greater potential for supporting a successful drive for world conquest than the crescent, which was not immune to invasion by land-based military power (Jordan 1973). Mackinder's ideas were challenged by Spykman (1944) who, although agreeing that Eurasia was the key continent, felt that more importance should be attached to the marginal crescent or rimland, which included nearly all the peoples and states of Europe. Although criticized by numerous people (see Cohen 1964; Blacksell 1977), the outdated theories of Mackinder and Spykman still form a basis for political argument.

In more recent years, geographers have applied the core–periphery concept to Europe itself and to divisions within the continent. Jordan (1973), for example, identified a distinct European core and periphery, based on the twelve 'European traits' listed in Table 1.1. Countries exhibiting all twelve characteristics formed a contiguous core area, centred on West Germany, northern France, Benelux, England, southern Scandinavia, and northern Italy. The number of characteristics declined in a fairly regular manner in all directions away from this core, with areas such as southern Italy, Ireland, Spain, and Yugoslavia forming the European periphery. A similar core–periphery relationship exists if the scale of analysis is reduced and western Europe only is analysed; evidence of this relationship is given in subsequent chapters of this book.

King (1982) and Ilbery (1984) have both drawn attention to the importance of the core–periphery concept in understanding the geography of Europe. A highly developed inner core of countries, the 'mega-core' (Seers 1979) is the focus of economic, political, and social dynamism and is characterized by innovation, concentration, and an agglomeration of capital and investment. This mega-core is surrounded by a periphery of less developed states, especially to the south and east, and the gap between core and periphery is reinforced by the process of cumulative causation and the dominance of 'backwash' effects over 'spread' effects (Mabogunje 1980). Economic recession in the late 1970s helped to exaggerate these differences, as inflation, unemployment, and declining industrial activity affected peripheral areas most severely.

It was Friedmann (1973) who developed the formal core–periphery model. He discerned an autonomy/dependency pattern of development, with the core dominating the periphery in most political, economic, and social respects. Therefore, the core controls itself and the periphery in a kind of colonial relationship. However, Friedmann anticipated the eventual breakdown of core–periphery relations, as political and social tensions between core and periphery emerged. These tensions are apparent in Europe, but there is little empirical evidence of convergence between core and periphery.

Although characterized by unifying features, Europe is far from homogeneous and contains many divisions. Geographers have for a long time divided Europe into western, eastern, central, northern, and south-eastern regions and the broad contrasts between east and west will be systematically analysed in Chapter 11. The possibility of European unity was eroded less than twenty years after the Second World War by the formation of trading blocs (the European Free Trade Association (EFTA) and the European Community (EC) in western Europe and the Council for Mutual Economic Assistance (COMECON) in eastern Europe); the two affecting western Europe are analysed in some detail in Chapter 7. The fragmented nature of Europe has also been highlighted by the demands of territorial minorities within individual nation states to have greater powers for the management of their own affairs (Williams 1980). Greater regional autonomy has been demanded by such linguistic groups as the Basques, Welsh, Scots, Bretons, Walloons, Flemings, and Catalans (Stephens 1976), who feel that national political systems are unable to cope with the economic, social, and cultural problems of particular peripheral regions.

Europe is characterized by unequal economic development and a proper understanding of present-day patterns of human activity can only be obtained if a historical perspective is developed. This is especially true in most west European countries where, as a result of long traditions and invasions, the past has survived into the present in many areas. Geographers must keep the past in mind if they are going to understand the processes which are helping to create the present patterns and problems. It is this balance between old and new which makes Europe so difficult to understand.

1.2 The physical landscape

Man has modified and been influenced by the physical characteristics of Europe. Differences in relief and climate partially explain the existing variations in human activity and thus a brief survey of the physical landscape follows.

Both Pounds (1966) and Hoffman (1969) divided Europe into four physiographic units, which correspond closely with the 'structural-geological' divisions of Europe suggested by Ager (1975), emphasizing the importance of underlying geological structures to the morphology of the physical landscape. Moving in a north–south direction these are:

1. *The north-west highlands*. This region contains most of Scandinavia, Iceland, Scotland, and parts of Ireland and Wales, and falls within Ager's 'Palaeo-Europa' geological division (Fig. 1.1). The mountains have experienced pronounced glacial erosion and the rock surfaces have been smoothed and the soils removed and deposited over the lower lands to the south. This has resulted in an area of thin, infertile soils, unsuitable for highly productive agriculture, and low population densities. However, the highlands contain minerals, notably iron-ore, and this has led to much mining activity.

2. *The north European plain*. This low-lying area, mostly less than 165 m above sea level, includes most of England and Denmark, southern Sweden, Belgium, the Netherlands, half of France, and north-west Germany, before it broadens eastwards to include most of Poland, European Russia, and parts of Romania. This plain is part of the 'Eo-Europa' geological division and can be subdivided into a glaciated northern section and an unglaciated southern section. Deposits of moraine and boulder clay make the soils infertile in the northern section, as the clays are too heavy and the sands too light for productive cultivation. The unglaciated section consists of low, flat-topped hills and broad valleys. Deposits of loess cover the area and consist of glacial outwash and silt blown out of river valleys. Soils are thus highly productive and agriculture has flourished. High rural population densities encouraged rural–urban migration, industrialization, and urbanization. The western part of this unglaciated area coincides with the 'west European core'.

3. *The central uplands and plateaux*. Situated south of the north European plain and unaffected by glaciation, this region consists of plateau-like highlands separated by deep valleys and plains which form natural routeways. The area forms part of Ager's 'Meso-Europa' and includes such notable masses as the Central Massif of France, the Vosges Mountains, the Black Forest, and the Bohemian Massif. Soils on the steeper slopes are thin and infertile and the higher surfaces are amongst the least populated in Europe. Population densities are higher where the rocks contain minerals and coal measures, but with the decline in importance of coal as a source of power, these

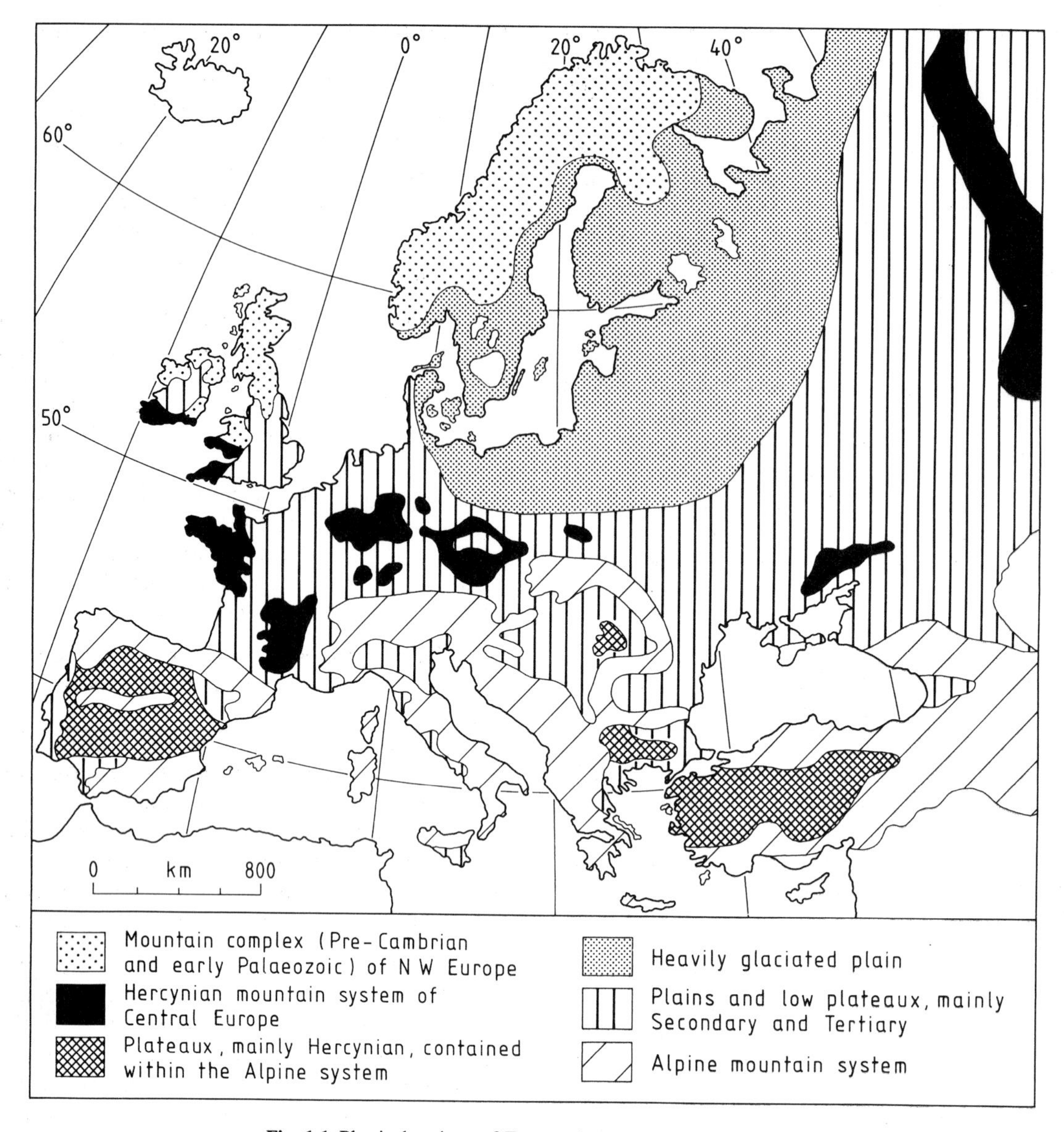

Fig. 1.1 Physical regions of Europe (*source*: Pounds 1966, p. 5)

particular areas are now over-populated in relation to resources.

4. *The southern Alpine system.* This chain of high and comparatively young rocks, which incorporates geologically old fragments, extends from Spain through France, Switzerland, and the Balkan countries and continues eastwards across Asia. The mountains in this 'Neo-Europa' geological division are discontinuous and separated by deep valleys, plains and plateaux; the north Italian plain and the Spanish tableland are good examples. However, the Alps of France, Switzerland, and northern Italy are continuous and act as an almost continuous barrier to communications. Rail and road tunnels have been cut through these mountains to link the peripheral regions with the core areas of western Europe. Agriculture and settlement have accordingly avoided the higher parts and are essentially concentrated in the valleys and on the plains.

Climate

Climatically, Europe is well suited to man: it does

not suffer from extremes of heat, except locally in the south and east, or from extremes of cold, except in the east and north-east (Pounds 1966). However, Europe has a far from uniform climate and it is the 'relative size and strength of the several different types of air-mass that in their seasonal variations give Europe its climate and the day-to-day variation that is weather' (Mellor and Smith 1979, p. 15). Marine air-masses from the west bring moist and humid climatic conditions, whereas further into Eurasia continental air-masses bring drier and colder weather in the winter months. Southern parts of Europe are also influenced by tropical air-masses from North Africa and this results in little or no rainfall during the summer months. The remainder of Europe receives rainfall at all seasons, although the amount decreases as one moves eastwards. In western Europe winter rain exceeds summer rain, whilst in eastern parts summer is the wettest period. Other factors affecting climatic elements within Europe, especially temperatures, include altitude and the surrounding seas. Temperatures are modified by altitudinal variations within the physiographic divisions and the seas moderate the extremes of temperature by having a warming influence on the land in the winter and a cooling influence in the summer.

Broadly speaking, Europe can be divided into three major climatic regions: first, in the south a Mediterranean region of hot, dry summers and mild, wet winters; secondly, a region of cool temperate climate to the north of the Mediterranean region, which stretches from west to east across Europe and where rainfall decreases and temperatures become more extreme toward the east; and thirdly, a region of cold climate to the north, with cool summers and cold winters, which are slightly modified in the west by the surrounding seas.

There is evidence in Europe to suggest that man has modified both the physical landscape and the climate. This can be seen in the building of agricultural terraces, slag heaps, and polders, and in the effects pollution, industry, and urban areas are having on atmospheric conditions. A strong correlation exists in Europe between adverse environmental conditions on the one hand and low standards of living and low levels of technology on the other.

1.3 Variations in social and economic welfare

Europe is characterized by inequalities in both social and economic welfare both between and within member states. Social welfare can be defined as the enjoyment of health and prosperity and a composite measure of social well-being will be developed for each west European country in Chapter 9. Economic welfare can best be expressed by measures of levels of economic development. Overall contrasts exist, particularly between western and eastern Europe, and these variations will now be considered in more detail.

Western Europe as a whole exhibits levels of social welfare not achieved in the east European bloc. The choice of criteria upon which to measure social well-being will obviously condition the ultimate patterns of variation. In this case, four social 'indicators', for 1980, have been chosen in order to monitor the broad contrasts between western and eastern Europe:

1. The 'consumption' of newsprint per inhabitant; an indirect indicator of education and literacy.
2. The number of television receivers per 1000 inhabitants; an indicator of access to information and the importance attached to consumer goods.
3. The number of hospital beds per 1000 inhabitants; an indicator of health care.
4. The number of telephone receivers per 1000 inhabitants; an indicator of material well-being.

The patterns of variation displayed by the indicators are shown in Fig. 1.2. Such variations are at the national level only, although regional variations in welfare exist in all European countries. Nevertheless, Fig. 1.2 shows some discernible patterns and two broad observations can be made about these four indicators of welfare: first, there are wide disparities between the European countries; and secondly, core–periphery contrasts, although varying in magnitude, are evident.

More specifically, the 'consumption' of newsprint per person varies from 1.3 kg in Yugoslavia to 35.6 kg in Sweden (Fig. 1.2a); indeed, the highest consuming areas (Sweden, Netherlands, Finland, Denmark, and Switzerland) have values more than five times as great as the lowest consuming areas (Portugal, Spain, Poland, Yugoslavia, Romania, Albania, and Bulgaria). Therefore, contrasts in the standards of education and

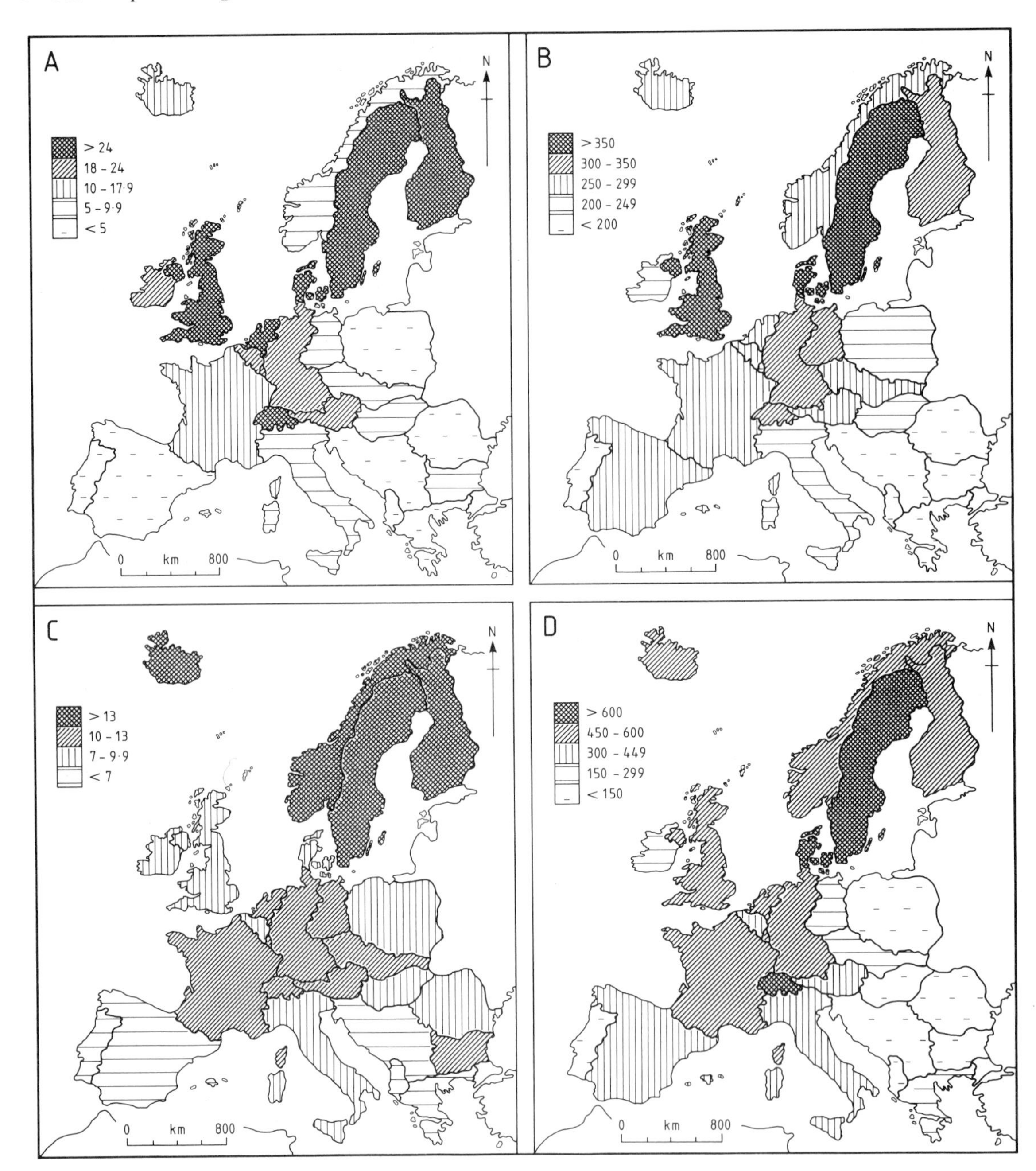

Fig. 1.2 Indicators of social welfare in Europe:
(a) consumption of newsprint per person in 1980 (kg);
(b) number of television receivers per 1000 people, 1980;
(c) number of hospital beds per 1000 people, 1980;
(d) number of telephone receivers per 1000 people, 1980

literacy, as gauged by this indicator, are apparent between the countries which form the European core and those which represent the European periphery.

A similar pattern emerges when the second indicator of welfare, the number of television receivers per 1000 inhabitants, is mapped (Fig. 1.2b). In 1980, a distinct core area centred on north-western Europe was in evidence, with four countries having in excess of 300 television receivers per 1000 people: the United Kingdom led the way with 394, followed by Sweden (374), Denmark (358), and West Germany (337). These countries have between three and four times as many television receivers per 1000 people as Portugal, Greece, and Albania, the only countries with a total below 150. In terms of this particular indicator, the European periphery is comprised of countries with less than 250 television receivers per 1000 inhabitants and includes Portugal, Albania, Italy, Ireland, Greece, Bulgaria, Romania, Yugoslavia, and Poland.

It is realistic to regard an increase in the number of hospital beds per 1000 people as representing an improvement in health care and hence social well-being. The provision of hospital beds varies spatially in Europe (Fig. 1.2c) and once again a core–periphery relationship exists. The core is larger in extent than those produced by the previous indicators and is truly centred on Scandinavia and Iceland where, with the exception of Denmark, the number of hospital beds is relatively favourable, at over 14 per 1000 people. Portugal and Spain in the south-west and Greece, Yugoslavia, and Albania in the south-east represent the other extreme, where the ratio falls below 6.5 per 1000 people. An important feature of this indicator is the inclusion, for the only time in these four examples, of Denmark, Belgium, and the United Kingdom in the semi-periphery of western Europe.

When the final indicator of social welfare, the number of telephone receivers per 1000 inhabitants, is mapped (Fig. 1.2d), the familiar contrast between an affluent west European 'core' and a poorer Atlantic and Mediterranean 'periphery' is distinguishable (Clout *et al.* 1985). Sweden (796) and Switzerland (727) easily head the list and are followed by Denmark (641) Luxembourg (547), and the Netherlands (509). At the other extreme, all east European countries, together with Ireland and Portugal, have less than 300 receivers per 1000 inhabitants.

Two important qualifications need to be made in relation to indicators such as those portrayed in Fig. 1.2 (Clout *et al.* 1985). First, the differences identified are 'spatially-packaged expressions of social differences with regard to employment, class and access to financial resources, educational capital and political power' (p. 3). Inequalities in social and economic development are the result of deeply-rooted social conditions which have grown in each specific political economy. Secondly, many indicators of social well-being have undergone a reversal since the mid-1970s, reflecting the uncertain economic future of western Europe.

Levels of economic welfare also differ between the countries and regions of Europe. One indicator of economic welfare is income levels, which can be measured in terms of per capita gross national product, per capita gross domestic product, or per capita money income. None is totally reliable, since all are based on data collected in different ways in different countries, and they reveal nothing of the ways in which wealth is distributed within a nation (Coates *et al.* 1977). Hence national figures are not strictly comparable, but the distribution of gross domestic product shows wide disparities between regions (Fig. 1.3). Some general observations can be made from this map.

In Europe generally, the level of economic welfare tends to be lowest in regions furthest removed from the powerhouse of industrial activity, which embraces a triangular area bounded by the Ruhr, Paris, and Birmingham (see Chapter 6). Thus the poor regions in individual countries are situated in the peripheral parts of Europe; for example, south-east Hungary, eastern Poland, and the west and north-western regions of the British Isles and Scandinavia. This familiar core–periphery pattern is the most noticeable feature if the scale of analysis is reduced and western Europe only is examined. In this situation, the core includes the Ruhr, Belgium, northern France, and lowland Britain, while the periphery includes southern Italy, Brittany, the greater part of Portugal, Wales, Ireland, and Scotland.

The wealthiest regions (southern Sweden, Paris, Brussels, Hamburg) have gross domestic product values more than three times as high as

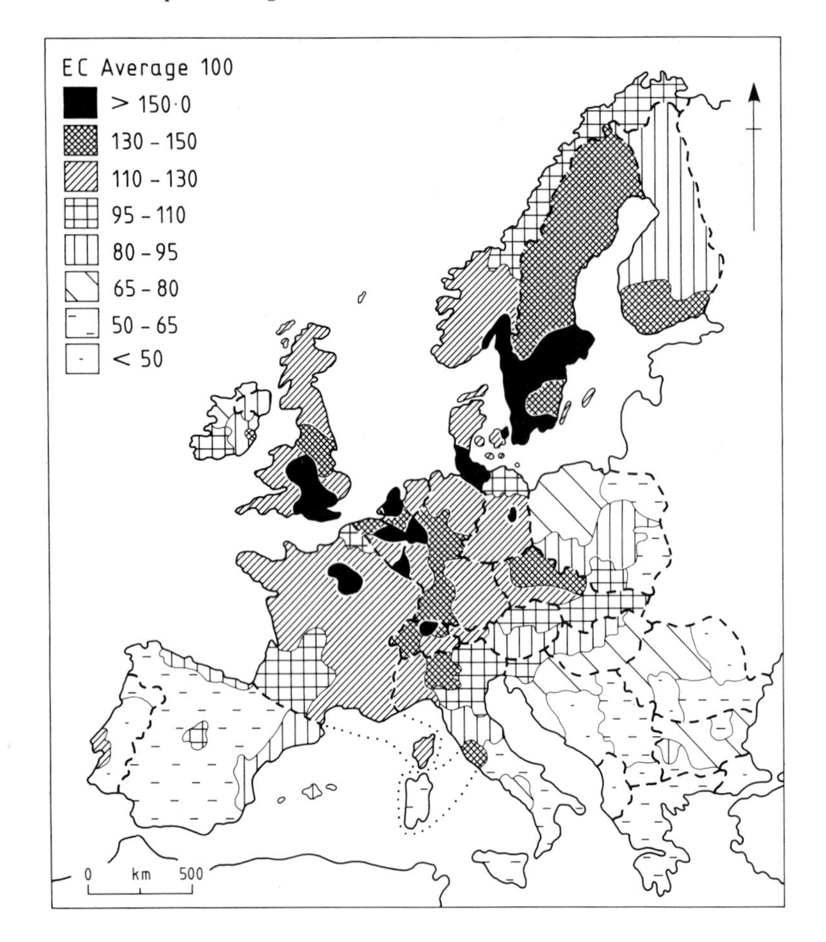

Fig. 1.3 Gross domestic product per person in Europe

the poorest (central Portugal, north-west Ireland, and southern Italy). In general terms, two broad correlations exist (Sant 1974). First, the distribution of wealth corresponds with that of population density (Fig. 2.1) which is highest in the triangular area between the Ruhr, Paris, and the English Midlands, with scattered outliers in a number of predominantly urban areas. The exception to the rule is Scandinavia, and notably Sweden, where high incomes have resulted from a good resource base, skilled technology, capital-intensive industry, and a consistent economic policy. Secondly, there is a relationship between incomes and the percentage of the workforce employed in agriculture and industry; the wealthier areas tend to have small shares of their economically active population employed in agriculture and large shares in manufacturing. Anomalies occur however, and, relative to their manufacturing sectors, regions in Belgium and the United Kingdom are badly off, compared with those in France and Denmark. Fig. 1.3 also identifies a number of metropolitan areas which have incomes well in excess of their surrounding areas; Paris, Lisbon, Madrid, and Rome are good examples.

Regional variations are often striking within individual countries and in France and Italy, for example, the gap in economic welfare between regions has if anything increased (Holland 1976; Armstrong 1978). Manufacturing employment in these countries was, up to the mid-1970s, especially expanded in the more industrialized areas and agricultural productivity remained low in the poorer regions and showed little sign of progress. However, there are contrasts between Italy and France regarding the factors which have contributed to regional disparities in economic development. In the former, natural population increase is high in the poor southern regions, but mi-

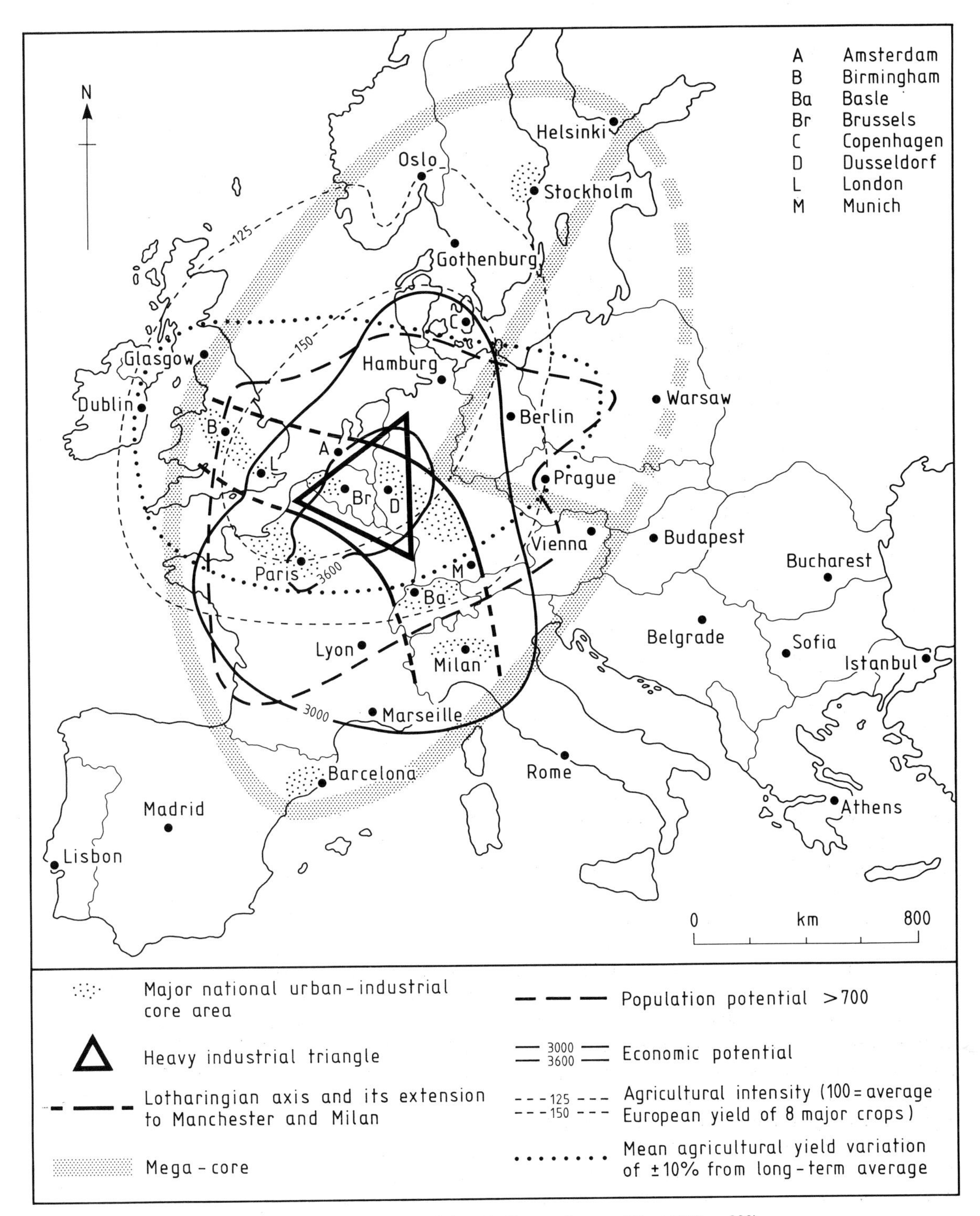

Fig. 1.4 Core and periphery in Europe (*source*: King 1982, p. 223)

gration to the north is low because of limited employment opportunities. In France, natural population increase is low and a favourable employment situation in the richer areas has led to large-scale migrations, especially of the younger age groups. This has led to depopulation and an unbalanced population structure in the poorer rural areas.

Similar regional variations can be found in most other European countries and in Norway, for example, gross domestic product decreases with movement north and coincides with declining population densities and an increasing dependence on primary production. A general trend in most countries, especially in the least developed, is for the capital city to dominate the regional pattern of economic welfare. Exceptions to this trend include West Germany and Switzerland which have various centres of similar importance.

Many other criteria have been used to delimit the core of western Europe. These have been summarized by King (1982) and are mapped in Fig. 1.4. As well as Seer's 'mega-core', agricultural, economic, and population potential have been used to provide isopleths of equal potential. Similarly, the yolk of the core has been delimited in terms of the heavy industrial triangle, the Lotharingian axis and the 3600 economic potential line. All of these measures and 'areas' will be referred to in subsequent chapters.

Whichever criterion is employed the distinction between core and periphery is clear. Indeed, the peripheral areas of Europe have increasingly become the focus of attention (Seers 1979; Lewis and Williams 1981; King 1982; Williams 1984). Numerous reasons for the lower levels of economic development in the peripheral regions can be forwarded (Friedmann and Alonso 1975), including the following:

1. The location of economic activity is still partly governed by historical patterns which emerged when proximity to coalfields and navigable rivers was vital. Although industry has a more flexible choice of locations today, geographical inertia has prevented major shifts in the distribution of industry.
2. The physical distance from the main industrial centre of Europe has itself been a powerful locating factor, making the regions nearer the centre more and those further away less attractive to potential industrialists.
3. In the low-income regions of Europe, a large share of the working population is employed in agriculture and productivity is low in comparison with the more prosperous parts.
4. The poorer countries are usually dependent on the capital city, which dominates the surrounding hinterlands.
5. Particular systems of land ownership are associated with the lack of agricultural development; the large estates (latifundia) in southern Italy and the inheritance laws in many parts of Europe, where land was divided equally among heirs on the death of the owner, are good examples.
6. The market for products is often external and subject to wide fluctuations. There has been a tendency in these areas to export food, forest products, and raw materials to the richer regions and to import manufactures and services. There is little control over the few resources that do exist, as major decisions are taken in the core.
7. The poorer countries and regions are lacking in local innovations and technologies. This has not been helped by the selective out-migration of labour involving the younger and more able people.
8. The climate tends to be unfavourable in the regions of the European periphery in comparison with those in the European core.

With such wide disparities in living standards between and, more especially, within countries, various planning measures aimed at reducing the gap between regions have been formulated. Regional planning and the associated regional policies will be discussed in Chapter 10.

From this introductory survey, it is clear that Europe is characterized by core–periphery contrasts and regional disparities in human activity. Western Europe emerges as an ambivalent region, but nevertheless characterized by a human geography which distinguishes it from its east European neighbours. Subsequent chapters will consider in detail the broad human contrasts manifest upon the landscape, together with the processes which have created such a dynamic entity.

References

Ager, D. V. (1975). 'The geological evolution of Europe'. *Proceedings of the Geologists Association*, **86**, 127–54.

Anderson, J. G. C. (1978). *The structure of western Europe*. Pergamon, Oxford.

Armstrong, H. W. (1978). 'Community regional policy: a survey and critique'. *Regional Studies*, **12**, 511–18.

Blacksell, M. (1977). *Post-war Europe: a political geography*. Dawson, Folkestone.

Coates, B. E., Johnston, R. J. and Knox, P. L. (1977). *Geography and inequality*. Oxford University Press, Oxford.

Clout, H., Blacksell, M., King, R. and Pinder, D. (1985). *Western Europe: geographical perspectives*. Longman, London.

Cohen, S. B. (1964). *Geography and politics in a divided world*. Methuen, London.

Friedmann, J. (1973). *Urbanisation, planning and national development*. Sage, Beverly Hills.

Friedmann, J. and Alonso, W. (1975). *Regional policy: readings in theory and application*. MIT Press, Cambridge, Mass.

Gottman, J. (1969). *A geography of Europe*, 4th edn. Holt, Rinehart and Winston, London.

Hamilton, W. (1970). 'The Uralides and motion of the Russian and Siberian platforms'. *Bulletin of the Geological Society of America*, **81**, 2553–76.

Hoffman, G. W. (ed.) (1969). *Geography of Europe: including Asiatic USSR*. Ronald Press, New York.

Holland, S. (1976). *The regional problem*. Macmillan, London.

Ilbery, B. W. (1984). 'Core–periphery contrasts in European social well-being'. *Geography*, **69**, 289–302.

Jordan, T. G. (1973). *The European culture area*. Harper and Row, London.

King, R. L. (1982). 'Southern Europe: dependency or development?' *Geography*, **67**, 221–34.

Lewis, J. R. and Williams, A. M. (1981). 'Regional uneven development on the European periphery: the case of Portugal'. *Tijdschrift voor Economische en Sociale Geografie*, **72**, 81–98.

Mabogunje, A. (1980). *The development process: a spatial perspective*. Hutchinson, London.

Mackinder, H. J. (1919). *Democratic ideals and reality*. Holt, Rinehart and Winston, New York.

Mead, W. R. (1982). 'The discovery of Europe'. *Geography*, **67**, 193–202.

Mellor, R. E. H. and Smith, E. A. (1979). *Europe: a geographical survey of the continent*. Macmillan, London.

Pounds, N. J. G. (1966). *Europe and the Soviet Union*, 2nd edn. McGraw-Hill, London.

Sant, M. E. C. (1974). *Regional disparities*. Macmillan, London.

Seers, D. (1979). 'The periphery of Europe'. In, Seers, D., Schaffer, B., and Kiljunen, M. L. (eds), *Underdeveloped Europe: studies in core–periphery relations*. Harvester Press, Sussex.

Spykman, N. J. (1944). *The geography of peace*. Harcourt Brace Jovanovich, New York.

Stephens, M. (1976). *Linguistic minorities in western Europe*. Gower, Llandysul.

Williams, A. M. (ed.) (1984). *Southern Europe transformed*. Harper and Row, London.

Williams, C. H. (1980). 'Ethnic separatism in western Europe'. *Tijdschrift voor Economische en Sociale Geografie*, **71**, 142–58.

2

Population

2.1 Demographic characteristics

In 1981, Europe contained 10.76 per cent of the world's population, even though it occupied just 3.6 per cent of the land surface. Thus the people of Europe live at a comparatively high density of 98 per km^2, which is 3.1 times the global average (30 per km^2). The density varies between and within the main regions and individual countries, and although western Europe has a similar population density to Europe generally, it accounts for 72 per cent of the total population. Within western Europe, the highest population densities are found in the European Community (EC). The Community contains 77 per cent of the west European population, living at a density of 164 per km^2, which is nearly four times as great as that of non-EC countries (Table 2.1).

Population distribution and density

Western Europe is typified by a densely populated core and a sparsely settled periphery. High densities are maintained by the industrially developed economies of the central and north-western parts of Europe: the Netherlands (346 per km^2), Belgium (323), West Germany (248), and the United Kingdom (231) are followed by a group of countries comprising Italy (190), Switzerland (156), and Luxembourg (141). Together they form a core of high population density. In contrast, exceptionally low densities are found in the northern countries of Iceland (2), Norway (13), Finland (14), and Sweden (20).

Comparisons of this nature are necessarily generalized and hide notable regional variations (Fig. 2.1). Kosinski (1970), Pressat *et al.* (1973), Salt and Clout (1976), and Knox (1984) have all identified a zone of high population density, extending from lowland Great Britain to north-east France, Belgium, and north-west Germany where it then divides, with one axis moving southwards into Switzerland and northern Italy and the other moving eastwards from the Ruhr to Hanover, Saxony, Silesia, and into the Soviet Union. In the same density category are isolated pockets such as Paris, Madrid, Barcelona, Lisbon, the Po Valley, Athens, and the lower Seine. At the opposite extreme, most of Scandinavia, the Scottish Highlands and Islands, mid-Wales, and parts of the Iberian meseta and the Massif Central are Europe's 'empty lands', with a density of less than 20 per km^2

Table 2.1 The population of Europe

	Population 1981 ('000 000)	Density/ km^2 1981	Surface area ('000 km^2)	Projected population 1990 ('000 000)	Project annual population growth 1975–90 (%)
World	4508	32	135 897	5020	1.7
Europe	485	98	4937	499	0.5
Western Europe	350	99	3594	358	0.3
EC	270	164	1656	275	0.2
Non-EC	80	46	1938	83	0.7

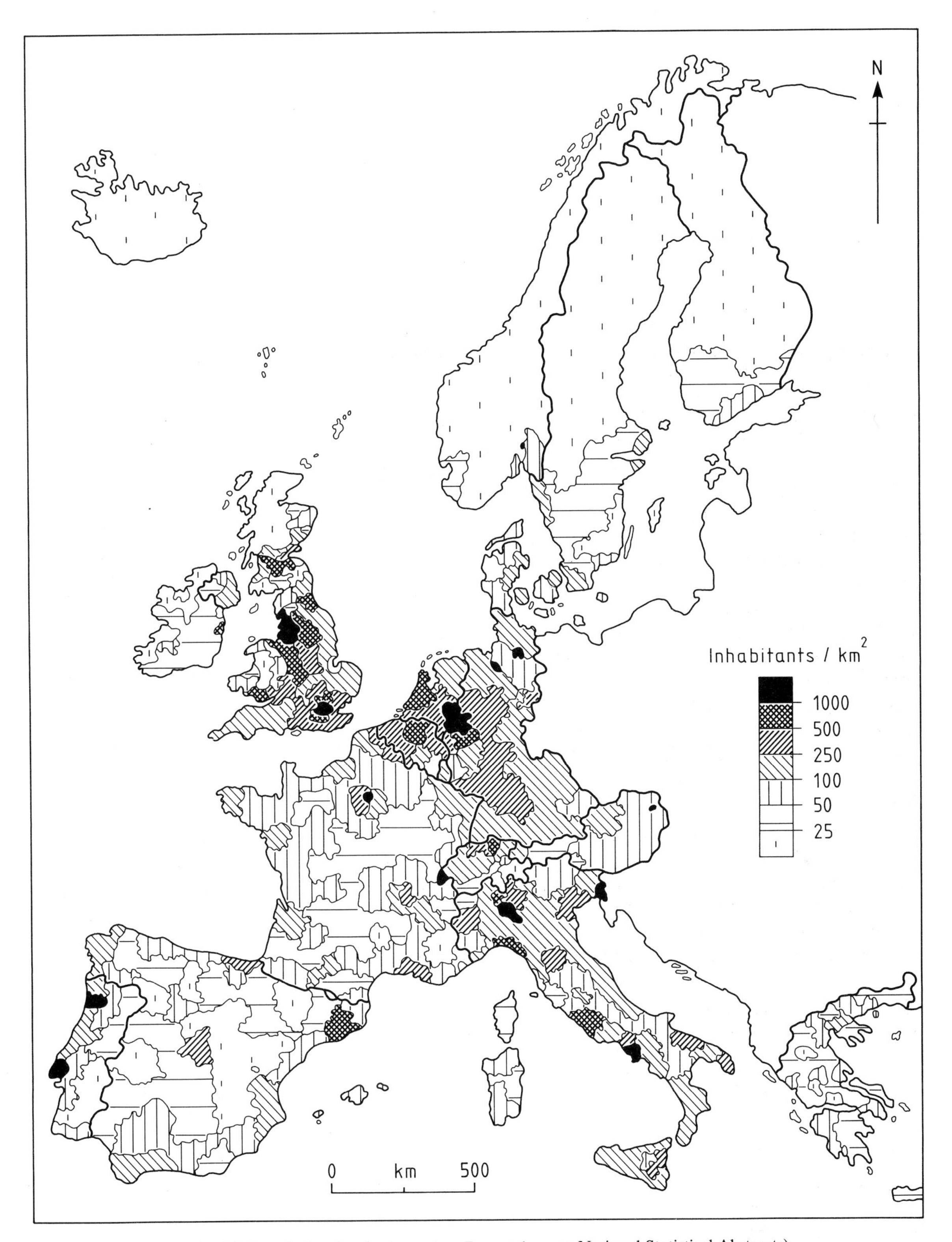

Fig. 2.1 Population density in western Europe (*source:* National Statistical Abstracts)

These differences are the result of spatial disparities in both physical and human factors. Environmental factors, such as mountainous terrain and climate, partially explain the low population densities in the peripheral areas. The distribution of natural resources, especially deposits of coal, has also affected the pattern of population density. Exploitation of coal resources gave rise to industrialization, which in turn led to urbanization, and together these two processes help to explain the growth of urban agglomerations in areas like the Rhine–Ruhr, the Sambre–Meuse trough, and the English Midlands. Other factors of an economic nature include transport and trade, agricultural potential, and the role of the capital city. Many centres owe their early development to favourable positions on trade routes; this is especially true of Milan, Turin, and Genoa, which form an industrial triangle in northern Italy. Concentrations of people in areas such as Catalonia, the Netherlands, and the Po Valley are also partly attributable to fertile soils and intensive forms of land-use. The dominance of capital cities like Paris, Athens, Madrid, and Copenhagen accounts for the increasing concentration of people in relatively small areas.

Cultural factors have also helped to shape the pattern of population distribution and Jordan (1973) believes they are most important. Inheritance laws represent one such factor and Europe was traditionally divided into northern and southern zones on the basis of legal systems. In the south, divided inheritance was dominant and has only been restricted in recent years in many of the romance-language nations. Under this system, landholdings and other possessions were divided equally among all heirs and farms were continuously subdivided and so became too small to provide an adequate living. Consequently, this so-called Roman law led to an increasing farm population and rural overpopulation. In the north, Germanic law and its English common law offspring supported the principle of undivided inheritance (primogeniture) where land was passed to one child only, usually the eldest; remaining children received compensation, but this was often inadequate and had the effect of either encouraging migration to the cities or emigration. Nineteenth-century Germany provides a good example of both systems: in the south, for example in Baden-Württemberg, farm division reached a critical level by the 1840s and rural over-population was a major problem; in the north, rural population densities were much lower as the area was a stronghold of Germanic law.

Another cultural factor affecting the past and present pattern of population is voluntary and forced migration. In the nineteenth century, there was a mass emigration from Scandinavia and Ireland, to escape poverty and famine, and as it was mainly the young who left, this aggravated the low economic development of these areas. Cousens (1960), in a study of emigration during the great Irish famine, has shown that 800 000 people starved and over one million emigrated between 1846 and 1851. This had a marked impact on the growth of population, so much so that over 100 years later the population of Ireland was just 50 per cent of the 1841 figure. Between 1920 and 1955, over 30 million people were forced to leave their homelands in Europe and prevented from returning (Jordan 1973). These forced movements, including the transfer of linguistic minorities and the flight to escape zones of military activity, had a more noticeable effect on the distribution of population in eastern Europe, but Finland, for example, was forced to house nearly 500 000 displaced Finns when the country conceded its eastern territories to the Soviet Union.

Population growth

Western Europe's population is characterized by a slackening rate of increase, compared with other areas of the world. By the year 2000, the population will have reached 420 million, less than seven per cent of the world's total. Low birth and death rates are characteristic and between 1970 and 1980, western Europe had an average annual growth in population of just 0.8 per cent, compared with a world figure of 1.9 per cent.

Western Europe's population grew by 14.8 per cent from 304.7 million in 1960 to 349.8 million in 1981. This average figure hides a wide range amongst the individual countries, from 29.5 per cent in Iceland and 25.0 per cent in Spain to just 6.6 per cent in the United Kingdom, 6.8 per cent in Austria, and 7.6 per cent in Belgium. The highest growth rates have not been restricted to either 'core' or 'peripheral' areas and are found in Iceland, Switzerland, Spain, Ireland, and the Netherlands (Fig. 2.2a). The bulk of this growth

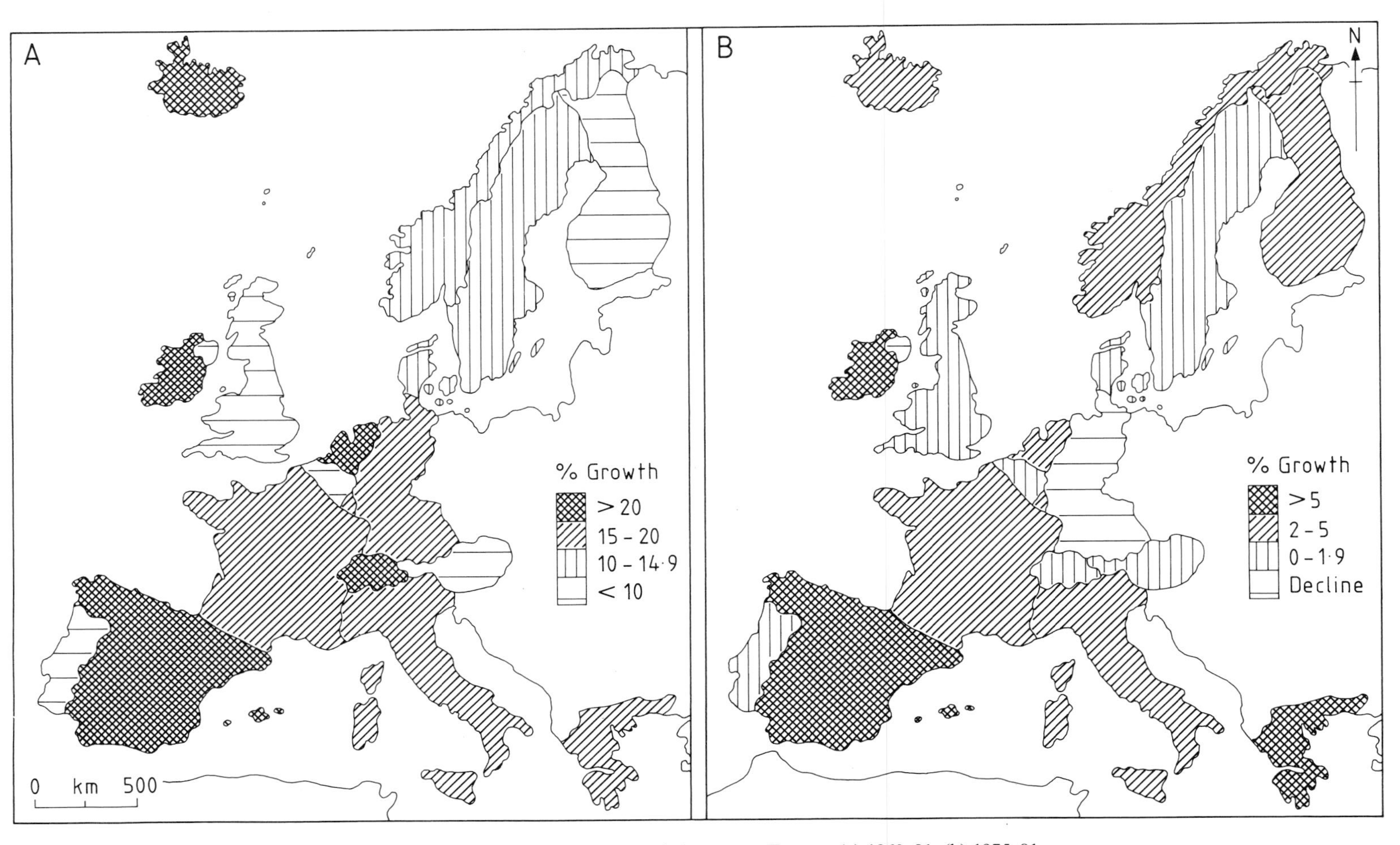

Fig. 2.2 National population growth in western Europe: (a) 1960–81; (b) 1975–81

is accounted for by natural increase, except in Switzerland, where immigration has been very important, and in France and the Netherlands, where repatriation has played an important role. Annual rates of increase tended to fall towards the latter part of the 1960–81 period (Fig. 2.2b), except in Ireland, Spain, and Greece. One of the most remarkable changes occurred in Switzerland, a country which had been experiencing a high rate of population growth. Here the average annual growth rate declined from 1.74 per cent in the 1960s to less than 0.1 per cent between 1975 and 1980. This can be accounted for by a combination of decreasing birth rates, associated with the spread of the contraceptive 'pill', and government policy designed to stabilize the immigrant population.

Analysis of national trends, although important, hides notable regional variations in the rates of population growth. The latter were examined in a study of western Europe between 1961 and 1971 by Pressat *et al.* (1973), and it is apparent from their work that the highest rates of increase (over 30 per cent) occurred not in the core regions but around the urban areas of Madrid, Barcelona, Bilbao, Copenhagen, and Oslo. Rates were also high in the familiar axial belt running from south-east England through the Rhinelands to northern Italy. In contrast, certain peripheral and predominantly rural areas lost population, including parts of Scandinavia, Ireland, Iberia, southern Italy, Wales, and Scotland. Pressat *et al.* (1973) identified additional pockets of population loss, amongst which the Massif Central and the Po valley, areas of strong out-migration, stood out.

Western Europe has not always been characterized by low levels of natural population increase. Member countries have undergone a pronounced demographic development, which has been conceptualized into the demographic transition theory, or the transition different societies make as they follow the path of socio-economic development. Basically, four phases of demographic development can be identified:

1. High stationary phase, characterized by high birth and death rates. Hence natural increase is low.
2. Early expanding phase, characterized by high birth rates and declining death rates, and therefore a rapid rate of natural increase. Decreasing death rates are the result of an improvement in social and economic development, in terms of improved hygiene and medicine.
3. Late expanding phase, characterized by a decline in birth rates as well as a continual decline in death rates. Natural increase is thus reduced, reflecting an urban society and better methods of birth control.
4. Low stationary phase, characterized by low birth and death rates.

Almost all west European countries have progressed through the four stages of the demographic transition, and it was among these countries that this regularity was first observed and the concept developed. However, the various countries did not undergo this type of demographic development simultaneously and, in general terms, the 'demographic revolution' began in north-western Europe in the late eighteenth century and gradually diffused to central, eastern, and finally southern Europe.

Natural population change

The pattern of peopling is the result of the interaction of the twin processes of natural change and migration (Salt and Clout 1976). Migration will be discussed in Section 2.2, but the components of natural change, birth and death rates, will now be considered.

Whilst all countries in western Europe have experienced a cut in their death rate, to less than 12 per 1000 inhabitants, certain countries and notably those in the European periphery are still associated with relatively high birth rates. In the 1920s, a clear distinction existed between Italy, Portugal, Spain, and the Netherlands on the one hand (over 30 births per 1000 inhabitants), and the industrial nations such as France, Germany, Sweden, and the United Kingdom on the other (less than 20 births per 1000 inhabitants). By the early 1970s, this distinction was less clear with virtually all countries having entered the fourth stage of their demographic development. If one compares figures for 1961 and 1975 (Table 2.2), a continual decline in birth rates was evident for most countries, although in the Netherlands and West Germany the rate of decline was more rapid, at over 8 per 1000 inhabitants. The one exception was Ireland, where the rate remained stationary

Table 2.2 Natural population change in western Europe

	Deaths/1000			Births/1000			Natural population change/1000			Infant mortality/ 1000 live births		
	1961	1975	1981	1961	1975	1981	1961	1975	1981	1961	1975	1981
West Germany	11.0	12.1	11.7	18.3	9.7	10.1	7.3	−2.4	−1.6	31.7	19.7	11.6
France	10.8	10.5	10.3	18.2	13.6	14.9	7.4	3.1	4.6	25.6	13.6	9.7
Italy	9.4	9.9	9.6	18.6	14.8	11.0	9.2	4.9	1.4	40.7	20.9	14.3
Netherlands	7.6	8.3	8.1	21.3	13.0	12.5	13.7	4.7	4.4	17.0	10.6	8.3
Belgium	11.6	12.7	11.2	17.3	12.2	12.7	5.7	−0.5	1.5	28.1	14.6	11.7
Luxembourg	11.4	12.2	11.2	16.1	11.2	12.1	4.7	−1.0	0.9	26.2	14.8	13.8
United Kingdom	12.0	11.8	11.8	17.9	12.4	13.0	5.9	0.6	1.2	22.1	16.0	11.1
Ireland	12.3	10.7	9.4	21.2	21.6	21.0	8.9	10.9	11.6	30.5	18.4	10.6
Denmark	9.4	10.0	11.0	16.6	14.2	10.4	7.2	4.2	−0.6	21.8	10.4	7.9
Greece	7.6	8.9	8.9	17.9	15.6	14.4	10.3	6.7	5.5	39.8	24.1	16.3
Norway	9.2	9.9	10.2	17.3	14.0	12.4	8.1	4.1	2.2	17.9	10.5	7.5
Sweden	9.8	10.8	11.1	13.9	12.6	11.7	4.1	1.8	0.6	15.8	8.3	6.9
Switzerland	9.3	8.8	9.3	18.1	12.4	11.5	8.8	3.6	2.2	21.0	12.5	7.6
Austria	12.1	12.7	12.3	18.6	12.3	12.4	6.5	−0.4	0.1	32.7	20.8	12.7
Finland	9.1	9.4	9.1	18.4	14.2	13.2	9.3	4.8	4.1	20.8	10.2	6.5
Portugal	11.2	11.0	9.8	24.5	19.6	15.4	13.3	8.6	5.6	88.8	37.9	26.0
Spain	8.6	8.2	7.6	21.3	18.3	14.1	12.7	10.1	6.5	46.2	13.8	10.3
Iceland	7.0	6.9	6.8	25.5	20.6	19.9	18.5	13.7	13.1	19.5	11.1	11.3

Source: United Nations Statistical Yearbooks

at around 21.5 per 1000 inhabitants. In 1975, the crude birth rate was over 20 per 1000 inhabitants in just two countries, Ireland (21.5) and Iceland (20.6). The rates were lowest in the European core, with West Germany (9.7), Luxembourg (11.2), and Belgium (12.2) providing good examples. Reasons for such variations are individual and complex, but two general observations can be forwarded: first, the peripheral countries of western Europe are less industrialized and urbanized than those in the economic core, and birth rates tend to be higher in less developed areas; and secondly, higher birth rates are associated with lower standards of living, and living standards have been shown to vary substantially between European countries (Fig. 1.3).

Birth rates continued to decline in a majority of countries between 1975 and 1981, notably in Italy, Spain, and Portugal. However, in five 'core' countries—West Germany, France, Luxembourg, Belgium, and the United Kingdom—a reversal of the downward trend occurred and birth rates began to rise once again, although in the case of West Germany they are still the lowest in western Europe. According to Westoff (1983), this reversal is unlikely to herald the start of a rise in fertility similar to that of the early 1960s.

Post-war fertility trends in western Europe have been succinctly summarized by Westoff (1974 and 1983), who classified member countries into four groups:

1. The British Isles, France, Belgium, and the Netherlands. Following a temporary rise in fertility after the war, fertility rates declined up to the 1950s and then showed signs of increase, notably in England and Wales. This trend soon changed and the 1960–75 period was characterized by falling fertility. Ireland has experienced a unique pattern and fertility actually increased between 1955 and 1965. Since this date it has shown signs of declining, although the total fertility rate remains high at just below four births per woman.

2. Italy, Spain, Portugal, and Greece. Fertility patterns have followed a similar course to those countries in group one, although at a higher rate. In Portugal, the fertility rate has remained unusually stable, and it has also been fairly constant in Spain and Greece.

3. Austria, Switzerland, and West Germany. Since 1955, these countries have followed a similar pattern to those further west, although the decline in fertility has been quite rapid since the introduction of the 'pill' in the mid-1960s. The total fertility rate in West Germany is 1.5 births per woman, unquestionably the lowest in the world, and all three countries are significantly below the population replacement level.

4. Scandinavia. Apart from Sweden, which escaped the war, the countries of Scandinavia experienced a brief post-war boom in fertility and then followed a similar pattern of development to countries in the other three groups. However, Norway was anomalous in showing an increase in fertility for a decade beginning in the early 1950s. Overall, there has been a large decrease in fertility, as shown by a fall in the total fertility rate in Finland from 3.5 births per woman in the late 1940s to 1.8 births per woman in 1973. With the exception of Norway, fertility is below replacement level in Scandinavia.

The general decline in fertility is the result of a complex interrelationship of factors which can be attributed to changing societal attitudes. Amongst the more important factors are better birth control methods, a greater acceptance and availability of family planning and abortion, a decline in traditional rural societies, and a desire for smaller families as a result of the changing status of women and the wish to acquire extra material possessions. Westoff (1974), although agreeing that these factors play a role, felt that such casual explanations were inadequate. He considered declining fertility rates to be part of a longer historical process, which had been temporarily interrupted by the post-war baby boom.

Mortality rates are similar and comparatively low in all parts of western Europe, having declined with progression through the various stages of the demographic transition. This pattern of decline occurred as a result of earlier improvements in health conditions in the more urbanized countries (producing a marked ageing of population) and better medical care in the less developed parts. It is noticeable that some of the lowest death rates occur in the less industrialized countries such as Spain (7.6 per 1000 inhabitants), Iceland (6.8), and Greece (8.9). The relatively high rates of over 11 per 1000 inhabitants in the economically developed areas can be attributed to the higher proportion of old people. Between 1961 and 1975 the death rate actually increased in a majority of west European countries, including the peripheral ones, suggesting that the ageing of population is no longer a unique characteristic feature of north-west Europe. However, by 1981 this trend had reversed and, with the exceptions of Switzerland and the three Scandinavian countries of Norway, Sweden, and Denmark, death rates were lower than in 1975.

The general decline in mortality over the last 40 years is particularly attributable to large decreases in infant mortality (Table 2.2). These decreases are the result of better medical care and hygiene and rates of over 100 deaths in the first year of life per 1000 live births, which were common in many countries in the 1930s, had been reduced to rates of between 15 and 26 per 1000 in the early 1980s. The highest rates are found in Portugal (26.0 per 1000 live births), Greece (16.3), Italy (14.3), and Luxembourg (13.8), and these figures compare with very low rates in Finland (6.5), Sweden (6.9), Norway (7.5), Switzerland (7.6), and Denmark (7.9).

Infant mortality is one of the best indicators of variations in social well-being, as it reflects the quality of health care, public health services, housing, and income (Clout *et al.* 1985). The regional incidence of infant mortality in 1980 has been mapped by Knox (1984), who identified a 'north–south dimension, a core–periphery dimension, and something of an urban–rural dimension' (p. 20). A major contrast existed between the best region (Värmland, Sweden: 3.9) and the worst region (Trás-os-Montes/Alto Douro, Portugal: 43.1). As Fig. 2.3 clearly demonstrates, Scandinavia, Switzerland, north-west and south-east France, and most of the Netherlands emerge as having the most favourable rates of infant mortality. This compares with high infant mortality in many peripheral regions of peripheral countries, notably in Greece, southern Italy, and Iberia.

The overall rates of natural changes for 1961, 1975, and 1981 are listed in Table 2.2. In all cases except Ireland, there was a decline in the rates of natural increase between 1961 and 1975 and at the end of this period, deaths actually exceeded births in West Germany, Luxembourg, Belgium, and Austria. At a more localized level, high rates of

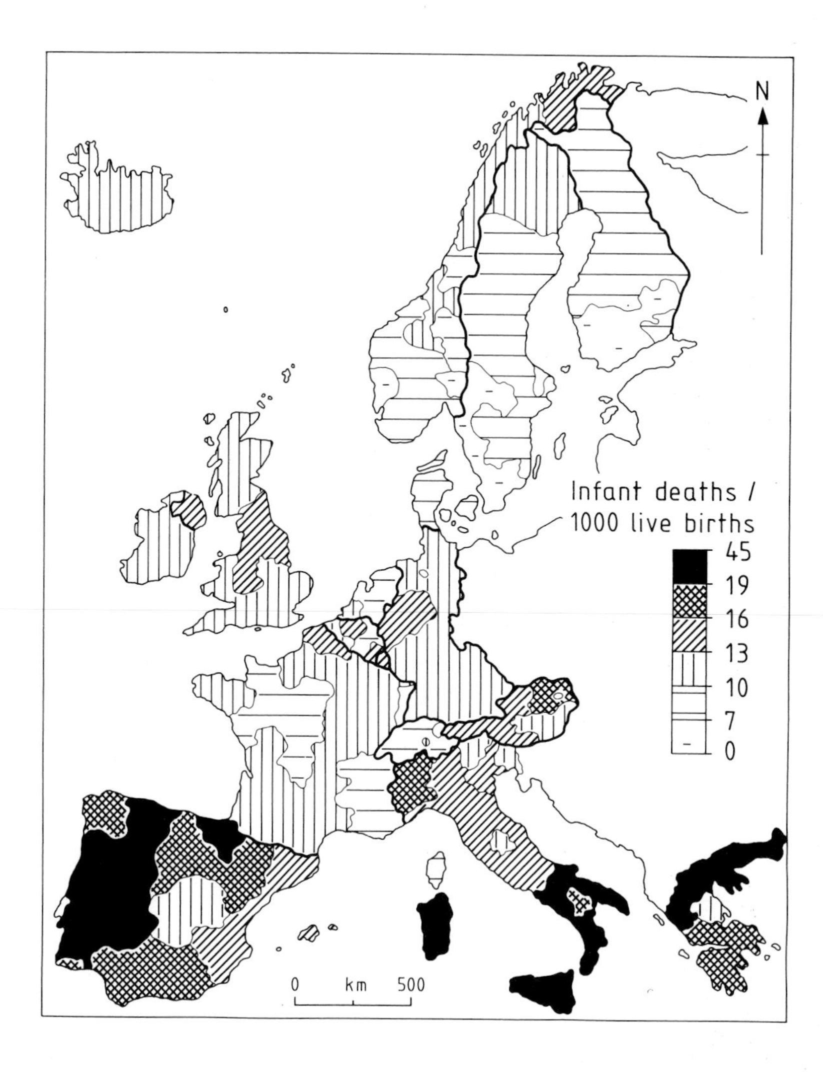

Fig. 2.3 Infant mortality in western Europe, 1980 (*source:* Knox 1984)

increase could be identified in the major industrial areas which attracted young people of employable and marriageable age from the provincial areas. Conversely, deaths exceeded births in two types of area: first, in severely depopulated rural areas like the Massif Central, the Pyrenees, and upland Wales; and secondly, in retirement areas such as parts of south-east England and the Alps-Maritimes (Salt and Clout 1976).

By 1981 deaths exceeded births only in West Germany and Denmark and there was an increase in the rates of natural population growth in seven countries. One of these was Ireland, the only country to experience a continual rise throughout the 1960s and 1970s; the remaining six were countries that had witnessed an upturn in birth rates during the late 1970s. The spatial pattern of natural change in 1983 is portrayed in Fig 2.4 and comparatively high rates are found in a peripheral zone which includes Iceland (12.2 per 1000 inhabitants), Ireland (9.7), Spain (6.0), and Portugal (5.1). In all cases except Finland there was a decline in the rates of natural change between 1981 and 1983, demonstrating the fluctuating nature of fertility and mortality trends in post-industrial western Europe.

Age and sex structure

In western Europe, there is a strong interrelationship between natural increase and the age–sex structure of the population. This relationship works both ways: mortality and fertility depend on the structure of the population, and the structure is conditioned by natural growth and its

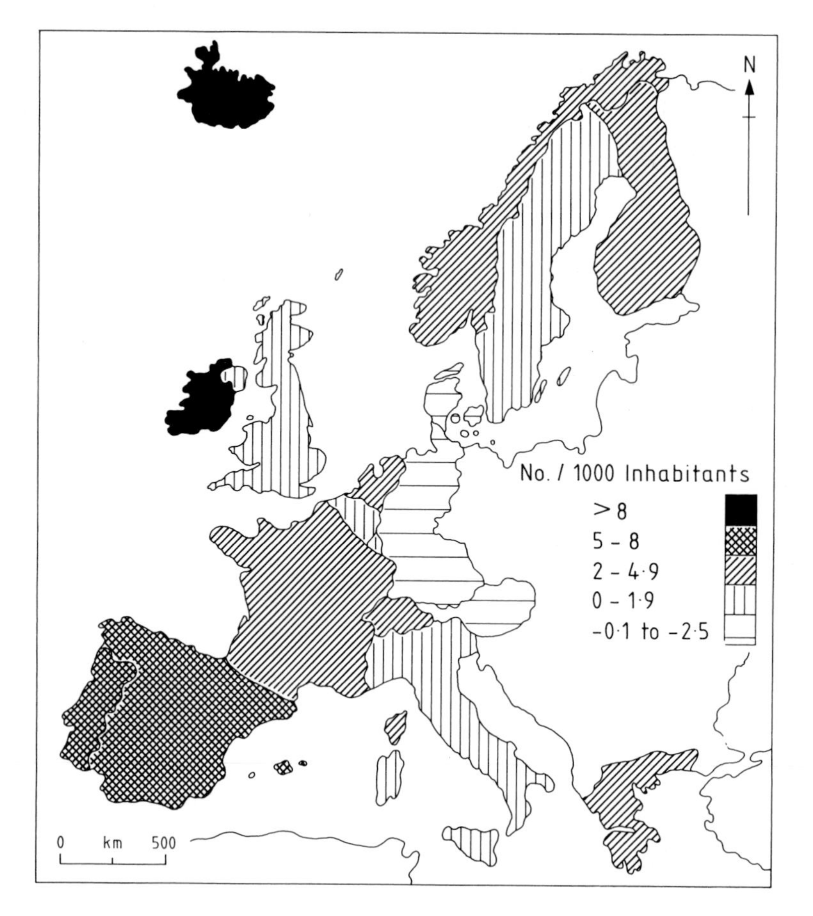

Fig. 2.4 Natural population change in western Europe, 1983

components. With plummeting fertility rates and an increase in average life expectation, western Europe is characterized by an ageing population and slow growth; in 1980 12.6 per cent of the population was over 65. The process of ageing is fairly well advanced and the proportion of young people is relatively lower and the proportion of old people much higher than in many other parts of the world. Consequently, the percentage of total population classed as productive has declined, causing fertility rates to fall near or below replacement levels. This in turn means that western Europe is rapidly approaching a situation of steady state or zero population growth, where birth rates equal death rates.

A potentially serious by-product of a stationary population with a high expectation of life is the nature of the age distribution which this implies (Boulding 1973). In traditional societies, a great deal of premature death results in a triangular age distribution, with large numbers of young people, smaller numbers of middle-aged, and very few old people. With the elimination of premature mortality in western Europe—up to say 70 years—the age–sex pyramids for individual countries have become sugarloaf-shaped (Mellor and Smith 1979) with almost equal numbers in each age group up to the age of about 70. This is characteristic of zero population growth and is exemplified in the age–sex pyramids for selected countries, shown in Fig. 2.5. With the exception of Portugal, where in 1980 35.2 per cent of the population was below 20 years of age (due to one of the highest rates of natural population growth in western Europe), the pyramids are characterized by ageing at the apex and a narrow base, indicative of a declining number of young people. The proportion of aged people, dominated by females, has continued to increase dramatically, a trend especially evident in Sweden, a country that has experienced very low rates of growth for a long period of time. More than 20 per cent of

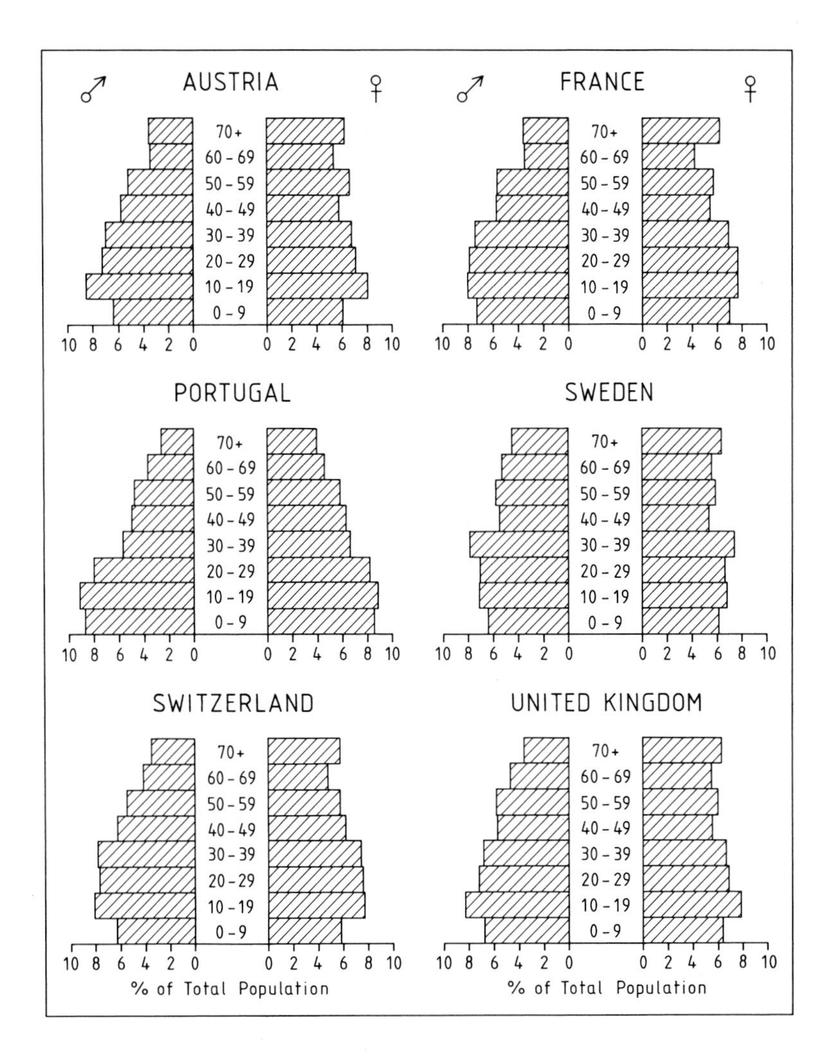

Fig. 2.5 Age–sex population pyramids for selected countries, 1980

Swedish people in 1980 were over 60 years of age with 5.6 per cent of these older than 75, the highest percentage in western Europe.

Although unlikely to continue to increase in the 1980s, ageing of the population has far-reaching social consequences. The proportion of old people is likely to reduce slightly by 2000, but absolute numbers will increase, leading to such problems as loneliness, changing consumption patterns in terms of food, housing, schools, hospitals, and pensions, an increasing death rate, and postponement or raising of retirement age. Older people accumulate knowledge and experience which, as a result of technological change and advancement, become less relevant. Inevitably, there is a greater reluctance to accept new ideas and adopt innovations, and the ageing of managerial groups, for example, can affect the expansion impulse of firms. Catering for the diverse needs of the elderly represents one of the most urgent problems in western Europe, especially as by 1990 half of those over 65 will in fact be older than 80 (Clout *et al.* 1985).

Population potential

As an index of relative accessibility, the population potential of an area is a measure of the nearness of people to that area or point. The technique is based on two variables—population and distance—and assumes that accessibility is inversely proportional to distance. The result is a generalized visual picture of a distribution in an area.

The measure was applied to each European

country and for ease of computation the population of each country was considered to be centred on its capital. In turn, the population of each country was divided by the distance of that country from every other country (i.e. the distance between the capitals) as well as from itself (the within-country distance). The population potential for each country is the sum of its relationship, derived as described above, to all countries (Cole and King 1968). A 'spot height' is therefore derived for each European capital and after being plotted on a map, isopleths or lines of equal potential (equipotential lines) can be interpolated. It is important to remember that the actual potential values need to be considered in a relative and not an absolute sense.

The results are shown in Fig. 2.6 and one can easily identify the areas of high and low potential. The map suggests a definite core–periphery relationship, similar to that outlined in Chapter 1. Circles of decreasing accessibility can be seen to radiate out from a distinct core area which includes the whole of Belgium, Luxembourg, West Germany and the Netherlands, a large proportion of France, the south-east tip of England, and small parts of Austria and Switzerland. Areas outside the 500 equipotential line are essentially the peripheral regions of western Europe and include Scandinavia, Greece, Portugal, Ireland, and the southern parts of Spain and Italy. Scandinavia is something of an anomaly, in so far as the population potential technique has placed most

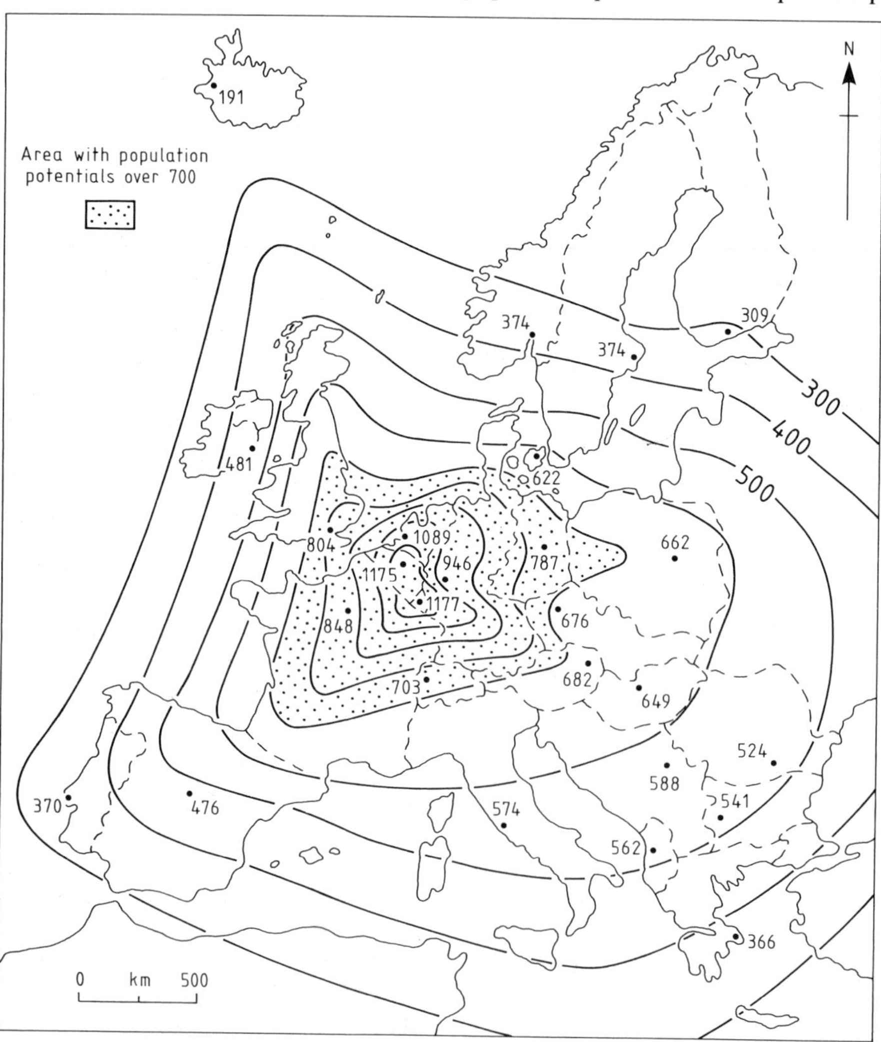

Fig. 2.6 Population potential in Europe

of Sweden, Norway, and Finland into the European periphery. This conflicts with the generally high standard of living in northern Europe, but as population potential is a measure of accessibility it highlights the low population densities and the relative isolation of these parts from the main core area of western Europe.

2.2 Migration in western Europe

The population of Europe has experienced a high degree of spatial mobility, both before and since 1945, and migration has become one of the key elements in the explanation of regional variations in population change. As most west European countries progressed through the demographic transition, the redistribution of population through migration replaced population growth as the main demographic feature. Indeed, patterns of migration are often a direct reflection of economic, social, and political conditions and affect such processes as urbanization, regional economic growth, and rural depopulation.

The direction and composition of European migration streams have varied considerably with time, and Salt and Clout (1976) recognized 'stages' in migratory evolution. These stages reflected the levels of economic and social development that had been reached in various parts of Europe. The dominant characteristic of most movements up to the mid-1970s was an outflow from regions of decline or stagnation to regions of relative prosperity and opportunity. Migration has not only been induced by economic factors, and both political situations and the desire to improve one's self socially must also be considered.

Migration patterns in western Europe can be recognized on two main levels; inter-continental and inter-European. Movement away from Europe was most important in the nineteenth and early twentieth centuries, but in more recent years moves between European countries and between regions within individual countries have become dominant. Studies of migration have suffered in the past from poor data sources and although greatly improved, it needs to be stated even today that sources vary in availability, reliability, and quality, and quite often relate to different time periods. Although it is possible to analyse general trends and patterns, this point should be remembered in the subsequent analysis.

Inter-continental migrations

Europe was for a long time characterized by emigration, although since the 1960s there has been a trend towards the immigration of people from former colonies like Algeria, India, Pakistan, and the West Indies into western Europe.

There have been considerable regional variations over time in the countries supplying emigrants and the centre of gravity has shifted since the nineteenth century from north-western Europe, to central and eastern Europe, and since 1945 to southern Europe. In more detail, five broad time-stages can be identified:

1. The mid-nineteenth century, when Britain accounted for 60 per cent of the total European emigrants.
2. The late nineteenth century, when peaks in emigration rates were reached in central and eastern Europe, especially in Austro-Hungary and Poland.
3. The early twentieth century (1911–20), when the countries of southern Europe accounted for 44 per cent of European emigrants. This area remained the most important source of emigrants during the inter-war years.
4. The early post-war period, when Italy experienced the highest rates of net emigration. This was partially a reflection of the country's low war losses, leaving a large number of young potential migrants, but also the poor living conditions in the Mezzogiorno area of southern Italy.
5. The 1950s to early 1960s period, when Spain, Portugal, Greece, and Malta experienced high rates of emigration, with Latin America for Spaniards and Portuguese, and Australia for Greeks and Maltese, being the primary destination areas. Movement from these areas, and Italy, to other continents had been reduced by the 1960s and the trend was for these people to migrate to other parts of western Europe.

Borrie (1965) estimated that between 1820 and 1840, 55–60 million people left Europe, with the American continent being the main destination area, receiving approximately 70 per cent of the total. Kosinski (1970) has given a further and more detailed breakdown of people moving to America from Europe up to 1960. Between 1820 and 1960, 40 million European emigrants entered the USA; a further 9.6 million and 7.5 million entered Canada and Argentina respectively between 1850 and 1960. Since 1945, emigration to former colonies has increased, at the expense of America, and 88 per cent of British post-war emigration, for example, has been to Australia (1.4 million). Other destinations for Europeans include New Zealand, South Africa, and Zimbabwe.

These marked regional responses in the source and destination areas of emigrants can be attributed to various 'push' and 'pull' factors. A historically important push factor was rapid population growth, which accompanied decreases in death rates as the countries of Europe entered the second phase of their demographic transition. This was felt at different times in different countries and the centre of gravity of emigration shifted accordingly. Pressure of population was first experienced in regions with an agriculturally-based economy, especially in those areas where farms were small and fragmented. The overcrowded countryside, often characterized by poverty and famine, produced large numbers of emigrants.

Industrialization and urbanization may have helped to check the outflow of people from western Europe, by accommodating the excess rural population. However, as there is a significant correlation between urbanization and industrialization on the one hand and emigration patterns on the other, it is highly likely that these processes actively encouraged emigration. Both occurred originally in the north-western parts of Europe, before spreading first, to the central parts and secondly, to the southern parts of western Europe.

Other push factors included the impact of war, notably World War II and special government influences. The peripheral regions of western Europe were slow to recover from the Second World War and a lack of opportunities led to migration overseas. Government influences were especially important in the Netherlands, where emigration was actively encouraged because it was felt that neither industrialization nor land reclamation could absorb the growing population. Special agencies were introduced and in 1952 an emigration quota of 60 000 per year was set. Between 1946 and 1963, approximately 500 000 people were assisted in emigration. However, the effect of these measures was offset by the collapse of the Dutch colonial empire and between 1952 and 1963 the Netherlands received 950 000 immigrants.

'Pull' factors tended to reflect economic opportunity. This was represented first, by an abundance of either free or very cheap land and later, by the demand for industrial employment. In reality, many of the later migrants found that these pull factors were no longer operative and as a result were forced to take jobs of a lower status. Developments in international transport and communications, enabling a more rapid spread of information, also helped to increase the mobility of people. Improvements in transport conditions, notably reduced costs, meant that long-distance migration became possible for a majority of west European people.

During the main post-war period of economic growth in western Europe (1950–70), large numbers of foreign workers were attracted to some of the more prosperous west Europen countries. The demand for labour could not be met by the indigenous population and immigrant workers, many from north Africa and former colonies of European countries, became vital, both in terms of their numbers and the jobs they performed. As Drewer stated (1974, p. 2), 'immigrant workers have been important in the sustained economic growth of western Europe'.

There has been much literature on international labour migration in western Europe and estimates of the numbers of foreign workers attracted have varied. Castles and Cosack (1973) thought there were nearly 11 million immigrants in western Europe in 1970, and Hume (1973) estimated the total in north-west Europe on the eve of the economic recession in 1973 to be 8 million (plus one million dependants). A more recent survey by Kane (1978) suggested that 12 million foreign workers and their families were living in western Europe in 1975. Whichever figure is accepted, it is clear that the United Nation's forecast of 22 million foreign workers by 1980 was not realized. Over 90 per cent of the immigrants live in just four countries—France, West Germany, the United Kingdom, and Switzerland—and in 1973, foreign workers represented 30 per cent of the labour force in both Switzerland and Luxembourg.

The flow of foreign workers was aided by the system of bilateral and multilateral labour agreements between industrial countries and those with labour surpluses. These agreements were developed in the 1950s and extended in the 1960s and although many were with countries in western Europe itself, some extended to overseas territories. France, for example, has sixteen agreements, eight of which are with black African states. However, the early 1970s witnessed a slowing down in the recruitment of foreign workers; this was associated with the problems of assimilation and poor living conditions experienced by the immigrants, as well as with political and economic reasons. The United Kingdom felt obliged to curtail the number of immigrants, mainly because the majority wanted to settle permanently, and this was achieved under the 1962 and 1968 Commonwealth Acts and the 1971 Immigration Act. West Germany imposed a ban on recruitment of foreign workers in 1973 and this was followed by similar bans in other west European countries, such as those in 1974 in Belgium, France, and Switzerland.

Inter-European migrations

The post-war decline in emigration overseas was reflected in the increase in economic labour migrations within western Europe and once again a definite movement, reflecting differential regional prosperity, can be traced. In the nineteenth century, migrations within Europe were limited, being either seasonal or temporary in nature, or balanced by counter-movements. By the inter-war period, large scale movements had occurred and Kirk (1946) suggested that these flows were from countries of lower levels of living and rural over-population, to countries of slower population growth and greater economic opportunities. The general direction of flow was westwards and France, for example, became a major receiving country. This particular country actively encouraged immigration as its unbalanced population structure meant a shortage of workers.

During the depression of the 1930s, there was a rapid decline in inter-European movements, suggesting that past migrations had been economically motivated. After 1945, the numbers again increased and Sweden and Switzerland, with less war damage than other west European countries, were the first to attract migrant labourers. Between 1946 and 1950, Sweden had an average of 10 000 immigrants per year, mainly from Finland, and by 1973 the country had a total of 250 000 foreign workers.

The recruitment of foreign workers became an important structural feature of the west European space economy between 1950 and 1970. Throughout this period there appeared to be a general flow of unskilled labour from the peripheral south to the core of north-western Europe (Fig. 2.7). For example, France attracted many immigrants from Spain, Portugal, and Algeria, whilst West Germany's main source was south-east Europe, in particular Yugoslavia, Turkey, and Greece (Clout *et al.* 1985). Italy sent large numbers to France, West Germany, and Switzerland. A smaller counter-flow of skilled labour from north to south also took place, owing to the inadequate educational and training structures in the south.

Labour

Post-war labour migration can be categorized into three main phases (Clout *et al.* 1985):

1. 1945 to 1961, when France, the United Kingdom, Belgium, Sweden, and Switzerland were the main destination countries, and Italy, Ireland, and Finland the major suppliers.

2. 1961 to 1973, when West Germany became the major host country for south European migrants. Italian migration to West Germany and France slowed as the home demand for labour increased and this 'stream' was replaced by Greek, Yugoslav, and Turkish migration to West Germany, and Spanish, Portuguese, and North African moves to France.

3. 1973 onwards, when flows to all major destinations were significantly reduced and return migration became an important phenomenon.

Labour migrants tended to come from over-populated, underemployed, and underdeveloped southern regions and Böhning (1972) forwarded the notion of a growing maturity of migration flows over time. Four stages in particular were identified:

1. Migrants are short-stay workers, young and single males. They originate from the more urbanized and industrialized parts of the source countries.

2. Migrant stream ages and married males become involved (leaving their wives at home). The length of stay increases and the area of origin

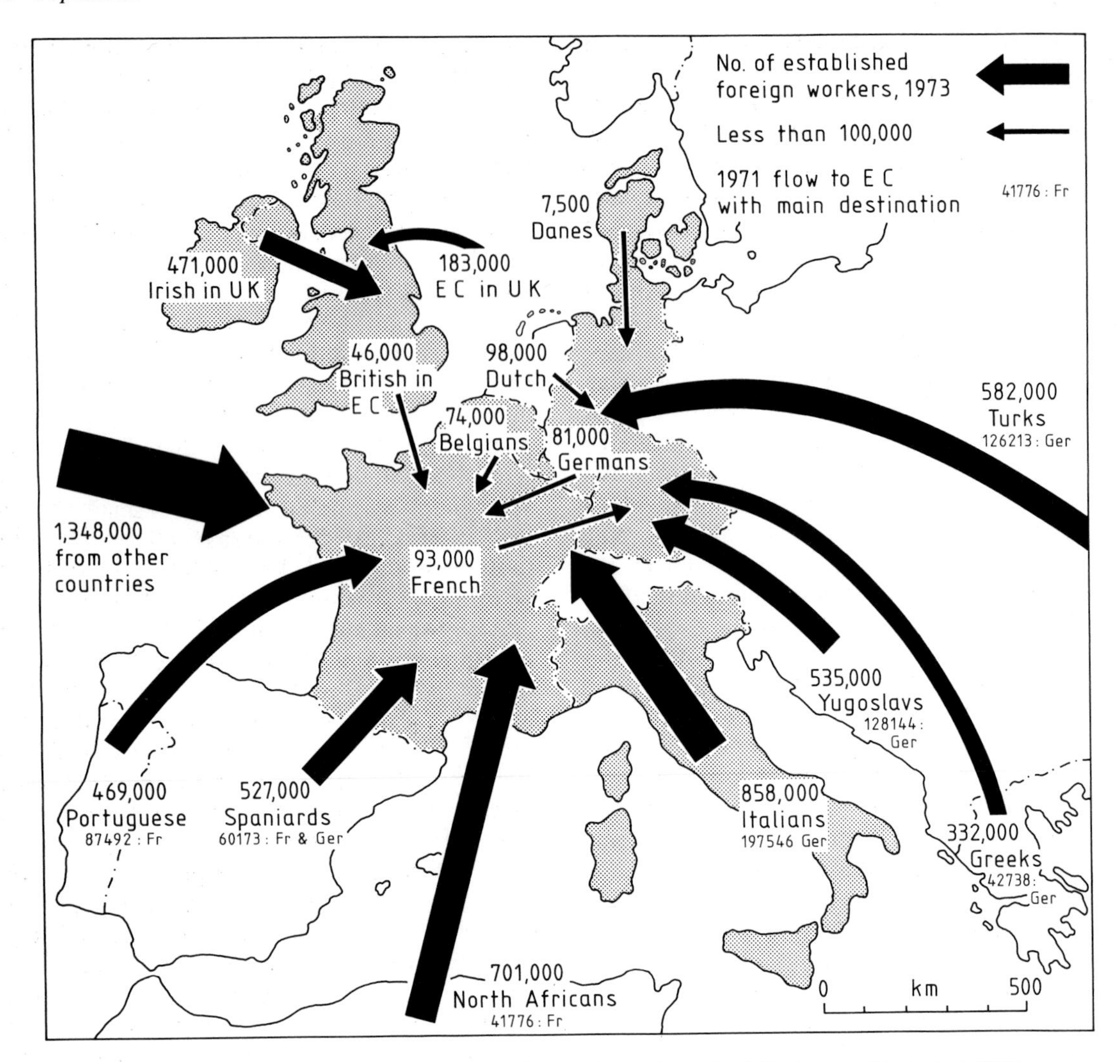

Fig. 2.7 Labour migration in the European Community in the early 1970s (*source:* Clout *et al.* 1985)

extends to small towns and surrounding rural areas.

3. Migration stream continues to age and wives and children are sent for. The length of stay increases further, as does the area of origin. Untrained rural workers begin to dominate the pattern of flow.

4. Migration process becomes self-feeding and creates its own demand for infrastructure and consumer goods. Return migration is limited.

Many migrant groups had reached stage four by the early 1960s and Clout *et al.* (1985) believe that Böhning's model helps to explain the regional differences in the intensity of out-migration within individual supply countries. The migrants tended to concentrate in the major cities and industrial districts of the host countries. For example, nearly half of the migrants in France lived in the Paris city region, with most of the rest in the industrial areas of Lyon, Marseille, and the north. This has led researchers to study the distribution and segregation of foreign workers within individual cities; Kinsey (1979) examined the distribution of Algerians in Marseille, whilst O'-Loughlin (1980 and 1984) provided details for German cities.

In comparison with the international migrations which characterized western Europe during the major post-war phase of economic growth, the movement of labour between member states of the EC has been of little significance. This could be a reflection of EC policy towards free movement between states. The establishment of common labour markets represents an attempt to allow unrestricted movement of member nationals and the Nordic Common Labour Market was the first to be formed, in 1954. This allowed workers from Norway, Sweden, Denmark, and Finland to travel freely within the market in search of employment; in practice Sweden became the main host and Finland the main supplier. Although the basis of an EC Common Labour Market was laid down in articles 48 and 49 of the Treaty of Rome, freedom of movement came in stages and it was not until 1968 that all restrictions were removed. Werner (1974) and King (1976) analysed the possible effects of these regulations and the main flows have been from Italy to France and West Germany. In reality, these effects have been negligible, for two reasons: first, migration between member countries actually declined over the period of free movement; and secondly, there were insufficient workers in the member countries to satisfy the labour demand and this led to recruitment from outside the Common Labour Market. For example, West Germany signed bilateral agreements with Spain, Greece, and Yugoslavia. However, one needs to emphasize the complexity of international labour migrations, as Italian emigration to Switzerland continued during the 1960s and more Italians were working in Switzerland than in the rest of the EC (except Italy). As King (1982, p. 227) has shown, labour migration is 'one of the most powerful links binding the economic core of Europe to an ever-widening periphery'.

The oil crisis of 1973 marked an important turning point in the pattern of European migratory movements and the 1970s became important for the process of return migration (Böhning 1979; King 1979; Clout *et al.* 1985). All sending countries, except Finland, experienced a decline in out-migration during the mid-1970s and the curtailment in recruitment of foreign workers was most dramatic in those countries that had previously attracted the largest migration streams. For example, between 1974 and 1978 the foreign workforce declined by 20.5 per cent in Austria, 19.8 in West Germany, 17.5 in Switzerland, and 9.2 in France (Clout *et al.* 1985). Böhning (1979) estimated that 1.5 million migrants returned to their Mediterranean homelands between 1974 and 1978, and King (1979) showed that an average of 22 000 labour migrants returned to Greece each year between 1969 and 1973, with approximately half of those coming from West Germany. Although return migration increased in the mid-1970s, it did not take place on a massive scale. Indeed, by the late 1970s labour flows into countries such as France and West Germany had increased again. However the volatile nature of labour migration is demonstrated by the fact that high unemployment in West Germany in the early 1980s led to further legislation to keep out foreign workers. The general trend of public opinion polls in that country, from 30 per cent in 1978 to 90 per cent in 1982, is to encourage guestworkers to return home (Clout *et al.* 1985). Therefore, it would appear that return migration will continue to be important but follow a downward trend.

King (1979) classified the factors responsible for return migration into the familiar push and pull categories. Amongst the push factors, King listed inadequate housing, the difficulty of occupational improvement, the strangeness of physical and social environments, and especially expulsion from the host country, due to the expiry of contracts or laws restricting the length of stay. For example, France introduced legislation in 1974 to restrict foreign workers; this was extended in 1975 and two years later the government offered a special financial bonus and one-way air ticket for returnees. Similar restrictions were imposed in West Germany, including legislation for the *Länder* in 1981 and financial incentives for families to return home in 1982. The most important pull factors were a desire to rejoin the family, a plan to invest accumulated savings in a house or business, anticipation of retirement in the place of birth, and the prospect of satisfactory employment back home.

Return migration has been commented upon in some detail, for as Böhning (1979, p. 404) stated 'the period since the oil crisis reveals starkly the inequity of the migration system, both at the level of the individual, where the recent migrants and the unskilled suffered most from unemployment and profited least from exemptions to the general recruitment freeze, and at the national level, where the rich and powerful countries shifted a burden onto the poor and weak countries which they should rightfully have borne themselves'.

Not all inter-European migrations have been economically motivated and politically induced movements, although small in comparison, have been taking place for a considerable time. However, it is sometimes difficult to separate these from economically motivated moves; refugees from East to West Germany, for example, were motivated to move by the greater prosperity in the latter as well as by political reasons. A succession of wars has led to continuous adjustments of population and Kosinski (1970) estimated that 7 700 000 people moved within Europe as a result of World War I alone. Changing political policies were responsible for much movement in the inter-war years too. More than 60 000 people left Italy after the establishment of the Fascist regime in 1924; approximately 300 000 people moved from Spain and France as a result of the Spanish Civil War; and there were large outflows of Jews from Poland, Germany, and Spain because of hardening political policies (Salt and Clout 1976). The Second World War led to further moves of over 25 million people and post-war Europe has witnessed additional inter-European migrations, following the redrawing of political boundaries.

Internal migration in western Europe has been examined by Fielding (1975) and Mayer (1975). A vast amount of inter-regional movement has taken place and in terms of numbers far exceeds inter-continental migration. However, it is a complex phenomenon and studies suffer from an absence of suitable comparative data. Inter-regional moves reflect regional shifts in the demand for labour and spatial differences in the levels of earnings and job opportunities (Salt and Clout 1976).

Fielding (1975) produced a map of internal migration rates for the 1960s, on a common scale (Fig. 2.8). On the basis of this, the countries of western Europe were divided into three types:

1. Less developed, mostly peripheral countries, characterized by a wide range of positive and negative migration rates. Areas of severe depopulation are dominant, but partially compensated by modes of rapid urbanization based on industry and tourism. The best examples are Spain and Portugal, but Ireland, Austria, and parts of France come within this category.
2. Countries where agglomeration has co-existed with localized decentralization of industry and population; for example, Sweden, Norway, Denmark, Switzerland, and on a smaller scale France.
3. Countries with regional industrial decline and metropolitan decentralization. The United Kingdom, West Germany, Belgium, and the Netherlands are good examples.

The actual pattern depicted in Fig. 2.8 has been described by Clout *et al.* (1985) who, as well as highlighting the major area of migration balance and gain in the industrial belt of western Europe, identified the areas with the highest net gains and losses. Amongst the former were favourable retirement areas (e.g. south-west England, Côte-d'Azur), expanding capital cities (e.g. Copenhagen, Oslo, Madrid) and newer industrial areas (e.g. western Switzerland, Barcelona, and south Paris Basin). The latter included remote upland areas of forestry and poor-quality farming (e.g. interior Spain, northern Scandinavia, and the Scottish highlands and islands) and to a lesser extent the older industrial areas of the central axis (e.g. Lorraine, French Nord, and Belgian Limburg). However, the same authors also emphasized the fact the Fielding's map was based on net migration and not on absolute numbers and flows; consequently the importance of areas like south-east England, which was characterized by high rates of gross mobility in the 1960s, has been understated. Western Europe in the 1980s is still characterized by high rates of inter-regional migration, but it is dominated by inter-urban movements.

Various types of inter-regional migration can be recognized and these are exemplified by the number of moves which have taken place in Great Britain. First, there was movement to the coalfields in the early and mid-nineteenth century, when coal was a prime resource. Secondly, a movement away from such areas occurred in the late nineteenth century as a result of over-specialization on just a few commodities. A general regional movement to south-east England resulted, labelled as the 'drift to the south'. A third type of movement was rural–urban migration as the country began to industrialize. This

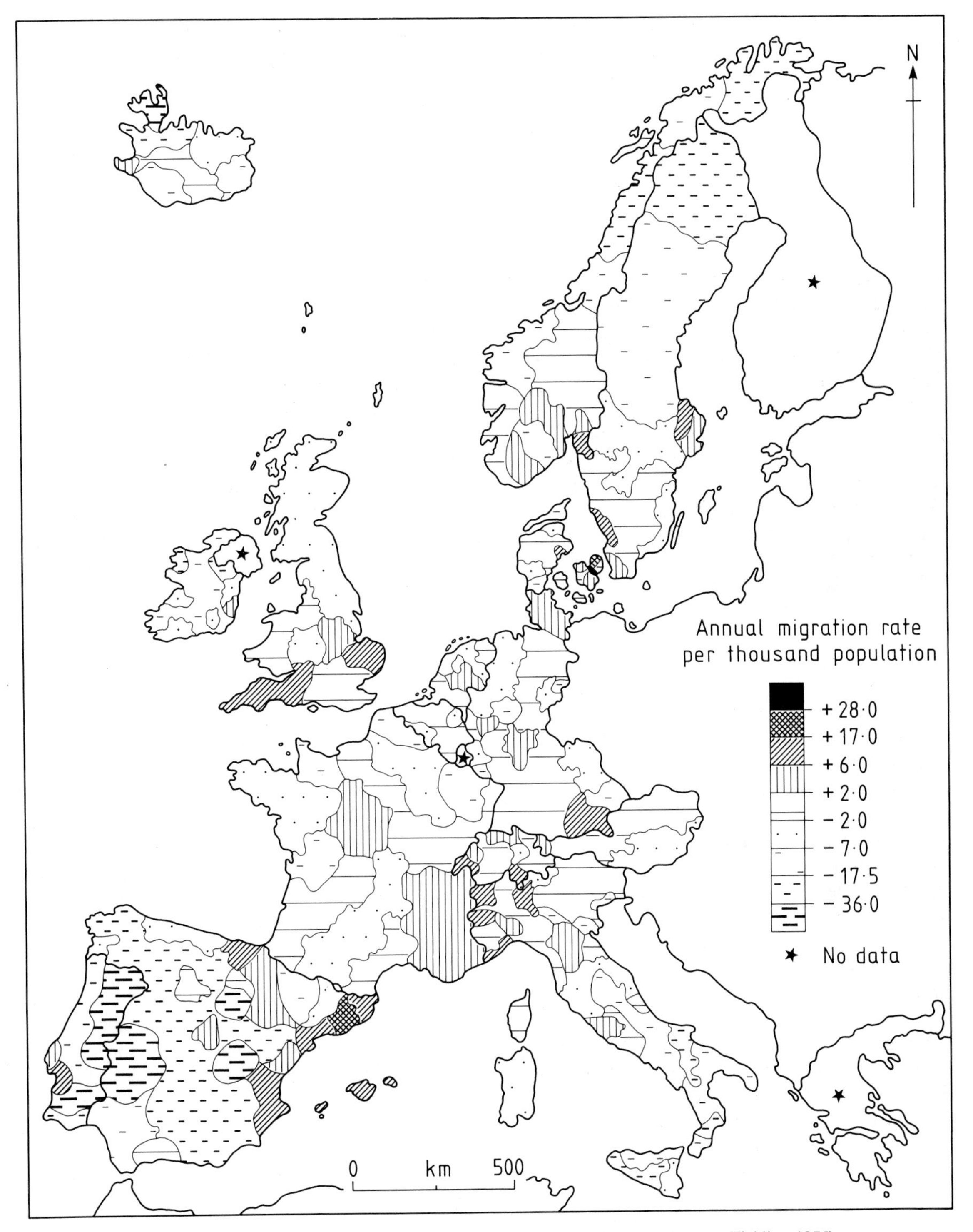

Fig. 2.8 Net internal migration in western Europe during the 1960s (*source:* Fielding 1975)

2.1 City centre redevelopment in Marseille (Photo Shop)

was caused by such 'push' factors as rationalization in agriculture and rural poverty, and the 'pull' of better wages and opportunities in towns. Rural–urban migration is one of the basic forces of urbanization and will be further discussed in Section 2.3. A fourth and more recent movement is the return of people to rural areas to live, reflecting the importance of such factors as increasing affluence, car ownership, and a desire for more space. Places of residence and work have thus been separated and definite 'journey to work' patterns have emerged. Post-industrial Britain is finally characterized by increasing rates of inter-urban migration.

Most inter-regional migrations within western Europe have taken place with little attempt by governments to channel movements along desired lines. However, migration policies have been adopted in most west European countries. These are closely related to regional policies, which attempt to counteract the spatial imbalance in job opportunities (see Chapter 10), but the emphasis has been on moving work to the workers and not on moving the workers themselves.

2.3 Urbanization

Western Europe has for a long time been highly urbanized and one of the chief forces behind this process is pronounced rural–urban migration of the people, representing a movement to relative wealth. Large numbers of farm labourers have been 'pushed' from the land by farm rationalization and improved agricultural technology, and 'pulled' to towns by better wages and opportunities and by the aura of an urban way of life. Much confusion exists over the term 'urbanization', but for present purposes it can be defined as the proportion of the total population living in urban defined areas. Urbanization and urban growth are not synonymous, as the former takes into account rural growth or decline as well as growth in cites.

Assessing the extent of urbanization in western Europe is not easy, as national definitions of what actually constitutes an urban settlement vary considerably, making comparisons very difficult. The various definitions of an 'urban population' in western Europe have been summarized by

Kosinski (1970) and Clarke (1972). Three basic approaches can be recognized:

1. The classification of administrative divisions, in which the populations are classified as either urban or rural according to:

(i) the type of local government (the United Kingdom);

(ii) the number of inhabitants in the minor administrative divisions (Austria, Belgium, West Germany, Spain, Switzerland, and the Netherlands); and

(iii) the size of the principal cluster of the minor administrative division (France, Luxembourg, and Greece).

2. The classification of agglomerations or population clusters, where the urban population is identified as the residents of centres above a given size. The minimum size varies greatly, from 200 in Denmark, Sweden, and Iceland to 20 000 in Portugal and Spain.

3. The differentiation of towns from rural centres by the presence of non-agricultural activities. This criterion is added to those above in certain cases: in the Netherlands, for example, urban areas have a minimum of 5000 people, of which not more than 20 per cent of the economically active males are engaged in agriculture; and in Italy the agricultural proportion should not exceed 50 per cent.

As a result of these varying definitions, the minimum size of urban areas in western Europe ranges from 200 to 20 000 people, although a more realistic range would be from 2000 to 7500. Therefore, in any discussion on the levels of urbanization the comparability of data is questionable.

If one uses the individual countries' definitions of 'urban', there are ten countries with over 70 per cent of their population living in urban settlements. Five of these—West Germany, Sweden, Denmark, Iceland, and the United Kingdom—are highly urbanized, with over four-fifths of their population living in towns and cities (Table 2.3).

The highest levels of urbanization, with the exceptions of Iceland and Sweden, are thus found in the countries that border the North Sea, reaching a peak in the United Kingdom (91 per cent), Denmark (84), and West Germany (85). Coun-

Table 2.3 Urbanization and millionaire cities in western Europe

	Total population 1981 ('000 000)	Projected population 1990 ('000 000)	% population urbanized 1982	Millionaire cities
West Germany	61.68	60.64	85	Hamburg (1.64); Munich (1.29); West Berlin (1.90)
France	53.97	56.14	78	Paris (8.61); Lyon (1.18); Marseille (1.08)
Belgium	9.85	9.89	72	Brussels (1.00)
Italy	57.20	57.26	69	Rome (2.91); Milan (1.67); Turin (1.18); Naples (1.22)
Luxembourg	0.36	0.37	79	
Netherlands	14.25	15.00	76	Rotterdam (1.02)
Denmark	5.12	5.10	84	Copenhagen (1.38)
United Kingdom	56.02	57.03	91	London (6.87); Birmingham (1.04); Glasgow (1.73)
Ireland	3.43	3.77	58	
Norway	4.10	4.27	48	
Sweden	8.32	8.40	83	Stockholm (1.38)
Switzerland	6.43	6.49	61	
Austria	7.56	7.65	54	Vienna (1.59)
Finland	4.81	4.92	60	
Portugal	9.79	10.53	32	Lisbon (1.61); Oporto (1.31)
Spain	37.65	39.69	75	Madrid (3.19); Barcelona (1.75)
Greece	9.73	9.88	64	Athens (2.90)
Iceland	0.23	0.25	88	

Source: United Nations Demographic Yearbook

tries that are least urbanized are situated in the European periphery and include Portugal, Greece, Austria, Ireland, and Norway.

Urbanization and industrialization are closely related in the western world. However, it is important to remember that industrialization does not lead to urbanization unless people actually settle at their place of work. Urbanization in western Europe has thus resulted from rural–urban migration. The problems of studying this process have been outlined by Salt and Clout (1976), but one can safely say that the movement of people from the land has taken place in most west European countries since 1945, having started far earlier in some. In France, for example, the proportion of the workforce engaged in agriculture has steadily decreased, from 45 per cent in 1906 to 27 per cent in 1954, 14 per cent in 1970, and 8.1 per cent in 1983; between 1960 and 1975 alone, 2.8 million people left farming. The figures for West Germany are even higher, with 4.3 million leaving agriculture over the same period. In Italy, the process started later and continued into the 1970s on a large scale. Over three million left the land between 1960 and 1975, especially in the rural south, and most migrated to Rome or the city regions in the north.

In a detailed survey of the west European urban system, Hall and Hay (1980) demonstrated that up to 1975 there was a tendency for population to move from rural to urban areas. The number of people living in metropolitan areas increased from 270.5 million in 1950 (86 per cent of total population) to 316.2 million in 1970 (88 per cent). Between these two years, seventeen 'megalopolitan' growth zones could be identified (Fig. 2.9). None of these was located in the outer peripheral regions and only one (Paris) was focused on a major city; instead the highest rates of growth in north-west Europe were found in medium-sized cities away from the old industrial conurbations. However, in southern Europe the growth zones did involve such major cities as Valencia, Barcelona, Milan, and Rome. Conse-

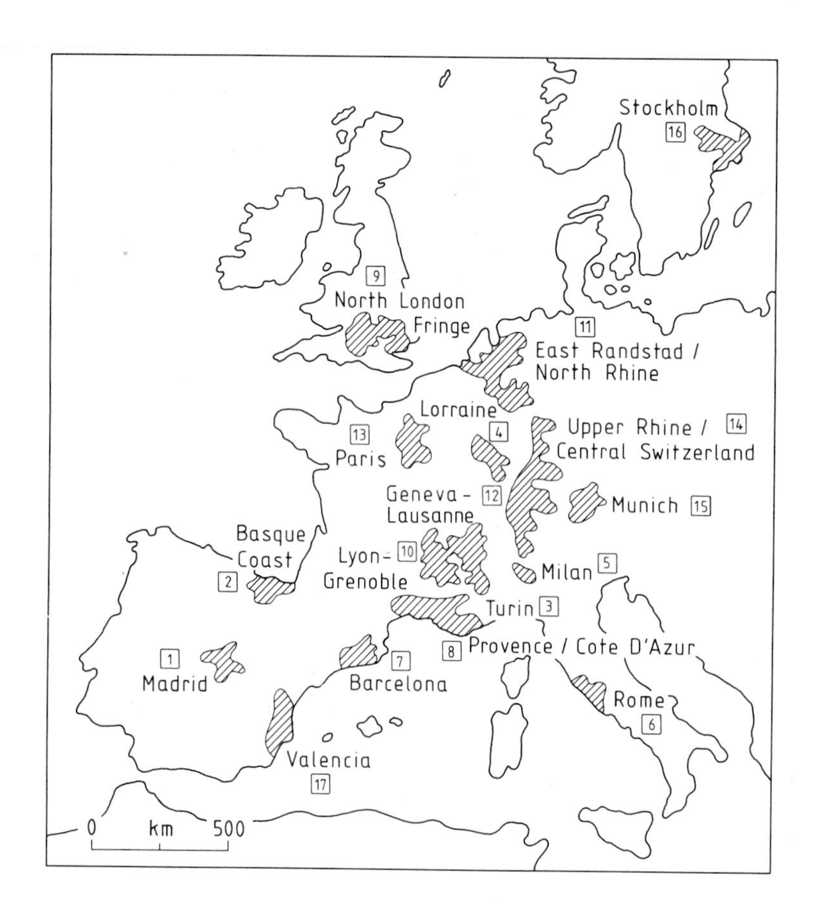

Fig. 2.9 Megalopolitan growth zones in western Europe, 1950–70 (*source:* Hall and Hay 1980)

quently, different parts of western Europe were experiencing the urbanization process in different ways.

It is possible to isolate a number of important features associated with the urbanization process in western Europe:

1. In general terms, the later a country began to urbanize the faster it took place. Urbanization in the peripheral regions of western Europe, and indeed in eastern Europe, advanced (and in some instances is still advancing) at a fairly rapid rate, compared with the process in the economic core areas. It is quite feasible that the south-eastern parts will witness even faster urban expansion than that experienced in the rest of western Europe.

2. A characteristic of urbanization in western Europe has been the growth of large cities. Clout (1975) has noted that the number of cities with more than 100 000 inhabitants apiece grew from 200 in 1950 to 300 in 1970 and accounted for 27 and 35 per cent respectively of the west European population. By 1980 there were 131 cities of more than 250 000 inhabitants each (Clout *et al.* 1985). Indeed, western Europe accounts for more than 80 per cent of the total number of European urban agglomerations with over 100 000 inhabitants. Cities over this size are especially important in three countries, accounting for 53 per cent of the population in the United Kingdom, 42 per cent in the Netherlands, and 41 per cent in France.

Capital cities became very important in western Europe as a destination for rural migrants. These cities, often exhibiting primate tendencies characteristic of developing countries, hold substantial proportions of their country's urban population. London, Paris, Stockholm, Copenhagen, Athens, and Madrid are good examples. This tendency helps to explain the growth of the 'millionaire city', the number of which has increased from an inter-war figure of 13 to an early 1980s figure of 25 (Table 2.3); if one accepts Rhine–Ruhr and Randstad Holland as world cities (Hall 1977 and Section 2.4), the total rises to 27. Over three-quarters of these urban agglomerations are to be found within the 'Euro 10' and 58 per cent in just four countries, the United Kingdom, Italy, France, and West Germany. With the accession of Spain and Portugal into the EC in 1986, only two millionaire cities—Stockholm and Vienna—will be outside the Community. Despite these figures, western Europe has only London (4th) and Paris (7th) in the world's twenty largest cities.

3. Another post-war characteristic of urbanization is the attempt by the governments of western Europe to limit the growth of the very large city region. These regions are expanding at the expense of rural areas and the cities of the older industrialization (Elkins 1973). Paris, London, and the north Italian cities provide good examples and government schemes exist to aid the decentralization of industry and services away from such areas. The French government attempted to curb the remarkable dominance of Paris in the country's urban hierarchy by demarcating a number of *métropoles d'équilibre* (see Chapter 10 and Fig. 10.3). The *métropoles*, including cities such as Marseille, Toulouse, and Lille, were designed to act as urban counter-magnets to Paris. Elkins (1973) notes that the policy was successful, in so far as population migration into Paris in the late 1960s was checked and industrial employment between 1962 and 1968 decreased by 1.1 per cent, when in the country as a whole it increased by 6.5 per cent.

4. In north-western Europe, the urbanization process seems to have reached saturation point. Industrialization bears little relationship to what is happening in cities and industrial development often takes the form of industrial estates in 'green-field' sites. People are not necessarily living at their place of work and suburbanization is increasing as lower housing densities on the outer rings of urban areas become the norm. The 1970s witnessed the process of counterurbanization (Fielding 1982), as more people chose to live in rural areas.

As Hall and Hay (1980) noted, a 'clean break' in urbanization occurred during the late 1960s and early 1970s, when decentralization to the suburbs and rural areas grew in popularity. Detailed research on counterurbanization in western Europe has been conducted by Fielding (1982) and more recently in France by Ogden (1985). Fielding's work demonstrated how the relationship between net migration and settlement size in western Europe had changed from a positive one in 1950 (urbanization) to a negative one in 1980 (counter-urbanization) (Fig. 2.10). By 1980 the largest

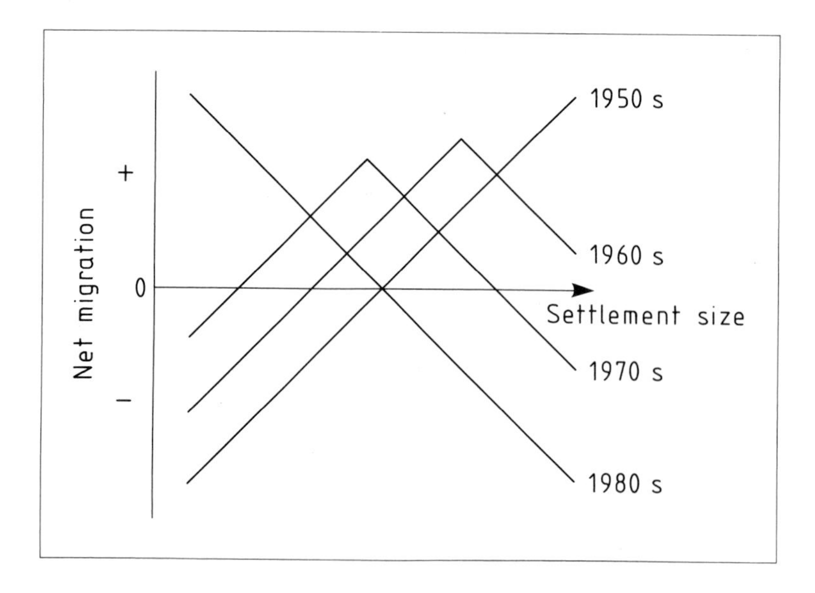

Fig. 2.10 Net migration and settlement size in western Europe, 1950s–1980s (*source:* Fielding 1982)

urban settlements were experiencing net migration loss and rates of gain were highest among small and medium-sized settlements. Therefore, counterurbanization represents a shift within the urban system rather than large-scale moves to rural regions (Knox 1984).

Various possible explanations for the changing relationship between settlement size and net migration have been forwarded (Fielding 1982), including individual life-style preferences, the economic recession, government regional policies, and the end of rural depopulation. However, a key factor has been the 'new geography' of production. Fielding advocated that the urban–rural shift in manufacturing industry (Chapter 6), away from the largest cities and older industrial regions and towards the smaller towns and rural areas, has created a new regional division of labour. The latter have benefited considerably from modern technological industries, branch plants, and R & D companies, with the more attractive surroundings being an added influence. Therefore, firms have acted as major agents of change in the distribution of population and been a prime generator of counterurbanization.

One important consequence of counterurbanization is that the total population living in urban-defined areas has fallen and this is most marked in the inner city. Problems of the inner city are threefold. First, they are losing population as a result of out-migration. This is exemplified by a study of the British major cities, where inner London lost 535 150 people between 1971 and 1981 and where the population of inner Manchester and Liverpool declined by 17.4 and 16.4 per cent respectively over the same decade. Out-migration is selective, involving those most able to move, and the inner city is left with elderly and unskilled workers who lack formal qualifications. Secondly, the inner cities often consist of old, substandard houses and as a result suffer from environmental deterioration. The situation is usually made worse by a mortgage famine in these areas. Thirdly, employment opportunities are declining, especially in manufacturing jobs. Even where employment is available, it is often the wrong type for the residents of the inner city.

2.4 The polycentred metropolises

The growth of larger cities in western Europe has led to the development of city regions, or the agglomeration of conurbations to produce large polycentred metropolises (Burtenshaw *et al.* 1981). One of the major factors aiding this movement towards regional entities is improved communications. As the European people became increasingly mobile, they were less place-bound and there was a tendency for progressive movements to the suburbs. Similarly, industry is now given a more flexible choice of locations.

Two of the greatest urban agglomerations to emerge in western Europe are the Dutch Randstad and the West German Ruhr (Fig. 2.11). A

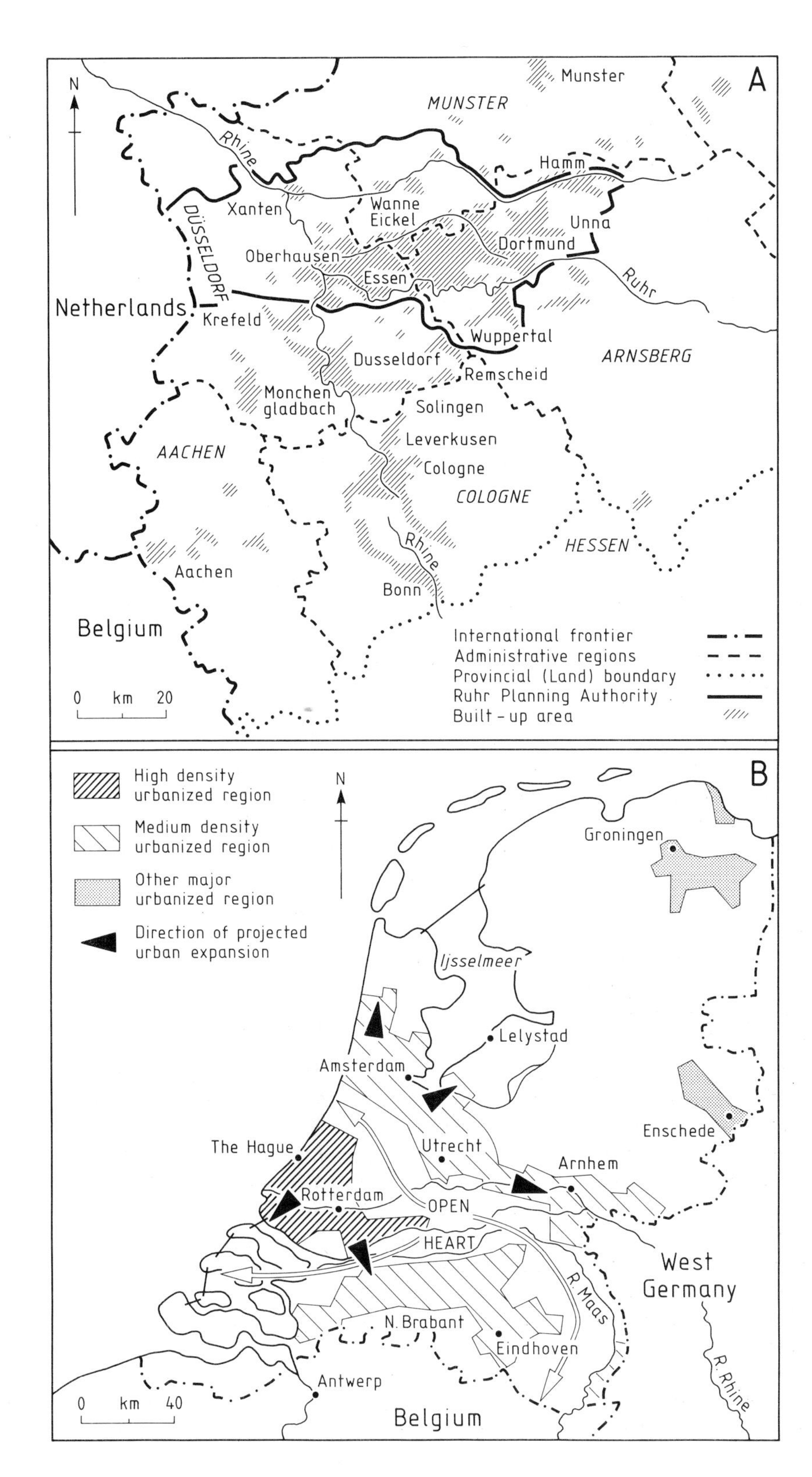

Fig. 2.11 The west European polycentred metropolises: (a) The Rhine–Ruhr (*source:* Hall 1977); (b) Randstad Holland (*source:* Burtenshaw *et al.* 1981)

number of distinguishing features can be identified in the two areas. First, these particular urban forms are naturally unique in a European situation and the scale of urban sprawl is not reproduced elsewhere. Unlike the unicentric development of Paris, London, and other large European cities, Randstad and the Rhine–Ruhr are polycentric in nature, with several cities competing for primacy.

Secondly, both areas consist essentially of a number of towns which have grown so close together that they form one city region. Consequently, areal definition is difficult and no general agreement exists as to their real extent (Burtenshaw 1974; Riley and Ashworth 1975; Hall 1977). According to Hall (1977), the Rhine–Ruhr encompasses an area 40 miles in radius and consists of eight contiguous metropolitan areas, which contain 22 cities with a population of over 100 000 each, and a total population of approximately 11 million. It is triangular in shape and bounded by the autobahn network on all sides, with apices at Bonn, Xantem, and Unna. Randstad is a horseshoe-shaped conurbation, with its open end pointing towards the south-east, and measures 100 kilometres from north to south and 80 kilometres from east to west (Riley and Ashworth 1975). This city region consists of six cities with a population of over 100 000 each, including Rotterdam, Amsterdam, and The Hague, and contains nearly 5 million people overall.

Thirdly, and possibly the most dramatic feature, the two urban sprawls are still expanding and growing closer together. At present there is approximately 70 miles of open land between the Ruhr and Randstad, but the two are connected by the E36 motorway and Dutch planners are concerned that by the end of the twentieth century the two metropolises will have coalesced into one gigantic European megalopolis (Hall 1977). This huge urban region would be 180 miles long, stretching along the Rhine from Bonn to the Hook of Holland.

Fourthly, a characteristic of megalopolitan development is the interdependence of urban centres, as cities begin to specialize in certain activities. This is largely an outcome of improved transport linkages and increased mobility within the urban structure. Whereas one centre would previously be associated with a cross-section of activities, such as administration, manufacturing, and services, the modern trend is for each of these functions to be concentrated in different parts of the polycentred metropolis. This is not well-marked in the cities of the Ruhr, although there is some differentiation into mining, manufacturing, and service trades towns. However, the cities of Randstad have to a large extent specialist functions: there is heavy industry and wholesaling in Rotterdam; light industry in Haarlem and Leiden; administration in The Hague; and banking, commercial, and cultural functions in Amsterdam. This specialization is likely to be reinforced if and when the European megalopolis is formed.

Finally, the scale of development in these polycentred metropolises is creating problems and handicapping regional development elsewhere in their respective states (Lawrence 1973; Hellen 1974; Hall 1977). Urban sprawl is an obvious problem and congestion in the Ruhr in particular has led to much conflict over space. The situation in Randstad is not so severe as there is more land to accommodate future expansion. Different problems result from the contrasting urban structures in these areas: the Ruhr is an older industrial area with obsolescent towns and an inadequate or worn out infrastructure, whereas the Dutch cities grew very slowly until the modern period, despite their early foundation on European trade routes. Physical planning schemes have been designed in both areas (Burtenshaw 1974; Riley and Ashworth 1975; Hall 1977; Burtenshaw *et al.* 1981), and different methods of approach are evident. The 1966 development plan for the Ruhr, for example, was centrally organized (Section 6.2) unlike the haphazard and uncoordinated provincial planning by municipal authorities in the Randstad. German planners have favoured green belts and zonal planning, whilst the Dutch have preferred to use green sectors or wedges in an attempt to provide open country in the towns. A re-appraisal of planning measures occurred in both areas in the 1970s (Burtenshaw *et al.* 1981). This included a need to plan for decline rather than growth, especially in the inner city, to increase awareness of the environment, and to have a more responsible attitude towards the exploitation of resources.

The different scales of urban growth are associated with different periods of development, but it is clear that the megalopolitan developments of the Randstad and Rhine–Ruhr are products of the late twentieth century. It is interesting to note that a new megalopolis is perhaps emerging in Britain, extending from Brighton to Liverpool and recently accentuated by the designation of Milton Keynes. Urbanization patterns in western Europe are constantly evolving and increasingly large-scale problems are developing as a result.

References

Böhning, W. R. (1972). *The migration of workers in the United Kingdom and the European Community*. Oxford University Press, Oxford.

Böhning, W. R. (1979). 'International migration in western Europe: reflections on the last five years'. *International Labour Review*, **118,** 401–14.

Borrie, W. D. (1965). *Trends and patterns in international migration since 1945*. United Nations World Population Conference, Belgrade.

Boulding, K. E. (1973). 'The shadow of the stationary state'. In Olson, M. and Landsberg, H. H. (eds), *The no-growth society*. The Woburn Press, London.

Burtenshaw, D. (1974). *Economic geography of West Germany*. Macmillan, London.

Burtenshaw, D., Bateman, M., and Ashworth, G. J. (1981). *The city in West Europe*. Wiley, Chichester.

Castles, S. and Kosack, G. (1973). *Immigrant workers and class structure in western Europe*. Oxford University Press, Oxford.

Clarke, J. I. (1972). *Population geography*. Pergamon, Oxford.

Clout, H. D. (ed.) (1975). *Regional development in western Europe*. Wiley, London.

Clout, H. D., Blacksell, M., King, R., and Pinder, D. (1985). *Western Europe: geographical perspectives*. Longman, London.

Cole, J. P. and King, C. A. M. (1968). *Quantitative geography*. Wiley, London.

Cousens, S. H. (1960). 'The regional pattern of emigration during the Great Irish Famine, 1846–51'. *Transactions of the Institute of British Geographers*, **28,** 119–34

Davis, K. (1969). *World urbanisation, 1950–70*, Volume One. University of California, Berkeley.

Drewer, S. (1974). 'The economic impact of immigrant workers in western Europe'. *European Studies*, **18,** 1–4.

Elkins, T. H. (1973). *The urban explosion*. Macmillan, London.

Fielding, A. J. (1975). Internal migration in western Europe. In Kosinski, L. and Prothero, R. M. (eds), *People on the move*. Methuen, London.

Fielding, A. J. (1982). 'Counterurbanisation in western Europe'. *Progress in Planning*, **17,** 1–52.

Hall, P. (1977). *The world cities*, 2nd edn. Weidenfeld and Nicolson, London.

Hall, P. G., and Hay, D. (1980). *Growth centres in the European urban system*. Heinemann, London.

Hellen, J. A. (1974). *North-Rhine Westphalia*. Oxford University Press, Oxford.

Hume, I. M. (1973). 'Migrant workers in Europe'. *Finance and Development*, **10,** 2–6.

Jordan, T. G. (1973). *The European culture area*. Harper and Row, London.

Kane, T. T. (1978). 'Social problems and ethnic change: Europe's guest workers'. *Intercom*, **6,** 7–9.

King, R. L. (1976). 'The evolution of international labour migration movements concerning the EEC'. *Tijdschrift voor Economische en Sociale Geografie*, **67,** 66–82.

King, R. L. (1979). 'Return migration: a review of some case studies from southern Europe'. *Mediterranean Studies*, **1,** 3–30.

King, R. L. (1982).'Southern Europe: dependency or development?' *Geography*, **67,** 221–34.

Kinsey, J. (1979). 'Algerian movement to Greater Marseilles', *Geography*, **64,** 338–42.

Kirk, D. (1946). *Europe's population in the inter-war years*. Gordon and Breach, New York.

Knox, P. L. (1984). *The geography of western Europe: a socio-economic survey*. Croom Helm, London.

Kosinski, L. (1970) *The population of Europe*. Longmans, London.

Lawrence, G. R. P. (1973). *Randstad Holland*. Oxford University Press, Oxford.

Mayer, K. B. (1975). 'Intra-European migration during the past twenty years'. *International Migration Review*, **9,** 441–7.

Mellor, R. E. H. and Smith, E. A. (1979). *Europe: a geographical survey of the continent*. Macmillan, London.

Ogden, P. (1985). 'Counterurbanisation in France: the results of the 1982 population census'. *Geography*, **70,** 24–35.

O'Loughlin, J. (1980). 'Distribution and migration of foreigners in German cities'. *Geographical Review*, **70,** 253–75.

O'Loughlin, J. (1984). 'Residential segregation of foreigners in German cities'. *Tijdschrift voor Economische en Sociale Geografie*, **75,** 273–84.

Pressat, R., Biraben, J. N., and Duhourcau, F. (1973). 'La conjoncture démographique: L'Europe'. *Population*, **28,** 1155–69.

Riley, R. C. and Ashworth, G. J. (1975). *Benelux: An economic geography of Belgium, the Netherlands and Luxembourg*. Chatto and Windus, London.

Salt, J. and Clout, H. D. (eds). (1976). *Migration in post-war Europe: geographical essays*. Oxford University Press, Oxford.

Werner, H. (1974). 'Migration and free movement of workers in western Europe'. *International Migration*, **12,** 311–27.

Westoff, C. F. (1974). 'The populations of the developed countries'. In Scientific American, *The human population.*

Westoff, C. F. (1983). 'Fertility decline in the west: causes and prospects'. *Population and Development Review*, **9**, 99–104.

3

Agriculture

3.1 Contemporary agriculture in western Europe

Western Europe is characterized by a complex and diversified system of farming and Table 3.1 demonstrates how the importance of agriculture varies quite remarkably between the member countries. In some 'peripheral' countries, agriculture accounts for over 10 per cent of the gross domestic product, reaching figures of 17 per cent in Eire, 16 per cent in Greece, and 13 per cent in Portugal. By way of contrast, less than 5 per cent of the gross domestic product comes from agriculture in the 'core' countries, with the United Kingdom and Belgium recording a very low 2 per cent.

Between 1960 and 1983, every country in western Europe experienced a decline in the number of people employed in agriculture. However, agriculture remains an important source of employment in certain countries, employing over 20 per cent of the economically active population in Portugal and Greece. At the other end of the range, agriculture in Belgium and the United Kingdom employs just 2.7 and 1.8 per cent respectively of the working population. However, as Minshull (1978) notes, agriculture in the

Table 3.1 Western Europe: agricultural statistics

	% GDP from agriculture 1980	% economically active population employed in agriculture				Land utilization, 1982		
		1960	1970	1975	1983	% Arable	% Permanent pasture	% Other
Denmark	5	18.1	11.2	9.8	6.1	62.6	3.3	34.1
Eire	17	36.5	26.5	24.3	19.3	14.1	70.4	15.5
UK	2	4.0	2.8	2.7	1.8	28.9	46.8	24.3
France	4	22.1	13.7	11.3	7.5	34.1	23.4	42.5
West Germany	2	14.2	7.5	7.3	3.3	30.5	19.1	50.4
Italy	6	30.8	18.8	15.8	9.5	42.2	17.4	40.4
Belgium	2	8.0	4.8	3.6	2.7	25.4	20.7	53.9
Luxembourg	3	15.4	7.7	6.2	5.0			
Netherlands	3	10.8	8.1	6.6	4.7	25.4	33.7	40.9
Spain	7	42.1	26.0	22.0	15.1	41.0	21.4	37.6
Portugal	13	44.1	33.3	28.2	24.2	38.7	5.8	55.5
Norway	5	19.7	11.9	10.2	6.7	2.7	0.3	97.0
Sweden	3	14.1	8.3	6.4	4.3	7.2	1.7	91.1
Finland	8	36.1	21.3	14.9	11.1	7.7	0.5	91.8
Switzerland	—	11.4	8.8	8.3	4.6	10.3	40.5	49.2
Austria	4	23.8	14.8	12.5	7.9	19.8	24.7	55.5
Greece	16	55.8	43.1	35.6	34.9	30.1	40.2	29.7
Iceland	—	25.7	17.7	13.8	10.2	0.1	22.7	77.2

Sources: FAO Production Yearbook; UN Monthly Bulletin of Statistics; UN Statistical Yearbook

United Kingdom is more important than its manpower suggests, as UK farmers produce about 55 per cent of the food consumed in the country. These figures indicate a contrast between the predominantly agricultural and rural nature of Eire, Iberia, and Greece and the more industrialized nature of countries such as West Germany, Belgium, the Netherlands, and the United Kingdom.

Land utilization figures also vary between European countries. This time it is the Scandinavian countries of Norway, Sweden, and Finland that stand out, as less than 10 per cent of their total land area is devoted to permanent pasture and arable activities; the remainder is given over to woodland or, because of physical restrictions, is unsuitable for agriculture. Indeed, it is a combination of physical factors (topography, climate, and soils) and population density that has helped to determine the amount of land devoted to farming in individual countries. Arable land-uses are dominant in Denmark, Portugal, Spain, and Italy, whilst Eire, the United Kingdom, and Switzerland are predominantly pastoral. The only countries to have some kind of balance between arable and pastoral land are Belgium, Luxembourg, and Austria.

Although the physical environment contributes to the diversity of west European agriculture, the post-war era has been characterized by a process of modernization and structural reorganization. Government policies, of several forms, have been instrumental in shaping the pattern of farming over the past 30 years and Clout (1971) classified government measures into four main groups:

1. Those designed to expand output in the immediate post-war period. The situation today, however, is that of many Governments trying to control and regulate production rates and to integrate the farming systems more effectively to the needs of western Europe.
2. Those which have attempted to rationalize and consolidate farm buildings, in order to alleviate the problem of too many and too small farms.
3. Those which have tried to raise farm incomes, which continue to lag behind other sectors of the economy.
4. Those concerned with broader regional development schemes and which have attempted to view agricultural problems in relation to other regional problems.

Agricultural output since 1945

Attempts to expand output in the immediate post-war period were mainly successful. The recovery of agriculture was rapid and pre-war levels of production were reached in most European countries by 1950 (Clout 1971). Progress continued throughout the 1950s, as a result of government schemes to promote agricultural development and technological and biological advances. Measures designed to increase output were replaced by those which aimed at cutting back on certain types of production. The disposal of surpluses has been a central issue in the European Community and will be examined in more detail in Section 3.3.

Advances in output were accomplished with a declining labour input (Table 3.1) and an increasing machinery input, in the form of tractors and combine harvesters. It was the non-EC countries that witnessed a marked increase in the number of machines in operation, with Spain, for example, showing an increase in the number of tractors, from 22 443 for the 1952–6 period to 562 626 in 1982. This represents an increase of over 2400 per cent and was matched by a larger percentage increase in the number of combine harvesters, over the same time period. Within the European Community, the most noticeable increases occurred in Italy, for tractors and combine harvesters, and in Denmark, France, and West Germany, for combine harvesters.

Despite the adoption of improved methods of cultivation and the increased intensification of production, spatial differences in agricultural development are still apparent in western Europe. The member countries of the European Community predominate in the agricultural as well as in the other sectors and the role of the non-EC countries is peripheral in relation to that of the Ten. Only for certain products are they of substantial importance, as in the case of Norway for fish and Spain for wine and cereals. The latter is of particular significance and will be discussed in Section 3.3. A closer examination of trends in cereal production will serve as an example of spatial variations in output.

Cereal production in western Europe followed a steadily rising trend, with minor fluctuations, throughout the 1960s and 1970s (Fig. 3.1) and has since remained fairly stable. With the exception of the Netherlands, this trend is characteristic of every individual country. Whilst ten countries reduced the area they devoted to cereals between 1975 and 1983, notably France, West Germany, and Portugal, only three—Denmark, Sweden and Portugal—recorded a decline in cereal yields over that time period. Clearly, the output of cereals per hectare has increased, with overall production much higher on a smaller area than in 1975. However, western Europe accounts for less than 15 per cent of the world grain production and, with the exception of Spain, the major producers are from the EC. France is by far the largest producer with an output twice as high as West Germany, the second major producer.

The growth in production is the result of two main factors: first, plant breeding and improved husbandry practices have resulted in an improvement in yields in most countries; and secondly, government policy, in the form of support policies, has encouraged farmers to grow cereals. In the 1960s and early 1970s, the original members of the European Community were assured of prices above those on world markets, as a result of minimum import and internal market prices. Outside the Community, similar measures in Spain, the United Kingdom, Sweden, and Denmark guaranteed prices and encouraged self-sufficiency. Such an intervention system provided a limitless market for all major agricultural products (Fennell 1979).

Average yields vary considerably and the main improvements have occurred in countries such as Italy, France, and West Germany, where yields were relatively low in the late 1950s. In contrast. the increase has been more modest in the United Kingdom, Denmark, and the Low Countries, although the Netherlands has maintained its position of producing the highest cereal yields in Europe; in 1983 this stood at 6347 kg/ha, compared to a figure of 5424 for the UK, in second place (Fig. 3.2). A distinct pattern emerges in the yield of cereals, which declines in a fairly regular progression in all directions from the Netherlands. The composition of cereal production also varies widely from country to country. France is the leading producer of the three principal grains

(ECC)

3.1 Intensive agriculture in Flanders: a field of begonias

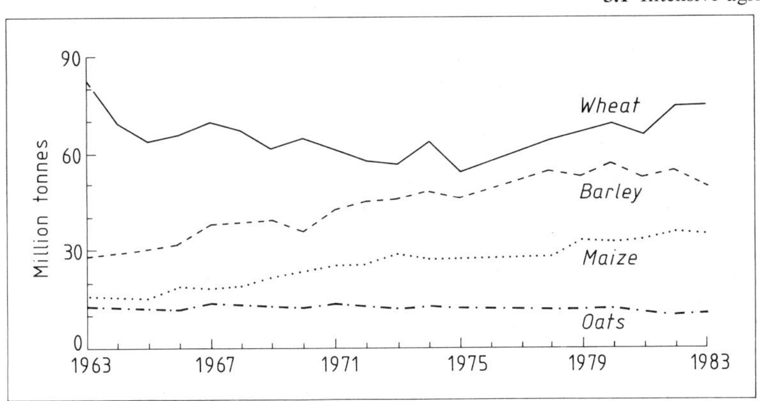

Fig. 3.1 Production of principal cereals in western Europe, 1963–83

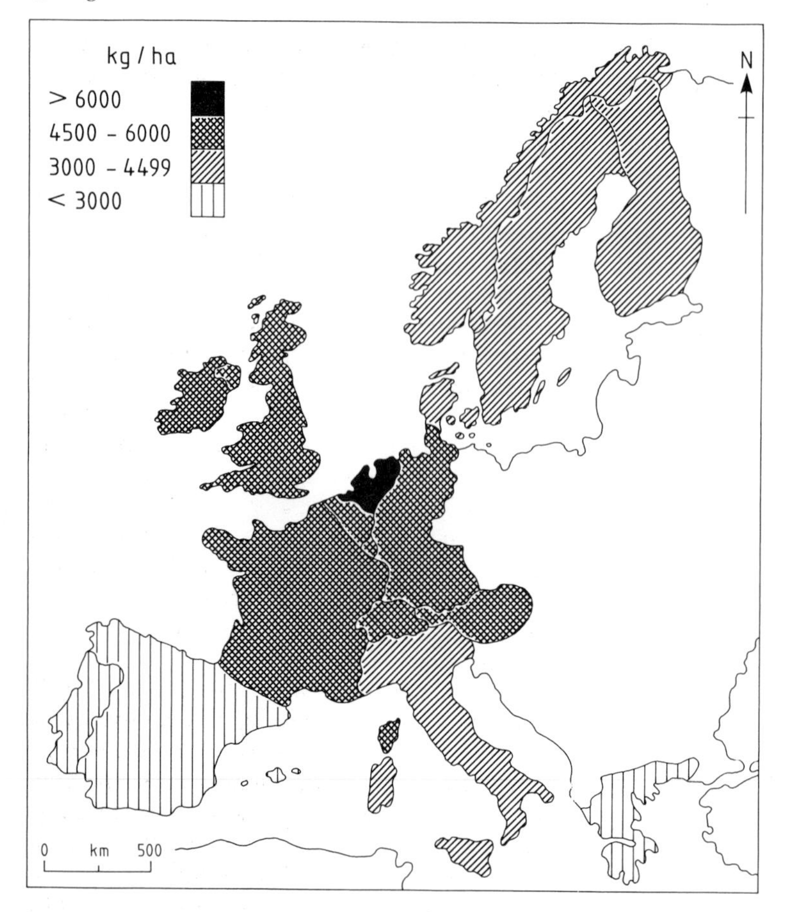

Fig. 3.2 Intensity of cereals production in western Europe, 1983

of wheat, barley, and maize; Italy is the second producer of maize and wheat; and the United Kingdom of barley. Wheat output increased in all the main producing countries except Italy, where improvements in yields were balanced by the withdrawal from cultivation of substantial areas of hill lands unsuited to cereal growing. France, with an output of 24.8 million tonnes in 1983 was a long way ahead of the United Kingdom and West Germany, the next two largest wheat producers with 10.9 and 9 million tonnes respectively. Five countries recorded outputs of less than one million tonnes, with Norway only managing to produce 0.09 million. Whilst wheat is important in all countries except Ireland and Scandinavia, barley is the dominant cereal in Denmark, just as maize predominates in Portugal and Norway, and oats in Sweden and Finland.

Spatial variations in yields are characteristic of many agricultural products in western Europe and various authors (e.g. Chisholm 1968; Minshull 1978) have paid attention to the model of agricultural intensity developed by Van Valkenburg (1960). Intensity of production, based on milk yields, production values for arable land, and yields for selected crops, was shown to decrease in a regular manner from a core area centred on the Rhine delta and adjacent areas. Thus, in terms of agricultural production, a distinct core–periphery relationship exists in western Europe (Fig. 3.3), demonstrating the strong influence of urban markets on agricultural intensity.

The increasing intensification of production in western Europe relates to both government policy and the rising level of purchased non-farm inputs in agriculture (capitalization), as well as to developments in biotechnology. The average growth in production per hectare has varied widely in western Europe and Bowler (1985) has shown that the regional pattern of growth is not core–periphery, as expected by Belding (1981) or shown in Fig.

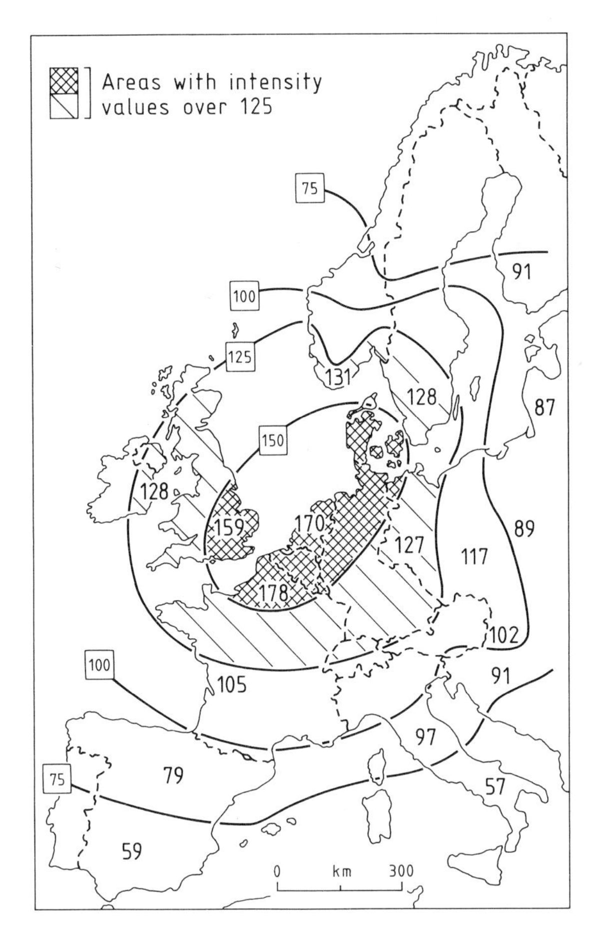

Fig. 3.3 Intensity of agriculture in western Europe (*source:* Chisholm 1968)

3.3. Instead, the pattern within the EC reflects the interaction between farm type and farm-size. In the first instance, some products, notably intensive livestock, have been more susceptible to intensification. Factors affecting the distribution of intensive livestock include large urban markets, access to major ports like Rotterdam for the import of feedstuffs, and proximity to major cereal growing regions. Consequently, the Netherlands, Belgium, north-west Germany, and northern Italy have developed into important intensive livestock areas. In the second instance, the pattern reflects a relationship between increasing intensity and farm-size. Occupiers of small farms have increased intensity, in order to survive the price-cost squeeze of post-war years, and thus areas like southern Germany, southern Italy, and Ireland compare favourably with the lower levels of increasing intensity in areas of large farm-size structures, such as central France and eastern England.

Structural changes in agriculture

In 1945 a great deal of farmland in western Europe was fragmented and in need of consolidation (King and Burton 1982). In many countries the land was divided into a large number of small-holdings and individual farms were further subdivided into strips. This situation arose as a result of two main factors: first, the fossilization of field patterns inherited from the open-field system; and secondly, inheritance laws in certain parts of Europe, where land was equally divided among heirs. The severity of the problem varied from country to country: in Denmark and Sweden, approximately 5 per cent of the farmland was in need of consolidation, compared to 50 per cent in West Germany and Spain, and 60 per cent in Portugal (Clout 1971).

To help overcome this problem, many countries adopted farm consolidation measures (King and Burton 1983). The schemes represented an attempt to adapt peasant agriculture to modern and highly commercial systems of farming. Schemes have existed in all west European countries, varying in purpose from simple attempts to increase farm-size to being just one part of a much wider plan of rural development. As Lambert (1963) pointed out, it is difficult to draw an overall picture of western Europe's need for and progress in farm consolidation because of differing government policies. However, it can be stated that the effects of consolidation have been slow and very little has been done in Portugal, Italy, or Greece. By the late 1960s, the average farm-size in the six EC countries was only 11.7 ha, and even in 1982, 78.6 per cent of agricultural holdings in the 'Ten' were less than 20 ha in size.

The amount of government involvement in consolidation schemes in western Europe has varied. In Spain, Switzerland, Greece, and France, consolidation only took place when the majority of local farmers requested reform, and this was in strict contrast to West Germany and the Netherlands where there was direct government intervention. A more detailed analysis of reform in France and West Germany is given in the next section, but the 1954 Consolidation Act in the Netherlands was probably one of the most

3.2 A typical Walloon farm (ECC)

successful attempts in western Europe to sort out the problems of land fragmentation. Following earlier acts of 1924 and 1938, this drew together schemes for the consolidation of 1.8 million ha of land, or approximately three-quarters of the total agricultural area (Riley and Ashworth 1975). The schemes were concentrated in the less-intensive agricultural areas and Zeeland, with 115 schemes covering 67 000 ha, has had the largest area of land consolidated. Despite an annual rate of consolidation of approximately 50 000 ha, progress in this go-ahead country has been far too slow and some of the earliest consolidated land is in need of reconsolidation.

This situation was characteristic of western Europe generally and in Belgium, for example, the need for consolidation was not recognized until 1957. In that year, a government act envisaged the consolidation of 600 000 ha, or one-third of the total farmland. Progress was remarkably slow and by 1960 requests for consolidation had been made for only the equivalent of 120 000 ha. Land reform in Italy and Greece was quite exceptional and had the effect of reducing, rather than increasing, average farm-size. This was the result of government attempts to provide land for the landless agricultural workers and to increase the size of smallholdings, by subdividing the large estates or latifundia.

One country where land consolidation has made reasonable progress in recent years is Spain (Guedes 1981; Clout *et al.* 1985). The process peaked in the late 1960s and early 1970s, when an annual average of 350 000 to 400 000 ha were consolidated, and progress was most marked in the northern provinces. By 1978, 5 million ha of farmland had been consolidated, leaving 3 million ha in need of attention. Mean plot size rose from 0.35 ha to 2.52 ha and the average number of plots per holding for those affected was reduced from 14.6 to 1.9 (Guedes 1981). However, a lack of legislation to prevent continuing subdivision of farmland upon the death of the proprietor has restricted overall progress and consolidated farms are often less than 5 ha in size.

Therefore, consolidation does not necessarily lead to an increase in farm size and so farm amalgamation has also been encouraged. In West Germany, France, the Netherlands, Norway, and Sweden, a system was introduced whereby any farmland falling vacant could be used to enlarge neighbouring farms. For example, in Sweden a network of County Agricultural Boards was established in 1957 and given authority to intervene in the land market and buy and sell land (Whitby 1968). The objective was to produce a more rational pattern of land allocation. Other legislation was aimed at the old and young farmers alike. Grants and annuities were available to elderly farmers who retired from agriculture and released their land for future consolidation purposes, and financial assistance was provided to help retrain young farmers who were prepared to give up their land. Similarly, farm amalgamation grants were often available for those remaining in agriculture. Overall, the schemes have had little effect on average farm-size and until rapid reform is introduced, agriculture will remain overpopulated and inefficient.

Despite the limited success of consolidation and amalgamation measures, there has been increasing concentration in the distribution of farm-sizes (Bowler 1985). Fewer and larger farm businesses characterize the trend towards a more industrialized system of farming (Table 3.2). Within the EC, the United Kingdom displays the largest average farm-size which at nearly 70 ha, is more than twice that of the second country, Luxembourg. At the other extreme, Italy and Greece have an average farm-size of less than 10 ha. In the non-EC countries, farms are mainly small in Scandinavia, Austria, and Switzerland, whereas in Spain and Portugal polarization has occurred, whereby large farms account for most of the land but smallholdings account for most of the farm population (Clout *et al.* 1985). Between 1950 and 1980, the greatest decline in the number of farms occurred in the United Kingdom, Belgium, and Luxembourg; this compares with low figures for France and Ireland.

Increasingly, large farms are accounting for a majority of the land. For example, 2 per cent of farms in Italy account for one-third of the farmland, especially in the south and central parts of the country; in the United Kingdom, 19 per cent of farms control two-thirds of the land, particularly in the south-east of England and the highlands of Scotland (Bowler 1985). Accompanying this trend to farm-size concentration has been enterprise concentration, with livestock in particular becoming more concentrated, and specialization of production, as farms become involved

Table 3.2 Farm-size structure of EC countries, 1980 (% of farms in each country; 1970 in brackets)

Country	1–4.99	5–19.99	20–49.99	over 50	Average size (ha)	Annual rate of change 1970–76 in number of farms
United Kingdom	14 (19)	28 (29)	27 (26)	31 (27)	68.70	−1.7
Luxembourg	19 (21)	28 (37)	41 (38)	13 (4)	27.63	−4.2
France	20 (23)	37 (43)	31 (26)	13 (8)	25.41	−3.2
Denmark	12 (12)	46 (52)	34 (31)	9 (6)	24.96	−2.0
Ireland	15 (20)	47 (52)	30 (22)	9 (6)	22.52	−0.5
Netherlands	24 (33)	51 (51)	22 (15)	3 (1)	15.61	−2.5
Belgium	29 (34)	48 (51)	19 (13)	4 (2)	15.43	−4.0
West Germany	34 (37)	43 (46)	21 (15)	3 (2)	15.27	−3.2
Italy	69 (68)	26 (26)	4 (4)	2 (2)	7.42	−1.8
Greece	71 (—)	27 (—)	2 (—)	0.2 (—)	4.27	—
The Ten (The Nine)	46 (43)	33 (39)	15 (15)	6 (5)	18.00	—

Source: Bowler (1985)

in fewer enterprises. Together, intensification, concentration, and specialization are key factors in the modernization of west European agriculture, as is the growth of agribusiness, whereby agriculture is being drawn into the food processing chain and linked with industries both 'upstream' (e.g. chemicals, feedstuffs, and machinery) and 'downstream' (food processing) of farming itself.

Farm incomes

The individual countries of western Europe pursued separate policies in an attempt to raise farm incomes, although many have been subsumed within the Common Agricultural Policy (CAP) of the EC (Section 3.3). These measures fall into two main groups. Direct price support measures form the first group and include import restrictions and minimum internal prices for certain commodities. It is doubtful whether these measures have raised farm incomes sufficiently and they appear to have had adverse effects. In the first place, price support does very little for the smaller farmers with low incomes, of which there are many. These farmers aim to be self-sufficient and thus sell very little of their produce. Secondly, the measures have hindered the economic development of agriculture in western Europe by encouraging farmers to remain on the land, even though their holdings are not viable, and to continue with the existing methods of farming which may be inefficient.

The second group of measures attempts to raise the efficiency of existing farms and includes grants and loans for improvements, and advisory services. Unfortunately, the effects are biased towards the larger and usually more educated farmers, who know how to take full advantage of such schemes. The smaller farmers in the poorer areas do not benefit and, as with the first group of measures, the income of the majority of west European farmers is not raised and the gap in incomes between large and small farmers increases.

The desire to raise farm incomes is a social policy and can only be achieved by distorting economic forces. Consequently, the CAP is attempting to reverse the tide of economic development, which of course it is failing to do (Hill 1984). Increasingly, it is being realized that the major determinant of farm income is farm structure and thus the long-term solution to poor farm incomes can only come from such restructuring schemes as land consolidation, farm enlargement, and land reform and, as already discussed, these have been making little progress.

Rural management schemes

Schemes of this nature indicate a change in government attitude towards agriculture. Attempts to view the agricultural problems of western Europe in isolation had not been too successful and a different plan of attack was to consider agricultural problems in the context of the rural economy generally. This led to the implementation of integrated plans for rural management in Spain, southern France, southern Italy, the highlands and islands of Scotland, and Greece.

Two examples of such plans are the Spanish Badajoz Plan, for the middle section of the Guadiana river, and the French SOMIVAL scheme, for the Massif Central. Both schemes had similar objectives and were concerned with forestry and rural industrial employment, as well as with agriculture. The Badajoz Plan included the construction of dams, which enabled 130 000 ha of land to be irrigated and 40 settlements to be established. In-coming farmers were specially trained in irrigation and up-to-date farming techniques and the whole agricultural programme was coordinated with schemes for afforestation and industrial development. Although creating over 12 000 jobs between the early 1950s and 1971, the Badajoz Plan could not arrest the high rates of out-migration and over 200 000 people left the province during this period (Naylon 1975). This was partly due to the failure of the scheme to attract adequate industry and commerce, but also to the insufficient support given to the newly settled farmers in terms of marketing networks and pricing policies.

The SOMIVAL* has been similarly concerned with the modernization of agriculture, through irrigation schemes, animal rearing, consolidation of holdings, and co-operative schemes for livestock herds. However, the Massif Central is not an ideal agricultural area and in many parts coordinated afforestation schemes are very important. In addition, the SOMIVAL has encouraged tourism, and holiday villages and estates of

*Société pour la Mise en Valeur de l'Auvergne-Limousin

second homes have been built. These are part of a larger recreational scheme which includes parks, reservoirs, lakes, and camping grounds. Indeed, the Massif Central has experienced numerous rural planning measures since the 1960s (Clout 1984), culminating in the European Community's integrated development programme for the *département* of the Lozère (Section 3.3). Although certain *départements* within the Massif area continue to lose population, there were 800 more people in the region in 1982 than in 1975 and high on the list of achievements is the improvement to rail and especially to road communications.

A rather different and radical scheme of rural management exists in Sweden, where land is classified according to its estimated use in the future. Grants for farm rationalization are available only for land with a potential agricultural future, whereas poor land is eligible for grants towards the cost of afforestation.

This section has demonstrated that government intervention—in the form of a whole range of measures designed to increase output, reorganize farm structures, raise farm incomes, and develop integrated plans for rural development—is a key factor in the agricultural geography of western Europe. Its importance is further emphasized in the following two sections.

3.2 Post-war intervention in French and West German agriculture

Agriculture in France and West Germany after the Second World War was inefficient and reflected an outmoded structure. Farms were too small, holdings were fragmented, and rural areas were deficient in basic services. To keep pace with the technological and biological advances being made in west European agriculture, government intervention was necessary. This took the form of direct intervention in West Germany, whereas in France reform only took place if requested by a majority of local farmers. In this section, the individual policies of France and West Germany will be outlined and only those policies introduced before 1968 are discussed; more recent intervention will be examined in Section 3.3., under the umbrella of the European Community's common agricultural policy.

France

There was a drastic need for post-war structural change in France as land was severely fragmented. This had been the case for many years and as long ago as 1891, the cultivated area of France consisted of 151 million parcels of land, averaging just 0.36 ha in size (Baker 1961). Such extreme partitioning of land not only made a very large labour force inevitable, but it impeded mechanization and the introduction of new crops and methods of cultivation.

Three types of post-war policy were introduced to help remedy these problems:

1. Regional policies.
2. Technical modernization.
3. Improved market practices.

Regional policies represented the main form of government involvement and subsequent discussion will concentrate on such policies. The basis of modern consolidation in France was the adoption of the *Remembrement* policy in 1941, when 14 million ha of land were in need of immediate reorganization (King and Burton 1983). Under this policy, a special commission investigated the possibilities of consolidation in any commune where 75 per cent of the farmers demanded change. The actual task of regrouping was undertaken by the *Service du Remembrement*, which had as its main aim the improvement of land cultivation by the introduction of a reorganization of land into larger parcels, more suitable for machinery and more easily accessible.

The *Service du Remembrement* had three main principles:

1. All land was to be given a value, based on the productivity of the soil. Under any regrouping which took place, landowners would be given land of the same productive value as their former property.
2. New regroupings had to lead to the elimination of all enclaves and of course to consolidation.
3. A new network of roads was to be drawn up, so that each parcel of land had access to a road.

Remembrement was faced with problems from the outset and complete consolidation into single blocks was impossible. One reason for this was that with local variations in soil productivity, farmers did not want to work land from just one 'zone' but wanted to farm land from each 'zone' of soil. In addition, many farmhouses were located in the villages and in an attempt to equalize travelling time for the farmers, it was necessary to allocate land in various zones of distance from the villages. Another problem of *Remembrement* was the amount of time taken to carry through the necessary reform once it had been initiated in a commune. Reform was slow for numerous reasons, including the procedure involved, the difficulty of evaluating land potential, and the time taken on road-building schemes and the removal of hedges and banks.

As a result of these problems, *Remembrement* took 28 years to consolidate 43 per cent of the land in need of treatment and had only achieved one half of its target by the mid-1970s (Hirsch and Maunder 1978). However, *Remembrement* had certain geographical effects. At a local scale, Baker (1961) studied four communes in Beauce and found that *Remembrement* had led to a decline of 75–80 per cent in the number of farms in each commune. In the commune of Vézelise, *Remembrement* was completed by 1960 and the number of fields had been reduced from 2500 to 358. At the regional scale, there were notable variations in the overall success of this policy (Fig. 3.4A). Most progress was made in areas already supporting the larger, more modernized farms and where there was a high level of tenancy and a large number of young farmers. Consequently, the open-field zone of northern France benefited, with 19 *départements* in the Paris Basin each having over 100 000 ha of land restructured (Clout 1972a). In contrast, least progress was made in areas such as Brittany, Acquitaine, and the Massif Central, where smaller units, with a high level of owner-occupation, traditional farming systems, and older farmers, were characteristic.

Areas where structural reform was most urgent got left behind and *Remembrement* had the effect of increasing rather than reducing regional disparities in agriculture. As a result, the Government intervened and in 1960, 29 regional SAFERs* were created, designed to complement the work of *Remembrement*. Their objective was simple: they were set up to buy land offered on the open market and use it to enlarge farms that were too small for efficient production. The SAFERs could also acquire abandoned land, improve it, and use it for the same purpose, or to create new viable units. At the same time they had the rather contradictory task of safeguarding the family character of French farming (Clout 1972a). A limited budget reduced the scope of activity of the SAFER organizations, which appeared to work against *Remembrement* in that there was a tendency to split up large farms in order to increase the size of smaller units. Indeed, policy seemed directed towards making the family farm more viable.

Three main factors determined the regional impact of the SAFERs: the amount of land coming up for sale, the complexity of tenancy agreements, and the price of land (Clout 1968). As a result, the effect of the SAFERs was most marked in southern France, where land tenure is simple, land is comparatively cheap and rural out-migration is highest (Fig. 3.4B). Higher prices for land in Brittany and the north constricted the

3.3 Specialized tractor for the cultivation of grapes in France (ECC)

*Sociétés d'Amenagement et d'Establissement Rural

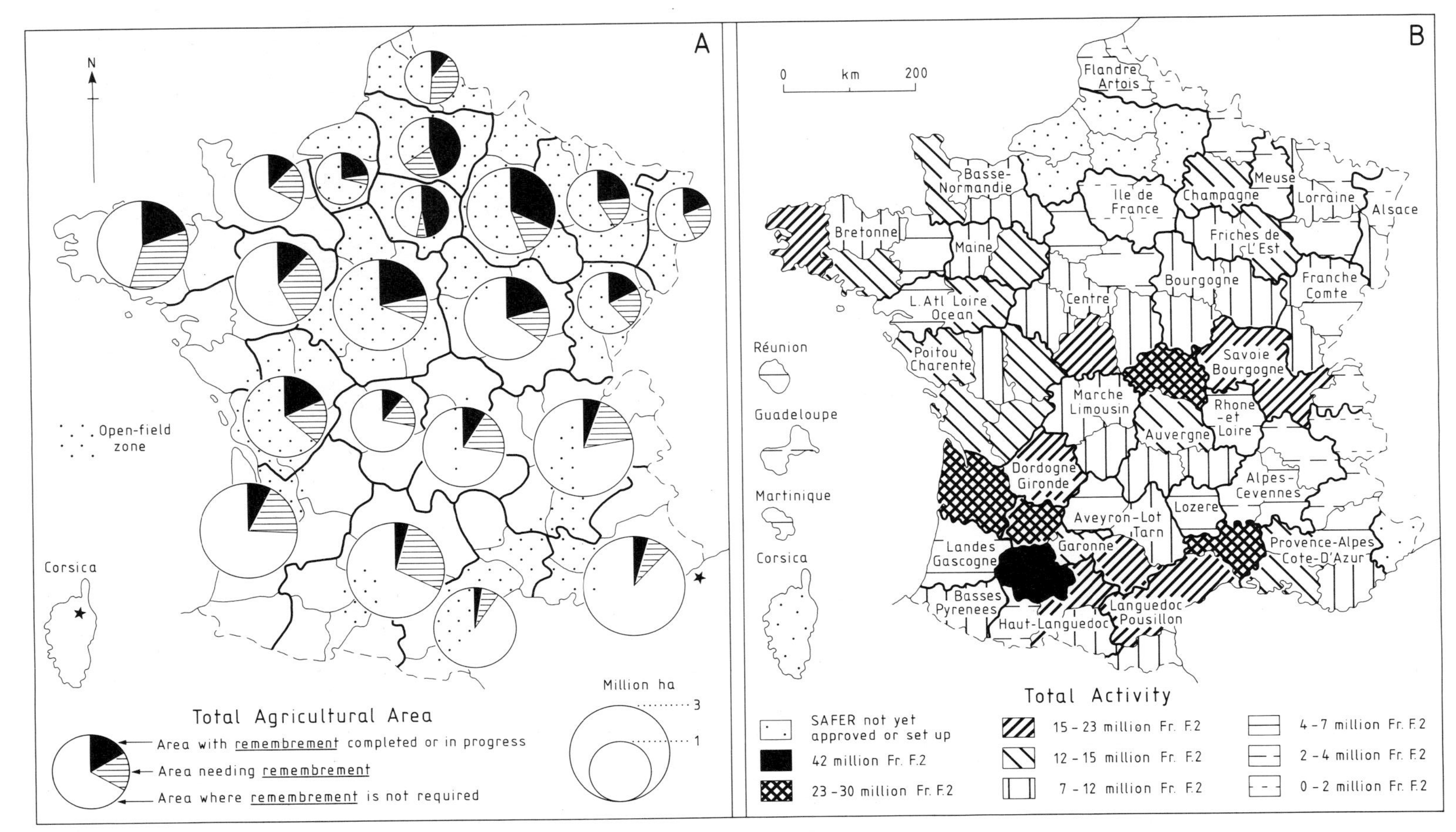

Fig. 3.4 Structural change in French agriculture: (A) *Remembrement* operations (*source:* Clout 1972a); (B) SAFER activity in 1970: value of acquisitions, assignments, and improvement work (*source:* OECD 1972)

actions of the SAFERs, and in the north-east their effect was minimal as only small parcels of land came onto the market. From this viewpoint, the regional SAFERs did complement *Remembrement*, by aiding areas such as the south which benefited least under the *Remembrement* scheme.

Further incentives for farm enlargement came in 1962 with the introduction of FASASA*. Under this scheme, various funds were established, including one designed to help retrain farm operators who were prepared to leave non-viable holdings. Similarly, installation grants were available to those farmers prepared to move from overcrowded 'departure' zones in the north to designated 'reception' zones in central and southern France, where there was a shortage of young farmers. Possibly the most widely adopted measure from FASASA was the retirement pension scheme (IVDs). If, at the age of 65, farmers voluntarily retired and vacated their land they would receive an annuity to supplement the state old-age pension. The vacated land could be used for the installation of farmers under the age of 45 on the same unit, for farm enlargement, or for transfer to the SAFER; the retiring farmer could retain a small subsistence plot (Naylor 1982).

As a result of its limited uptake in the mid-1960s, numerous amendments were made to the IVD scheme. First, an early retirement premium was available from 1968 for farmers between 60 and 65 years of age, in such problem areas as Brittany and the Massif Central, and the within-family transfer of land was allowed by leasing. In the same year, farm enlargement was further encouraged through a restructuring grant, which was given to retiring farmers who created an amalgamated unit larger than one and half times the minimum settlement area. Finally, in 1974 the early retirement premium became available throughout France and the additional restructuring grant was replaced by a supplementary payment, independent of the IVD scheme, given to farmers of any age transferring their land to the SAFER, for farm enlargement or the installation of suitably qualified young farmers (Naylor 1982).

Special annuities were paid to over 123 000 elderly farmers between 1963 and 1968 and the maximum number of IVDs were granted in 1969. Since then they have declined in attraction, due primarily to the increase in land values. The IVDs have been most popular among tenant farmers, but less effective in areas with a high proportion of part-time farming. One-third of the country's agricultural land has been transferred through the scheme, which has been most successful in central and south-western areas, where the small farm problem is particularly acute. Local customs, the age of farmers, the impact of advisers, and opportunities for selling land to industry, housing, and tourist interests are all factors affecting the spatial pattern of IVD adoption (Ilbery 1985). As with *Remembrement* and the SAFERs, FASASA has been restricted by a lack of funds and criticized for merely financing farmer retirement that would have taken place anyway. The scheme has had little real impact and only a very slight increase in farm-size has resulted.

By the end of 1982, 35 per cent of all French farmland had been consolidated; however, a further 10 million ha are in need of reallocation, some for the second time (Clout 1984). At the present rate of consolidation, structural reform in French agriculture will be incomplete by the end of the century and the problem is intensified by spatial inequalities in achievements, which favour northern rather than southern districts.

The second and third types of government policy in France were associated with technical modernization and improved market practices. An overseeing body in agriculture and rural development, the *Service du Génie Rural et de L'Hydrauligne*, acted as a coordinating body for *Remembrement*, regional agricultural development programmes, rural electrification, and water supply schemes. This body was also concerned with the promotion of mechanization and the general improvement of the rural environment. Despite such efforts, development in the backward areas of France was inhibited by a vicious circle effect: development in any area needs capital investment and the accumulation of capital requires high productivity, which in turn demands modern practices, and such practices were not found in many parts of France. The Government also attempted to adapt agricultural production to changing market conditions, especially to the demands of the European Community, by setting up a network of markets (*Marchés d'In-*

*Fonds d'Action Social L'Amenagement des Structures Agraries

térét National) which was concerned with the improvement of distribution and marketing in France.

West Germany

The traditional form of agriculture in West Germany is the family farm and this has resulted in an inefficient system of farming, with far too many small farms. Government intervention has been designed to improve the structure of agriculture, whilst at the same time maintaining the family farm as the basic unit of production (Burtenshaw 1974). Strong emphasis has been placed on the integrated agricultural development plan, with measures for the enlargement of farms, the consolidation of land, the relocation of farm buildings, and the improvement of buildings, drainage, and roads. The legal basis for these measures was the Land Consolidation Act of 1953 and financial support began in 1954, when DM50 million were allocated for structural improvement. The Federal Agricultural Act of 1955 reinforced government intervention and proposed the production of annual Green Plans, to allow a constant review of the agricultural reform programme. The first Green Plan was produced in 1956.

Intervention in West German agriculture has been directed in four major ways (Mayhew 1970 and 1971):

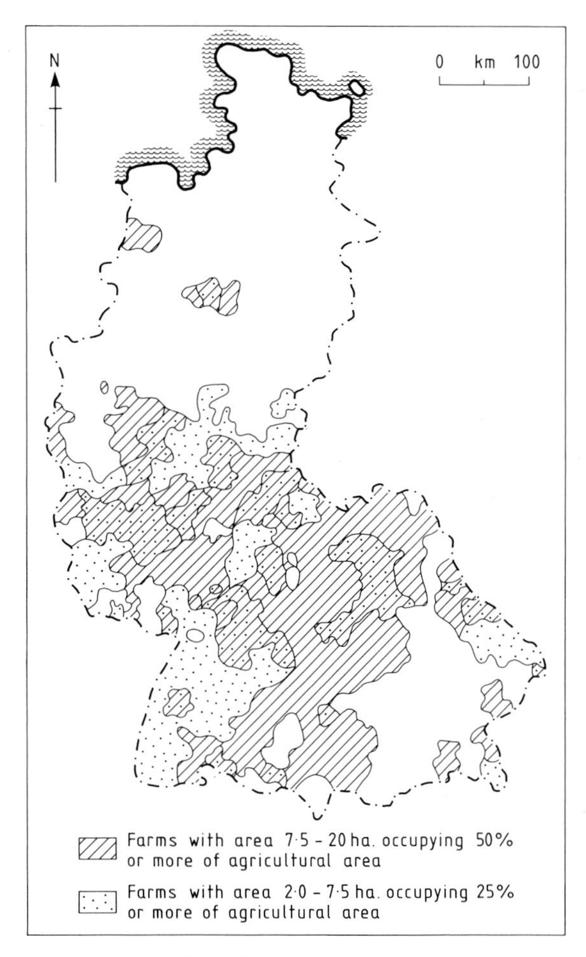

Fig. 3.5 Areas in West Germany dominated by small farms (*source:* Burtenshaw 1974)

1. *Farm size*. Government policy was designed to reduce the number of farms and to increase the proportion of large family farms. This was partly successful in that the number of farms with less than 10 ha decreased by 626 000 between 1949 and 1968, whilst at the same time the number of farms with more than 10 ha increased by 63 000. However, by 1967 only 13.1 per cent of the farms were greater than 20 ha and the process of enlargement appeared to be slowing down. Indeed, between 1967 and 1968 the number of farms between 2 and 10 ha decreased by only 3 per cent. By 1982, 50.3 per cent of farms were still below 10 ha in size.

The problem of small agricultural units was, and still is, most acute in the south, where divided inheritance was common. Rheinland-Pfalz, Hessen, Baden-Württemberg, and western Bayern suffered most (Fig. 3.5), and in Rheinland-Pfalz and Baden-Württemberg, 80 per cent of all farms had less than 10 ha of agricultural land in 1968. In fact the 1966 structural plan for agriculture in Rheinland-Pfalz saw farms of between 12 and 20 ha as being adequate and only recently has West Germany realized that the family farm of the future needs to be well in excess of 30 ha. In 1967, the only *Land* in which resettled farms averaged more than 30 ha was Schleswig-Holstein.

2. *Land consolidation*. The fragmentation of land is a legacy of open-field agriculture and the system of divided inheritance, both of which operated most severely in south-west Germany. To help overcome this problem, two main land consolidation schemes were introduced: the voluntary exchange of strips between farmers, and planned consolidation measures. The former was

supported by federal aid and was suitable for small-scale consolidation only. In 1966, 4460 ha were consolidated in this way and in 1969 the figure was 4441; such small figures made little impact on the overall rate of consolidation in West Germany. Planned consolidation measures included the accelerated consolidation plan, and the more common integrated structural reform plan. With the pace of consolidation needing to be increased, the first plan, which simply attempted to accelerate the rate of consolidation, gained new importance in the late 1960s. Nevertheless, in 1967, 244 298 ha were consolidated under the integrated plan, compared to only 39 736 ha under the accelerated plan.

Despite these measures, the problem was still severe and Burtenshaw (1974) noted that in 1968 5.8 million ha in West Germany were still in need of consolidation, with a further 2.8 million ha needing to be reconsolidated. The average rate of consolidation in the 1960s was 290 000 ha per year and at this rate it would take over 30 years to consolidate all the agricultural land (Mayhew 1970). Consequently, in 1976 a new Farm Consolidation Act was passed which combined consolidation with such aspects of rural planning as recreation and environmental preservation (Clout *et al.* 1985). However, by the early 1980s nearly one-third of the target still remained to be achieved. Even if this is achieved by 2000, large areas will be in need of reconsolidation.

3. *Farm resettlement and farm buildings.* Mayhew (1970) states that in 1961, in approximately 80 per cent of the agricultural parishes of West Germany, more than three-quarters of all housing was deficient in structure, in water supply, or in drainage. Yet again the problem was most acute in the south-west, especially in Rheinland-Pfalz and Baden-Württemberg. Housing grants and federal aid for the modernization of buildings became available. Where the cost of construction or conversion exceeded DM80 000, loans of up to DM50 000 could be obtained and were repayable at the very low rate of interest of one per cent.

Resettlement of farmers was also necessary and encouraged. Many agricultural villages were congested and farmers were moved mainly to hamlets, in preference to isolated settlements. The resettlement of farmers in groups helped to overcome the problems of service provision to isolated farmsteads, whilst preserving a 'community' spirit. Buildings in existing villages, vacated by resettled farmers, were taken over by the farmers who remained. However, the resettlement programme suffered from rapidly rising costs of constructing new buildings. Between 1958 and 1967, costs increased by 129 per cent (Mayhew 1970); in 1967, financial support was severely reduced and the Government began to question the sense of creating new farms, especially as they rarely exceeded 20 ha.

4. *Financial inducements to leave agriculture.* In an attempt to reduce the high proportion of old farmers in the agricultural labour force, financial aid was made available in 1968 to those farmers who were prepared to retire completely from agriculture. The vacated land was to be used once again for consolidation purposes. Further consolidation was encouraged by the introduction of retraining programmes, which were aimed more specifically at the younger farmers. Measures to reduce the number of elderly farmers were improved in 1971, when any farmer aged between 55 and 65, who agreed to leave agriculture, qualified for a minimum pension (Burtenshaw 1974). This went some way towards solving the problem of finding alternative employment so late in one's working life.

Mayhew (1970) notes that government policy has improved West German agriculture and yields have increased rapidly since the war. Nevertheless, the same author believes that this increase was the result of maintaining a high labour input and increasing the capital input, both of which were very costly. In the late 1960s, a number of problems remained, including the following:

(i) Basically, too many people remained in farming, particularly in the south-west. Support for the family farm was possibly too strong.

(ii) Although the farms were highly mechanized, they were inefficiently used. This was a reflection of farm-size and the technical education of the farmers.

(iii) Farm-size remained too small. In 1965, it was estimated that the 625 000 full-time farms would have to be rationalized to 400 000 by 1975. This was not and still has not been achieved.

(iv) Even if people left the land, there was often

a lack of alternative employment to soak up this surplus labour.

(v) Huge amounts of capital were required for consolidation purposes, livestock, and machinery. Approximately DM27 000 million were needed to carry out a complete programme of consolidation and even then it was highly likely that farm-size would still be too small.

(vi) Government reforms were taking place too slowly to be beneficial. An accelerated form of land consolidation was a necessity.

3.3 The European Community and agriculture

The Common Agricultural Policy (CAP) represents the most important sphere of government intervention in west European agriculture. Despite numerous problems and marked differences of opinion between member states, the CAP has been a forceful instrument of European integration and it has been comparatively successful when compared with rather feeble attempts to implement other forms of common policy. The Treaty of Rome (1957) singled out agriculture for special consideration and Article 39 laid down the objectives of a common agricultural policy: to increase agricultural productivity through technical progress and rational development; to ensure a fair standard of living for the agricultural community; to stabilize markets, assure the availability of supplies, and ensure that supplies reached consumers at reasonable prices. The agricultural sector needed to be strengthened; in 1958 it employed over 20 per cent of the working population in the Six, but accounted for only 8 per cent of the Community's gross product (Clout 1971).

The CAP was very slow in development and common price systems did not come into operation until ten years later, in 1968. A number of problems facing the CAP can be singled out:

1. The relative importance of agriculture varied between the member countries, from accounting for just 5.8 per cent of the national income in West Germany to 14.4 per cent in Italy.

2. The area given over to agricultural production varied enormously, from less than 50 per cent in Belgium and Luxembourg to over 66 per cent in Italy and the Netherlands.

3. The predominant type of farming varied between the countries. Of the total agricultural area in the Six, two-thirds was arable. On a national scale, significant differences existed; in Italy, approximately three-quarters of the area was under arable land-uses, whereas in the Benelux countries more than half was under pasture. In France and Germany, the ratio was approximately 60 : 40 in favour of arable.

4. The type of crops grown differed. In terms of cereal production, wheat was dominant and virtually the only commercially-exploited crop in Italy, whereas in the Netherlands and Belgium, barley, oats, and rye were just as important as wheat. West Germany had the smallest proportion (25 per cent) of its arable land devoted to wheat. In the case of fruit production, France and Italy had 98 per cent of the Community's fruit-producing land: one million ha in Italy and 40 000 ha in France.

5. Each country had introduced various price support measures and measures to improve farm structures. These had to be reconciled for a Community policy on prices, production levels, and farm structures.

6. The varying amounts of food being imported into the individual countries was another problem. Italy, for example, imported much of its food and exercised less intervention in home production than France, which was more self-sufficient. Each country protected its agricultural sector by tariffs and consequently it was very difficult to get farm produce to flow freely between member countries.

Successive enlargements of the EC, in 1973 and 1981 (and 1986 for Spain and Portugal) have only served to exaggerate the intensity of these problems and the diversity of west European agricultural systems; this point will be returned to.

Therefore, each country had different interests and the original members of the EC made slow progress in putting Community interests before national interests. As a result, positive steps towards a common agricultural policy were not made until 1962, when the Agricultural Guidance and Guarantee Fund (FEOGA) was established. This was designed to give price support for products when market prices fell below accepted fixed levels, and to work in coordination with a new levy system introduced by the Six. In an attempt to neutralize differences in the price levels of agricultural commodities between members and to replace former import duties, Community support systems were introduced. Hence, a system of price support represented the beginning of the CAP. Agreeing on a system of support and the level of prices was possibly the biggest problem in the 1960s; a problem which was to reappear with the formation of the Nine in 1973. The guaranteed prices set on products were relatively high and this had unfortunate consequences. Inefficient as well as efficient farmers were encouraged to remain in farming and there was overproduction of many agricultural commodities, leading to the now famous Community surpluses such as the butter and beef 'mountains'. It thus appeared that the policy of giving farmers assured markets and realistic prices, without burdening consumers with expensive food, had failed, as unmarketable surpluses and dear food soon became characteristic.

Many problems still faced the CAP. The major and continuing problem was that of surpluses. Policy seemed to work fairly well on storable products like cereals and sugar, but not on livestock products (Berendt 1974). Surplus goods had to be stored, exported at subsidized prices, or actually destroyed. In 1973, 200 000 tonnes of butter were exported to the USSR at knock-down prices, a move which promoted much public hostility within the Community. Surplus food continues to be sold off at a loss, often to east European countries.

The FEOGA itself had many problems. Initially, the fund was financed partly by national governments and partly from the proceeds of external levies. From 1969, a new arrangement was agreed and the Six started paying into a common Farm Fund. Finance for this Fund came from levies on food imports, custom duties from non-member countries and from part of the VAT revenue started in 1972. The problem here was that individual members did not benefit equally and the more agricultural countries were gaining at the expense of industrial countries such as West Germany. France, for example, has received over £600 million more out of the Farm Fund than it has contributed to it (Clout 1975).

The guidance and guarantee sections of FEOGA also appeared to have conflicting interests. The guidance section helps to improve agriculture and the overall economic structure of backward agricultural areas. Early projects to be financed included the improvement of farms and marketing facilities, especially in Italy and West Germany, and funds have also been used for the construction of dairies, processing plants, and irrigation works. However, all of this was undermined by the guarantee section and its system of high price support.

Little attempt to draft an overall coordinated and structured policy was forthcoming and the problems of too many farmers in western Europe and too much land in agricultural use remained. However, in 1968 the Vice-President of the European Community, Dr Sicco Mansholt, presented the Mansholt Memorandum, in which he suggested a number of measures 'to transform agriculture from its tradition-bound way of life into a modern business'. Mansholt pointed to certain weaknesses in agriculture. Although 500 000 people per year were leaving farming during the 1950s and 1960s, this was too little and the process was selective in that it was the young who were leaving agriculture. Consequently, the rural areas were left with old farmers and traditional farming systems. A further weakness was that 75 per cent of the farms in the Six were less than 10 ha and only 3 per cent were over 50 ha. Mansholt also felt that the system of guaranteed prices supported inefficient producers and kept marginal farmers on the land.

Mansholt proposed three objectives for west European agriculture by 1980:

1. To accelerate the drift from the land.
2. To change farm-size dramatically.
3. To balance out the supply and demand for farm products.

To achieve the first objective, Mansholt proposed retraining grants, more effective grants or annuities for the retirement of the elderly, and the diversion of children of agricultural families from the land. Mansholt envisaged an agricultural labour force of five million by 1980 (in the Six); just 25 per cent of the 1950 figure. However, the number of people leaving agriculture per year slowed down in the 1970s and this was because a certain minimum number of labourers was and always will be required. The problem of farm-size could be overcome by replacing peasant holdings with larger enterprises and Mansholt proposed certain optimum sizes of holdings, in which 80–120 ha of cereals were to be grown, or 40–80 dairy cows, 150–200 head of beef cattle, 450–600 pigs, or 100 000 head of poultry were to be raised annually. Concerning supply and demand, Mansholt suggested that guaranteed prices for over-produced products such as milk and sugar-beet should be cut. Production would thus decrease and marginal farmers would be forced to close. In the long-run, the cultivated area would decline by seven per cent, releasing land for afforestation and the creation of national parks.

Mansholt's proposals were not well received, especially by the more elderly farmers, as they would ultimately destroy the characteristic family farm. The various governments were uneasy, pointing to the high costs that would be involved, and many felt that surpluses would remain. The problem of finding alternative employment, notably in France and Italy, would be increased and the general feeling was that such rapid changes would be socially and politically undesirable.

Mansholt's reaction was to produce a modified and less dramatic memorandum in 1970, in which six proposals were set out for a five-year period:

1. Replacing the size criterion for the future farm, Mansholt suggested that a minimum gross output of $10 000–$20 000 per male farm worker per year should be retained. Only farmers, or groups working together, who submitted acceptable development plans should receive financial aid.
2. Farmers planning to switch from surplus products to beef and fodder grains should be given priority treatment.
3. Sums of over $1000 per year should be available for those willing to abandon farming and thus release land for future use.
4. Reclamation of land from the sea or woodland was to be discouraged and instead the Community would pay at least 80 per cent of afforestation costs.
5. Advice and training should be available for those willing to retrain for alternative employment.
6. The establishment of co-operatives and producer groups, to increase the scale of farming, was to be encouraged.

Mansholt's ideas have been dealt with in some detail as they are very important in the evolving CAP. More than anything else, the inefficient nature of west European agriculture was really driven home and Mansholt had demonstrated a clear and urgent need to revitalize the rural sector of society. It was shown that the problems of the countryside could not be solved solely by structural programmes in agriculture; instead integrated planning was a necessity. By 1971, headway was made in the direction of Mansholt's proposals and it was decided that the CAP would include social and structural measures to reform agriculture. This was followed in 1972 by further measures in line with Mansholt's ideas. Pensions were available to elderly farmers who were prepared to vacate their land, retraining facilities were available for younger farmers, and interest rebates would be provided for farmers who submitted acceptable farm modernization plans. The CAP was changing character and had for the first time separated price support measures from a set of structural measures aimed at solving the problems of the small farm sector.

The formation of the Nine, in 1973, created inevitable problems and a delay in the progress of the CAP. It also brought new ideas for agriculture. The British Government, for example, pressed throughout 1974 for changes in the Community's beef support system, advocating the introduction of deficiency payments in place of intervention. In that year, a beef-slaughter premium scheme was introduced as an alternative to intervention; this involved a direct payment on each beef animal going through the slaughterhouse, raising the farmer's return without pushing up the price of meat (Berendt 1974). Essentially, this was a recognition that deficiency payments* could play an important role in future agricultural policy. Berendt (1974) felt that the development of the CAP would certainly follow this path and indeed, deficiency payments have been incorporated into the Community's sheepmeat and beef support schemes.

The changing nature of agricultural policy in the European Community has been examined by Bowler (1976a and b). After outlining some of the structural measures mentioned earlier, such as the Pension and Amalgamation Scheme to encourage older farmers to give up their land and the Farm Development Scheme for farm modernization, Bowler demonstrated how the Community

*Deficiency payments cover the difference between the average market price and a guaranteed price

became more geographically sensitive with its adoption of the Less Favoured Areas Directive in 1975. Areas requiring priority aid from the FEOGA were identified and it was recognized that farm income problems were created by the nature of the areas and not necessarily by farm-size. Priority agricultural zones were those areas with an above Community average employment in agriculture, a per capita gross domestic product below the Community average, and a below average number of workers in manufacturing employment. The boundaries of three types of problem region were agreed (Fig. 3.6):

1. Areas with mountain and hill farming, where production is permanently handicapped by slopes, altitude, or soil type.
2. Areas with low population densities and severe depopulation.
3. Areas with specific problems, such as poor infrastructure or the need to maintain agriculture to protect the countryside or preserve tourist potential.

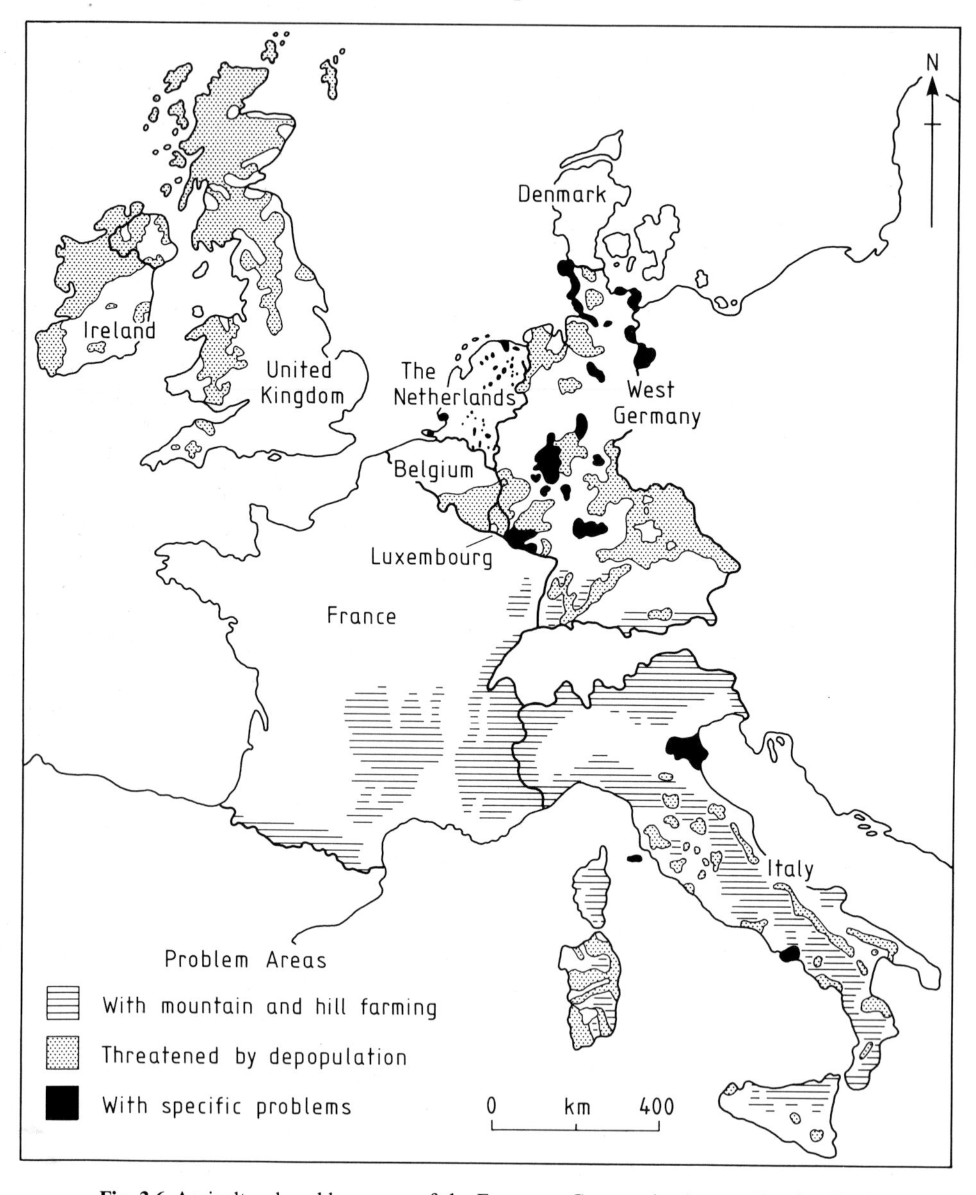

Fig. 3.6 Agricultural problem areas of the European Community (*source:* Bowler 1976b)

Ireland, upland Scotland, the northern parts of Denmark, the Netherlands and West Germany, southern and central Italy, and western France form the most extensive priority zones. An annual compensatory allowance is available in these areas to cover increasing costs of production, paid in terms of a grant per hectare or per head of livestock on a farm. In addition, favourable rates of aid are available for structural modernization and investment in non-agricultural enterprises such as tourist facilities. Between 1974 and 1976 1500 million units of account were budgeted for these regions, to aid agriculture and to provide a system of incentives to create non-agricultural employment opportunities for those leaving the land.

However, the effects of the post-1972 structural changes have been rather disappointing in comparison with the broad objectives of the Mansholt Plan. Clout *et al.* (1985) have catalogued the results of the different directives: 107 000 plans had been submitted under the farm modernization scheme by 1978, mainly in the UK, Belgium, Denmark, and the Netherlands; 46 000 farmers, essentially French, had taken annuities under the retirement scheme by the end of 1978; and the retraining initiative had very little uptake. The most successful directive would appear to be the Less Favoured Areas scheme and by 1981 over 550 000 farmers were receiving the compensatory allowance, accounting for approximately one-third of guidance payments from FEOGA. Most of these were in the UK, Ireland, France, West Germany, and more recently Greece; Italy was rather slow to benefit from this directive and only began to apply it in 1978.

The increasingly geographical nature of the CAP continued into the late 1970s with the 'Mediterranean package' in 1978 and the creation of integrated rural development programmes in three areas in 1979. In the former, specific types of improvement pertinent to Mediterranean farmers were supported, including irrigation measures in the Mezzogiorno, afforestation and improvement of rural infrastructure in the upland areas of southern Italy and southern France, and co-operative projects (Clout *et al.* 1985). With the accession of Greece to the EC in 1981 and the forthcoming membership of Spain and Portugal in 1986, the Mediterranean package will be a central plank of CAP policy in the 1980s, as demonstrated by the announcement in 1983 of a six-year integrated programme for Mediterranean regions. The three integrated rural development schemes in 1979 were for the western Isles of Scotland, the *département* of Lozère in France, and the province of Luxembourg in south-east Belgium. Under the programme, the CAP combined with both the regional and social funds of the EC (Chapter 10) to promote a mixture of agricultural and non-agricultural projects, demonstrating that policy was no longer solely agricultural.

A series of documents in the early 1980s laid down guidelines for European agriculture. Surpluses were identified as a major problem and it was recognized that prices should be fixed with regard to wider considerations than farm incomes alone; consumers' interests, market realities, and the degree of self-sufficiency were amongst the more important ones. Therefore, concern was expressed over the costs of achieving the objectives of the CAP and indeed, the share of the Community's budget spent on agriculture declined from over 70 per cent in 1980 to 65 per cent in 1983. March 1984 represented a major policy change for the CAP, when Ministers for the first time agreed to cut guaranteed prices for milk. New milk production quotas and prices were set for member states; price guarantees were restricted to 99.5 million tonnes of Community output, when 109 million tonnes were being produced and only 88 consumed. Whilst Ireland, Greece, and Luxembourg obtained favourable agreements, most members had their quotas reduced and this has caused considerable unrest among those farming communities that have little alternative to milk production. A similar cut in the guaranteed price of cereals could follow.

Therefore, the CAP has evolved in distinct stages since the early 1960s (Bowler 1976b) and has developed more quickly and further than any other Community policy. Nevertheless, in the eyes of many authors it has not achieved its objectives (Holland 1980; Hill 1984). Surpluses continue to mount, including a currant 'hill' in Greece (Clout *et al.* 1985), and despite the various structural measures for the poorer farming areas, the CAP raises revenues in proportion to output and so benefits better-off farmers in the richer regions. As a consequence, income disparities between rich and poor farmers have increased

(Fig. 3.7). During the 1970s, relatively declining incomes (in real terms) were evident in such rich agricultural regions as most of England and Wales, west-central France, southern parts of Belgium, and Hessen–Rheinland Pfalz in West Germany. Conversely, the increasingly rich agricultural regions included east Anglia, southern and central Scotland, northern Germany, the Netherlands, and north-east France (Bowler 1985). Of the poorer agricultural areas in the periphery of the EC, rising incomes characterized the eastern seaboard provinces of Italy from Veneto in the north to Abruzzi in the south.

Although attempting to raise farm incomes, the CAP is failing to halt the tide of economic development, a situation that will be intensified with the accession of Spain and Portugal in 1986. The enlargement of the EC from nine to twelve involves 'a 55 per cent increase in the farming population, a 57 per cent increase in the number of farms, a 49 per cent expansion of the agricultural area, and a 24 per cent rise in agricultural production' (Clout *et al.* 1985, p. 131). Greece, Spain, and Portugal produce the same range of products and will compete with Italy and southern France, and the sheer size and volume of production from Spain will have far-reaching ramifications for the CAP in the late 1980s. The future is uncertain and whilst the UK believes that radical reform is necessary, the farm lobby in Europe is very strong, which led Hill (1984, p. 137) to conclude that 'politicians have no real desire to reform the CAP'.

3.4 The worker-peasant phenomenon

The peasantry, the capitalist farmers, and the hired labourers were the basic elements of pre-war

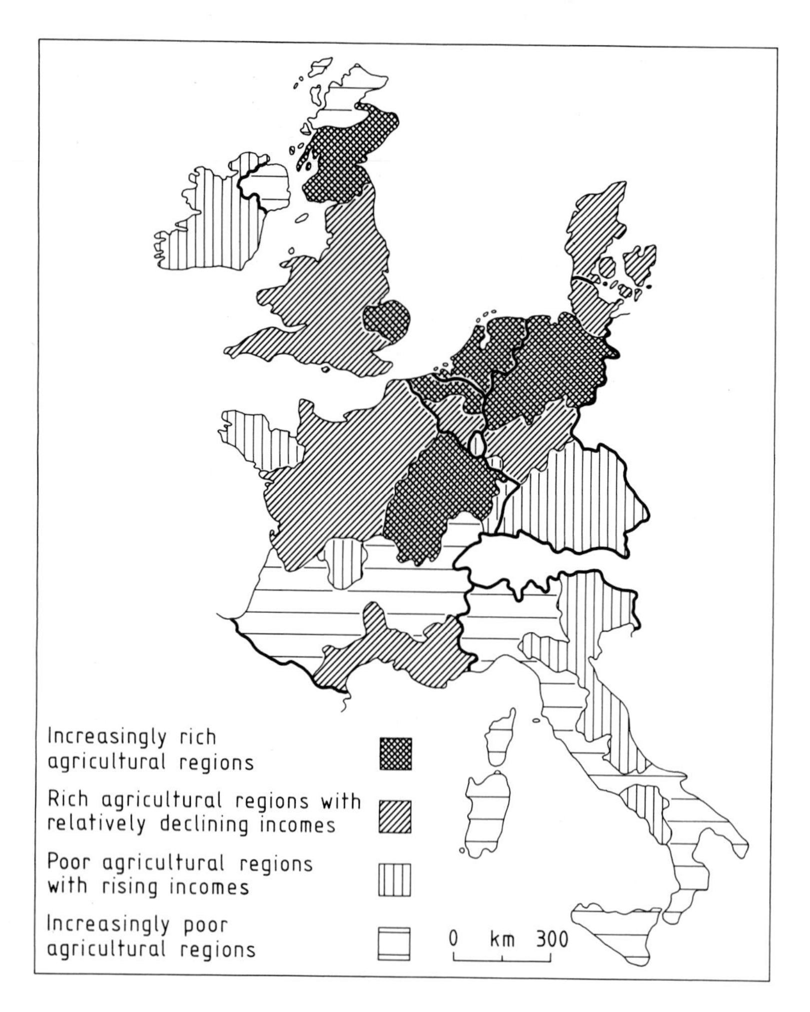

Fig. 3.7 Typology of regional trends in farm incomes, 1968/9–1976/7 (*source:* Bowler 1985)

rural society in Europe (Franklin 1969 and 1971). Although declining in numbers, the peasantry still forms the largest single category within the rural community throughout Europe and could number anything between 10 and 20 million. A true peasant society is characterized by a total labour commitment, no division of labour, no commercial objectives, and the working of the land to provide food, clothing, and shelter. The peasant's farm is the sum of the labours of the family and its generations, and the labour force consists basically of the peasant's kinship.

There are three types of peasant farmers:

1. The full-time farmers, where labour is supplied by family members on a full-time basis and where the main part of their income is derived from agricultural sources.
2. The marginal farmers, consisting in the main of elderly farmers who remain on the farm for the security it provides, but also including farmers who find it hard to make a living out of farming because of poor environmental and structural characteristics.
3. The part-time farmers, who have overcome the problem of inefficient income by adding a second, non-agricultural source of income.

It is this last group that witnessed a post-war expansion in numbers and their emergence was largely unassisted and independent of state policies or rural movements (Franklin 1971). Franklin sub-divided peasant part-time farmers into two groups:

1. Peasants who own a second business like a guest house, hotel, or some rural craft.
2. Peasants who also work in factories and are known as worker-peasants.

A worker-peasant is, therefore, a farmer who derives income from off-farm sources. In such communities, the agricultural population commute to urban industrial jobs each day, yet still manage to run their farms in the evenings and at weekends. These 'five o'clock' farmers are numerous and estimates suggest that the proportion of holdings operated by worker-peasants ranges from 15 per cent in France to 30 per cent in West Germany and even higher in Switzerland (Clout 1971). In West Germany, in the 1960s, there were approximately 617 000 worker-peasant holdings, compared with 322 000 part-time holdings and 511 000 full-time holdings (Franklin 1971). On a more localized scale, worker-peasants are heavily concentrated where industrial development and mining activity have been established and where public transport allows commuting from rural areas to urban factories. In West Germany, for example, 40 per cent of the worker-peasants around Kassel work in the factories of Volkswagen (Burtenshaw 1974), and over half the holdings in the Saarland and Baden-Württemberg are run by worker-peasants (Mrohs 1982).

The dual existence of the worker-peasant means that he and his family together work extremely long hours and it is surprising that these people continue to work the land. However, there is a strong sentimental attachment to family land and, apart from being able to produce their own food, land was seen to be a good investment. The action of authorities has been another factor encouraging worker-peasants to remain on the land. Whenever industrial unemployment was a threat, the authorities would sponsor peasant cultivation, thus releasing their industrial jobs for other people. Commuting from rural areas also helped to reduce the housing problem in urban areas and much post-war recovery of factory production was due to worker-peasant labour.

Clout (1972b) has investigated the worker-peasant phenomenon, listing a number of advantages and disadvantages which characterize these part-time farmers. The advantages are essentially fourfold:

1. It has been argued, mainly by Franklin, that a dual economy is useful; this is especially so during times of unemployment, when the worker-peasants can return to the land for a while.
2. The existence of worker-peasant communities in rural areas maintains a demand for local services, such as shops and schools.
3 The worker-peasants gain higher incomes than they would otherwise obtain from just farming or just industrial work. This extra money is normally spent on farm equipment and education for the children.
4. The mixing of worker-peasants in industrial areas leads to a diffusion of ideas into peasant societies.

However, the advantages are usually outweighed by numerous disadvantages:

1. Farm-size tends to be small and farms often become fragmented. This hinders consolidation plans in western Europe which would benefit the full-time farmers.

2. The farms tend to be over-mechanized. Tractors are often purchased, but infrequent use means they are uneconomic.

3. Worker-peasant farms are usually less productive because of limited labour inputs and a lack of formal training.

4. Much land becomes abandoned because of a lack of time. The land often returns to scrubby pasture and is sometimes planted with trees. Social fallow of this nature (*Sozialbrache*) is a wasted resource and hinders farm improvement schemes. The phenomenon is particularly acute in West Germany where in some districts up to 50 per cent of the land may be under social fallow (Franklin 1969; Künnecke 1974).

5. The worker-peasants are of a 'mixed' status; they are neither complete farmers nor complete industrial workers. The days are exceptionally long and many peasants remain unmarried and have no heirs to pass the land on to.

6. Franklin suggests that the peasants begin to compare their ideas with the modern industrial world. This leads to an erosion of traditional attitudes and, eventually, to the abandonment of farming and the disposal of the farm. However, this is a long process, as the peasants tend not to mix with their industrial colleagues and local ties remain exceptionally strong.

The 1960s and 1970s witnessed a decline in the number of worker-peasants in western Europe. One possible explanation for this trend is that the younger people were not prepared to become part-time farmers. A dual existence of this nature could lead to insecurity in both jobs, as well as providing little leisure time. The children of worker-peasants were receiving better education and being trained for industrial or service jobs. In all, this sentimental attachment of farmland is becoming weaker with successive generations. Another explanation is that the economic recession of the 1970s altered the position of the worker-peasant. Industries which had attracted local farmers began to shed labour as mechanization and automation became more common (Frank 1983). Rather than return to full-time farming, the peasants often abandoned it altogether, although they usually kept their land and so maintained the problem of *sozialbrache*. Despite this, a large number of worker-peasants remains, especially in Greece, southern Italy, south-central France, and the Iberian peninsula, representing a problem that affects the success of structural and consolidation policies in agriculture in western Europe.

Although the worker-peasant phenomenon is on the wane, there has been a general increase in part-time farming in western Europe and Frank (1983) estimated that 27 per cent of farmers in the EC had a gainful activity outside farming in 1980. For example, in the United Kingdom, the percentage of part-time farmers rose from 23 to 27 per cent between 1971 and 1979 (Gasson 1982). This growth has been most marked on the rural–urban fringe and relates in part to the desire of 'urbanities' to live in the country. These so-called 'hobby' farmers represent another tier in the hierarchy of part-time farming and many engage in such activities as horsiculture and farm tourism. Part-time farming, of different dimensions, would appear to be a permanent and not a transitional feature of the agricultural landscape of western Europe.

References

Baker, A. R. (1961). 'Le Remembrement Rural en France'. *Geography*, **46,** 60–2.

Belding, R. (1981). 'A test of von Thünen locational model of agricultural land-use with accountancy data from the European Economic Community'. *Transactions of the Institute of British Geographers*, **6,** 176–87.

Berendt, M. (1974). 'The Common Agricultural Policy I and II'. *European Studies*, **19** and **20,** 1–4.

Bowler, I. R. (1976a). 'The CAP and the space-economy of agriculture in the EEC'. In Lee R. and Ogden, P. E. (eds), *Economy and Society in the EEC*. Saxon House, Farnborough.

Bowler, I. R. (1976b). 'Recent developments in the agricultural policy of the EEC'. *Geography*, **61,** 28–31.

Bowler, I. R. (1985). 'Some consequences of the industrialisation of agriculture in the European Community'. In Healey, M. J. and Ilbery, B. W. (eds), *The industrialisation of the countryside*. GeoBooks, Norwich.

Burtenshaw, D. (1974). *Economic geography of West Germany*. Macmillan, London.

Chisholm, M. (1968). *Rural settlement and land-use*, 2nd edn. Hutchinson, London.
Clout, H. D. (1968). 'Planned and unplanned changes in French farm structures'. *Geography*, **53,** 311–15.
Clout, H. D. (1971). *Agriculture: Studies in contemporary Europe*. Macmillan, London.
Clout, H. D. (1972a). *Geography of post-war France: a social and economic approach*. Pergamon Press, Oxford.
Clout, H. D. (1972b). *Rural geography*. Pergamon, Oxford.
Clout, H. D. (ed.) (1975). *Regional development in western Europe*. Wiley, London.
Clout, H. D. (1984). *A rural policy for the EEC?* Methuen, London.
Clout, H. D., Blacksell, M., King, R. and Pinder, D. (1985). *Western Europe: geographical perspectives*. Longman, London.
Fennell, R. (1979). *The Common Agricultural Policy of the European Community*. Granada, London.
Frank, W. (1983). 'Part-time farming, underemployment and double activity of farmers in the EEC'. *Sociologia Ruralis*, **23,** 20–7.
Franklin, S. H. (1969). *The European peasantry*. Methuen, London.
Franklin, S. H. (1971). *Rural societies: Studies in contemporary Europe*. Macmillan, London.
Gasson, R. M. (1982). 'Part-time farming in Britain: research in progress'. *GeoJournal*, **6,** 355–8.
Guedes, M. (1981). 'Recent agricultural land policy in Spain'. *Oxford Agrarian Studies*, **10,** 26–43.
Hallett, G. (1968). 'Agricultural policy in West Germany'. *Journal of Agricultural Economics*, **19,** 18–95.
Hill. B. E. (1984). *The Common Agricultural Policy: past, present and future*. Methuen, London.
Hirsch, G. P. and Maunder, A. H. (1978). *Farm amalgamation in western Europe*. Saxon House, Farnborough.
Holland, S. H. (1980). *The uncommon market*. Macmillan, London.
Ilbery, B. W. (1985). *Agricultural geography: a social and economic analysis*. Oxford University Press, Oxford.
King, R. L. and Burton, S. (1982). 'Land fragmentation: notes on a fundamental rural spatial problem'. *Progress in Human Geography*, **6,** 476–94.
King, R. L. and Burton, S. (1983). 'Structural change in agriculture: the geography of land consolidation'. *Progress in Human Geography*, **7,** 471–501.
Künnecke, B. H. (1974). 'Sozialbrache—a phenomenon in the rural landscape of Germany'. *The Professional Geographer*, **26,** 412–15.
Lambert, A. (1963). 'Farm consolidation in western Europe'. *Geography*, **48,** 31–47.
Mayhew, A. (1970). 'Structural reform and the future of West German agriculture'. *Geographical Review*, **60,** 54–68.
Mayhew, A. (1971). 'Agrarian reform in West Germany'. *Transactions of the Institute of British Geographers*, **52,** 61–76.
Minshull, G. N. (1978). *The new Europe: An economic geography of the EEC*. Hodder and Stoughton, London.
Mrohs, E. (1982). 'Part-time farming in the Federal Republic of Germany'. *GeoJournal*, **6,** 327–30.
Naylon, J. (1975). *Andalusia*: Problem regions of Europe. Oxford University Press, Oxford.
Naylor, E. L. (1982). 'Retirement policy in French agriculture'. *Journal of Agricultural Economics*, **33,** 25–36.
Riley, R. C. and Ashworth, G. J. (1975). *Benelux: An economic geography of Belgium, the Netherlands and Luxembourg*. Chatto and Windus, London.
Van Valkenburg, S. (1960). 'An evaluation of the standard of land use in western Europe'. *Economic Geography*, **36,** 283–95.
Whitby, M. C. (1968). 'Lessons from Swedish farm structure policy'. *Journal of Agricultural Economics*, **19,** 279–99.

4

Energy

There has been a dramatic increase in the world's consumption of energy during the post-war period and the average annual growth rate of five per cent means a doubling of demand every fifteen years. Readily available supplies of relatively cheap energy are essential for industrial growth and economic development.

Between 1945 and 1973, western Europe witnessed tremendous growth patterns in energy consumption, with an average annual rate of increase of 10 per cent. This was accompanied by significant regional changes in the patterns of energy supply and demand. In the post-war period, the energy economy of western Europe has changed from a one-fuel economy to a competitive, multi-fuel economy (Jensen 1967). The decline of coal and the emergence of oil, and more recently natural gas and primary electricity as major energy sources is exemplified in Table 4.1. In a relatively short time-span, coal has declined in importance, from a situation of accounting for over three-quarters of the energy consumed in western Europe in 1950, to one of accounting for less than one-fifth in 1980. By the mid-1980s, oil was easily the major energy source, with natural gas still increasing and competing with coal as the second major source of energy. The last decade has also witnessed an ever-increasing role played by electricity generation and nuclear power in the west European energy scene; by the year 2000 primary electricity could account for one-fifth of all the energy consumed in western Europe.

Table 4.1 The balance of energy sources in western Europe, 1950–2000

	Percentage share of total energy from:			
	Solid fuel	Oil	Primary electricity	Natural gas
1950	83	10	7	—
1970	27	58	6	8
1975	21	57	9	13
1978	20	55	10	15
1982	23	50	11	16
2000	6	64	20	10

4.1 The energy economy, 1945–70

Industrial recovery after World War II was quite rapid in western Europe and the ten year period from 1945 to 1955 represented an age of cheap, indigenous coal. Pre-war levels of coal output were reached in most countries by 1947 and considerable government investment was placed in the coal industry, especially in the United Kingdom and the West German Ruhr. At this time there was very little use of hydro-electric power, natural gas, or indigenous oil and in France and the United Kingdom, the coal industry further benefited from nationalization. Coal output continued to increase in the early 1950s and the industry was given a boost in 1952 when the European Coal and Steel Community (ECSC) was set up, comprising the six countries which later formed the European Community. The ECSC represented a common market for coal and steel and one of its chief aims was to coordinate coal production in member countries and to reduce the internal cost of energy.

By the mid-1950s, the demands for energy had increased and the indigenous coal industry was unable to expand sufficiently to sustain the rate of economic growth which it had generated (Odell 1976). This was the beginning of a period when imported oil came to dominate investment and decision-making in western Europe. Between

4.1 The Wendel coal pit in Lorraine (ECC)

1950 and 1970, oil consumption increased by seventeen times and imported coal from America began to compete with local coal supplies. By the late 1950s, western Europe was a two-fuel economy and a net consumer of energy rather than a net producer. In 1950 western Europe was dependent on the rest of the world for 10 per cent of her energy needs; by 1968 this figure had risen to 57 per cent and by 1970 it stood at a very high 59.1 per cent. The figures for certain individual countries were even higher, reaching 71 per cent in France, 82 per cent in Italy, Belgium, and Ireland, and 99.6 per cent in Denmark. The energy economy had become dominated by foreign supplies and suppliers (Odell 1983).

The growth in importance of imported oil can be attributed to four main elements:

1. *A growth element*, whereby the total energy demands of western Europe increased and oil took its fair share of the growth.

2. *A structural element*, whereby many users of oil were themselves growing, such as the petrochemical and transportation industries.

3. *A competitive element*, whereby oil was much cheaper than European coal. Between 1950 and 1970, oil actually fell in price, as a result of new oil discoveries in the Middle East and new economies in oil transportation in the form of supertankers and oil pipelines.

4. *A cost element*, whereby the costs of mining coal increased as the more accessible seams became exhausted. Indigenous coal became uneconomic, compared to oil and imported American coal, and lost its commercial markets.

As a result of the rise of oil and the decline of coal, the ECSC was forced to play a broader role and its objectives were threefold:

1. To control and guide the inevitable contraction of the coal industry and to ensure that the process would cause as little disruption as possible.

2. To help plan output so that coal was mined in the most efficient way, at competitive prices. The objective was to maintain output of the specific types of coal that would satisfy the market.

3. To help ease the social and economic problems in the coalmining areas. Assistance from the ECSC came in the form of retraining schemes, direct grants for unemployed miners for one year, financial help in moving to new employment

areas, and money for the modernization of the coal industries. West Germany has received over 50 per cent of the ECSC's funds for modernization, of which 40 per cent has gone to the Ruhr industrial district.

Therefore, coal experienced a relative and absolute decline and output in the six main producing nations in western Europe (West Germany, the United Kingdom, France, Belgium, the Netherlands, and Spain) decreased from an output high of 467 million tonnes in 1955 to 279 million tonnes in 1972, and 267 million tonnes in 1975. A distinct pattern of decay was evident and between 1955 and 1970, the decline in coal output ranged from just 12 per cent in Spain and 23 per cent in West Germany, to 45 per cent in Belgium and 77 per cent in the Netherlands. Indeed, as Mellor and Smith (1979) note, the fall would have been more dramatic if it had not been for two factors: first, the increasing demand for electricity (i.e. coal-fired power stations); and secondly, the continued high demand for metallurgical coke for iron and steel production.

Hydro-electric power and natural gas accounted for less than 10 per cent of energy consumption in western Europe between 1950 and 1970, although this figure had risen to nearly 15 per cent by 1970, due almost entirely to the growing importance of natural gas. Specific physical conditions are required for the generation of hydro-electric power, such as high land, impervious rock, incised valleys, and large quantities of water, and this restricted its development in many parts of western Europe. However, this form of energy became important in Scandinavia and the Alpine countries, accounting for between 25 and 60 per cent of the total energy consumed in Sweden, Norway and Switzerland. Until the mid-1960s, natural gas production was on a small scale and was of importance locally only in south-west France and northern Italy. The discovery of large gas reserves in the Gröningen province of the Netherlands in 1964 changed the situation and Dutch gas began to enjoy a monopolistic position on the mainland.

Thus, this period saw a movement away from coal, of which western Europe had plenty, to oil, of which it had very little (Parker 1981). This was to have a marked impact on the regional patterns of energy supply and demand, and a movement of industry away from the coalfields towards coastal locations, to benefit from imported energy, was inevitable.

4.2 Energy in the 1970s and beyond

Since the mid-1970s, western Europe has been a four-fuel economy: oil, coal, natural gas, and primary electricity. The relative importance of each varies between countries and the geographical patterns of production and consumption are best analysed in relation to the four maps produced in Fig. 4.1. Total energy consumption in western Europe in 1982 amounted to 1062 million tonnes of oil equivalent (Mtoe) with 83 per cent of that being consumed by the ten EC members. In the same year, energy production totalled a mere 603 Mtoe, with 81 per cent being supplied by the European Community. Consequently, in 1982 western Europe was dependent on imported energy for 43 per cent of its total energy requirements; although high this figure compares favourably with the 57 per cent for 1974.

The largest producer of energy is the United Kingdom, with 220 Mtoe, followed by West Germany with 125, and the Netherlands with 64 (Fig. 4.1a). These three account for 66 per cent of the total energy produced in western Europe. The only non-EC countries to produce energy in any quantity are Norway (57 Mtoe), Spain (19), and Sweden (16). Oil is a rapidly increasing contributor to total energy production (24 per cent) and the largest single producer is the United Kingdom with 102 million tonnes (72 per cent of the west European total), followed by Norway (24.4 million), and West Germany (4.3 million). The European Community accounted for just 46 per cent of the oil produced in western Europe in 1975, but with the emergence of the United Kingdom as a major producer, this has since risen to over 80 per cent. Coal and lignite, natural gas, and primary electricity all make significant contributions to total energy output. Coal remains western Europe's most important indigenous source of energy, accounting for 32 per cent of total energy production. The countries of the Ten are responsible for over 90 per cent of the coal produced in western Europe, with West Germany and the United Kingdom playing leading roles. Natural gas and primary electricity account for 24 and 20 per cent of total energy production but whereas

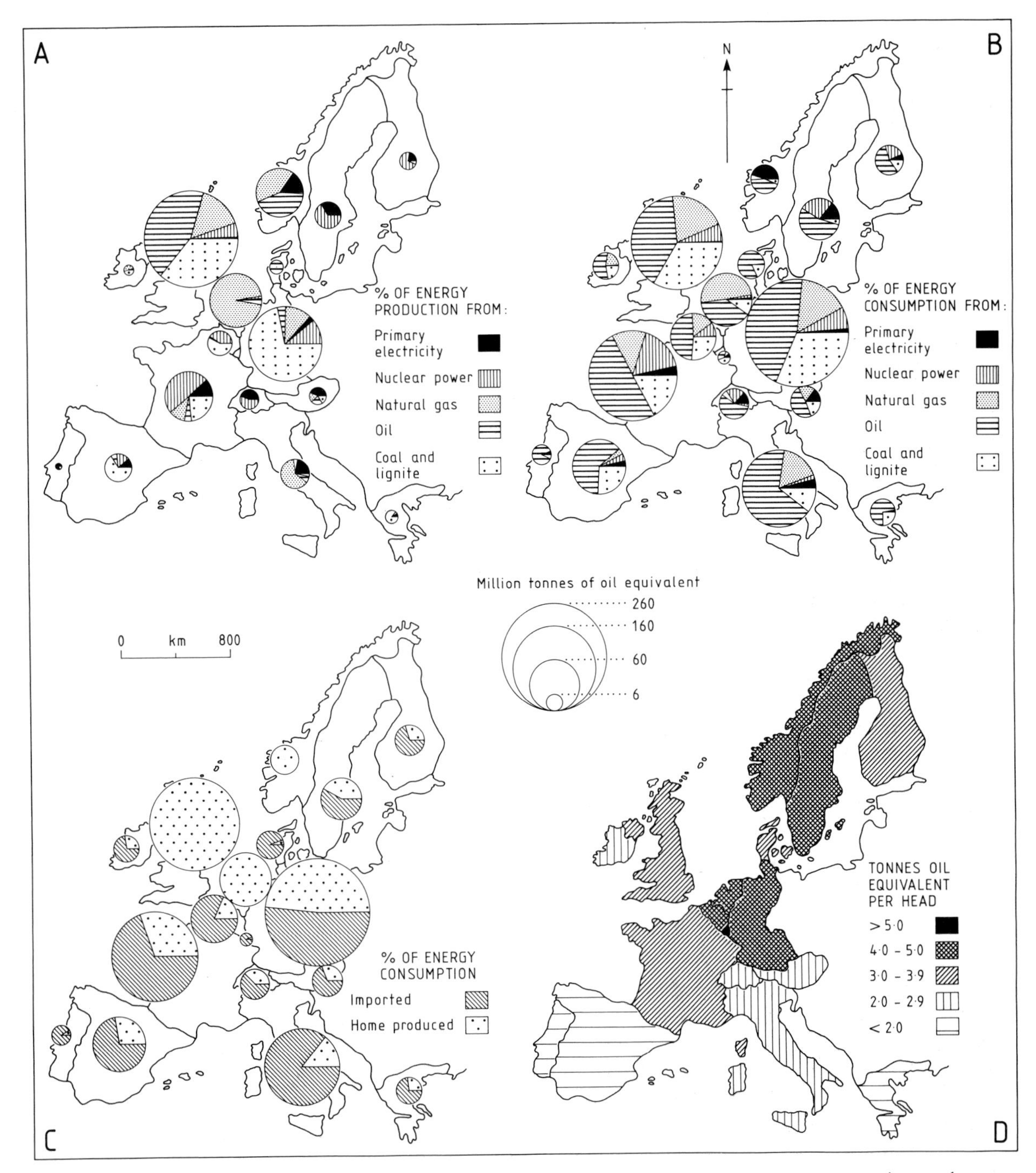

Fig. 4.1 Energy in western Europe, 1982: (a) energy production; (b) energy consumption; (c) dependence on imported energy; (d) consumption of energy per capita

the EC countries account for 82 per cent of western Europe's gas, notably from the Netherlands and the United Kingdom, they account for only 63 per cent of its primary electricity. Norway, Sweden, and Switzerland, with their good physical conditions for the production of hydroelectric power, have remained important producers of primary electricity, and Sweden is the fourth producer of nuclear power after France, West Germany, and the United Kingdom.

In terms of energy consumption, West Germany (248 Mtoe) and the United Kingdom (194 million) again lead the way, although France (175) and Italy (125) are heavy consumers too (Fig 4.1b). The non-EC countries are a long way behind, with Spain (62) and Sweden (36) the only significant consumers. The largest individual consumers of oil are West Germany, France, the United Kingdom, and Italy, and the Ten accounts for 81 per cent of all oil consumed in western Europe. In fact, oil, accounting for half of the total energy consumed in western Europe in 1982, is the largest energy source for all west European countries (with the exception of Luxembourg and the Netherlands) and in Denmark, Greece, and Portugal it totally dominates the energy economy. Coal is the second major source, accounting for 23 per cent of the total energy consumed in western Europe. The major consumers are again the members of the European Community, who use 88 per cent of the total energy produced from coal in western Europe. Natural gas is increasing in importance as a major energy source within the European Community, especially in the Benelux countries and the United Kingdom, and the Ten accounts for 95 per cent of total gas consumption in western Europe. It is now the major energy source in the Netherlands, although more is consumed in West Germany and the United Kingdom. Outside the EC, natural gas is insignificant as an energy source, and whilst Norway's sector of the North Sea has high production potential, most of its gas is exported to Community members. Overall, natural gas accounts for 16 per cent of total energy consumed in western Europe. Primary electricity is the least important source of energy, accounting for 11 per cent of total energy consumption, but nuclear power's share of this continues to increase.

If the figures for energy production and consumption in the countries of western Europe are compared, it becomes clear that there is an overwhelming dependence on imported energy (Fig. 4.1c). The only countries to produce more energy than they consume are the Netherlands, Norway, and the United Kingdom. Despite this, the Netherlands is still dependent on imported coal and oil, just as the United Kingdom is on gas, and Norway on coal. Two other countries—West Germany and Sweden—produce more than 40 per cent of their energy requirements. At the other extreme, Denmark, Luxembourg, and Portugal are over 90 per cent dependent on imported energy. In terms of the consumption of energy per inhabitant, there is little variation between the countries of western Europe, except for the two extremes of 8676 tonnes (oe) in Luxembourg and only 858 tonnes in Portugal (Fig. 4.1d). However, it is noticeable that consumption is lower in the more peripheral Mediterranean countries and Ireland.

Western Europe will continue to depend on imported energy during the 1980s, despite the growing importance of oil and natural gas supplies in the North Sea. With imported oil dominating the energy economy in the 1970s there was an expansion of industry in coastal locations and major inland centres near pipelines, and a huge increase in refining capacity, which in the peak year of 1976 stood at 1032 million tonnes. Italy had the largest refining capacity (221.9 million tonnes), followed by France (171.8), West Germany (153.9), and the United Kingdom (145.6). As early as 1972, there were 136 oil refineries in the European Community alone, with coastal locations dominant and ports such as Marseille, Antwerp, Rotterdam, Le Havre, and Genoa growing in importance. A network of crude-oil pipelines emerged, transporting oil from import terminals to inland centres (Fig. 4.2). Refineries in the Ruhr and Rhine Valley were being supplied from Rotterdam and northern Germany, whilst centres in Bavaria, France, and Switzerland received oil from the south European ports of Marseille, Genoa and Trieste. The development of transnational crude-oil pipelines epitomized the geographically integrated nature of the oil market in western Europe (Odell 1976). The pattern of oil refining is dominated by the massive agglomeration at Rotterdam–Europoort, and Rotterdam, the largest port in the world, handles over 130 million tonnes of crude oil annually and

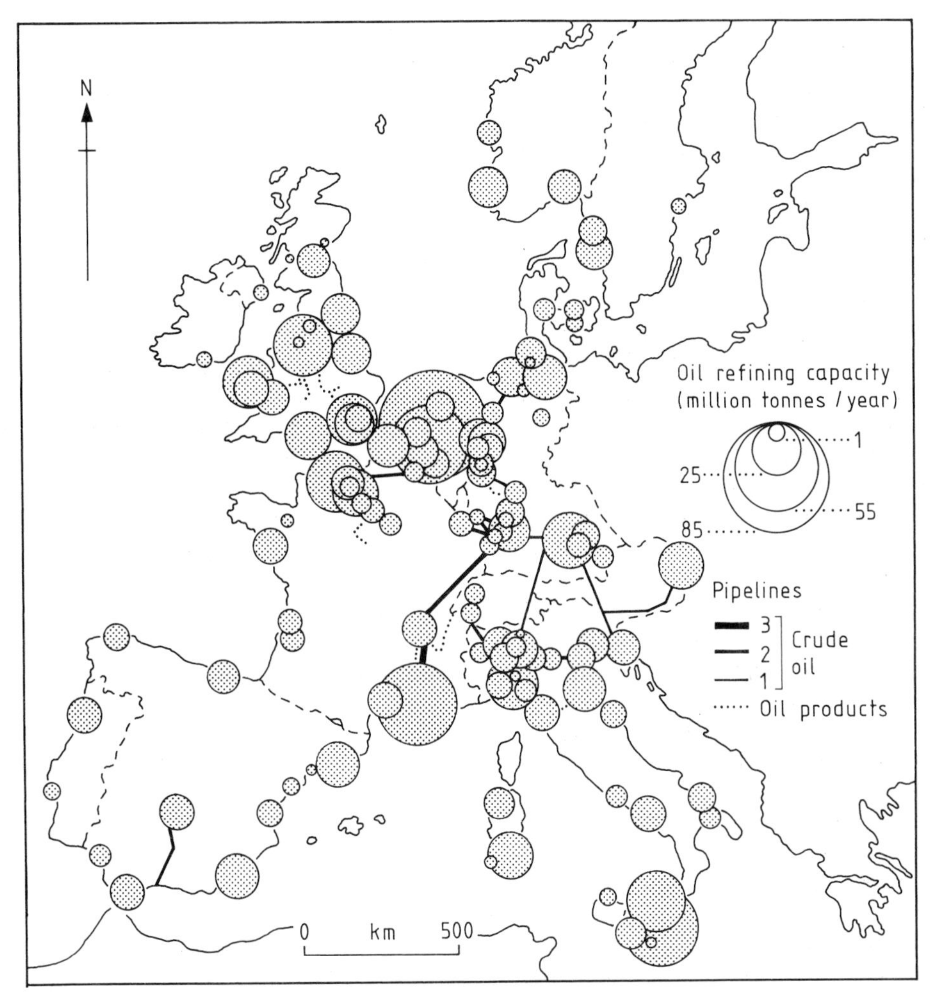

Fig. 4.2 The pattern of oil refining in western Europe, 1976 (*source:* Odell 1976)

has been termed the oil jugular of western Europe (Clout 1975) with pipelines supplying Belgium and West Germany. It supplies more than one-fifth of western Europe's daily oil needs and is ideally situated to receive supplies from the North Sea.

However, by the early 1980s a fall in the demand for oil products, together with cheap imports from big refineries built in the Middle East and North Africa, led to a surplus oil refining capacity of around 400 million tonnes. As a consequence, capacity declined by 25 per cent between 1980 and 1985. It was mainly the smaller refineries that were closed, although Clout *et al.* (1985) noted that BP's Isle of Grain refinery (10.3 million tonnes) and Texaco's Ghent refinery (9 million) were among the casualties. The decline in oil refining has not been spatially uniform and countries with small total capacity (Scandinavia, Portugal, and Greece) have suffered less than the 'core' countries, Spain and Italy. With falling oil prices in the 1980s, many suriving refining companies have been making losses on their operations and responded by incorporating modern cracking processes into their refineries. This transforms heavy oil products into more profitable lighter products and by 1986 more than '30 refinery-conversion projects will have been completed throughout the core and southern periphery, raising western Europe's total cracking capacity by 50 per cent' (Clout *et al.* 1985). However, the viability of such a conversion programme is dependent upon an increasing demand for lighter oils, which as will be shown cannot be guaranteed.

Indeed, oil imports continued to decline and by 1985 accounted for only 39 per cent of primary energy consumption. This, in part, reflects the rapid exploitation of North Sea oil reserves since 1975. In 1976 western Europe produced 22 million tonnes of oil; by 1979 this had risen to 150 million. A figure of 500 million could be realized by 1990, from the North Sea and coastal waters off Spain, Greece, and Italy, representing a major transformation in 15 years. Most of the oil is located in the British and Norwegian sectors of the northern basin of the North Sea and some large oilfields, such as Brent, Forties, and Statfjord, have been discovered. An extensive system of pipelines links these fields to the Shetland Isles, mainland UK, and Norway. North Sea oil has certainly tempered OPEC's power and oil prices in the cartel have been falling since 1983. However, estimates of the total reserves of North Sea oil are conflicting and supplies could be depleted by the early part of the twenty-first century. Consequently, the search for alternative energy sources continues.

The upsurge in oil prices associated with the 1973 crisis provided an ideal opportunity for the indigenous coal industry to reappraise its future, especially as western Europe's reserves of this fuel are huge when compared with other energy sources (Table 4.2). There is enough coal in western Europe to sustain all current energy needs for approximately 75 years, with West Germany and the United Kingdom particularly well endowed. Indeed, the United Kingdom's Plan for Coal (1974) suggested that production could increase from the 130 million tonnes in 1973 to between 135 and 150 million by 1985; similarly the 1977 'Coal for the future' forecast 170 million tonnes by 2000. In West Germany, similar plans for expansion saw output increasing from 129 million tonnes in 1975 to between 145 and 150 million by 2000 (Manners 1980). These figures have proved to be wildly optimistic and whilst the demand for coal has increased, production in the European Community continues to follow a downward path (Sill 1984), from 270 million tonnes in 1973 to 229 million in 1983. Output has become increasingly concentrated in West Germany and the United Kingdom, has declined significantly in France and Belgium, and ceased altogether in the Netherlands. Outside the EC, Spain has managed to increase its production of coal, from 10.6 millions tonnes in 1975 to 15.5 million in 1982.

Table 4.2 Energy reserves in western Europe, 1980 (proven recoverable reserves)

	Oil (million tonnes)	Gas (Thousand million cu. metres) (1982)	Bituminous coal (million tonnes)	Lignite and brown coal (million tonnes)
West Germany	42	176	23 991	35 150
France	8	73	550	60
Italy	51	172	—	31
Netherlands	9	1515	—	—
Belgium	—	—	440	—
Luxembourg	—	—	?	—
United Kingdom	1906	633	45 000	—
Ireland	—	—	55	—
Denmark	46	50	—	—
Greece	21	11	—	1550
Norway	550	1440	18	—
Sweden	—	—	—	1
Switzerland	—	—	—	—
Austria	18	10	—	65
Portugal	—	—	5	33
Finland	—	—	—	—
Spain	18	18	398	553

Sources: European Marketing Data and Statistics, 1984; Basic Statistics of the Community, 1983.

4.2 Veba Chemical Company's petroleum refinery and 1380 MW coal-fired Scholven power station at Gelsenkirchen in the Rhine-Ruhr (Presseamt Stadt Gelsenkirchen)

Although the demand for coal in power stations increased from 73 million tonnes in 1973 to 88 million in 1985, the rundown of coal capacity in the late 1960s and early 1970s left the industry unable to respond adequately to market opportunities. As a consequence, much of the increased demand for coal has been met by imports, which had the advantage of being cheaper than indigenous supplies. As early as 1973 45 million tonnes of coal were imported, mainly from the USA, Poland, and Russia, and by 1981 23.2 per cent of hard coal consumption was imported, this time essentially from the USA, South Africa, and Australia (Sill 1984). Manners (1980) has argued that coal will continue to serve a declining share of total energy requirements in the remainder of the 1980s. However, he feels that by the 1990s, the prospects for coal will be much better, due to a decline in oil supplies, but the problems of mining, waste disposal, and poor labour relations in western Europe will ensure further significant increases in imports.

Natural gas will continue to emerge as an important element in the European energy scene. As with oil, this energy source has benefited from large discoveries in the North Sea and from the network of gas pipelines which traverse Europe (Fig. 4.3). The main west European reserves are found in the Netherlands, the United Kingdom, and Norway. By the mid-1970s, gas production had increased in scale dramatically and the Soviet Union had become a major exporter of gas to western Europe, especially to Italy, West Germany, and France. The known reserves of the Dutch Gröningen field, together with the actual and potential supplies from the British and Norwegian sectors of the North Sea, mean that western Europe is comparatively well endowed with natural gas. However, it is difficult to evaluate the extent of gas reserves in the North Sea, as they depend in part on oil production. This is because gas in the northern part of the North Sea is found in association with oil although in the southern part it is found on its own. With the possibility of other coastal reserves in the Mediterranean, Adriatic, and Celtic Seas, together with the large

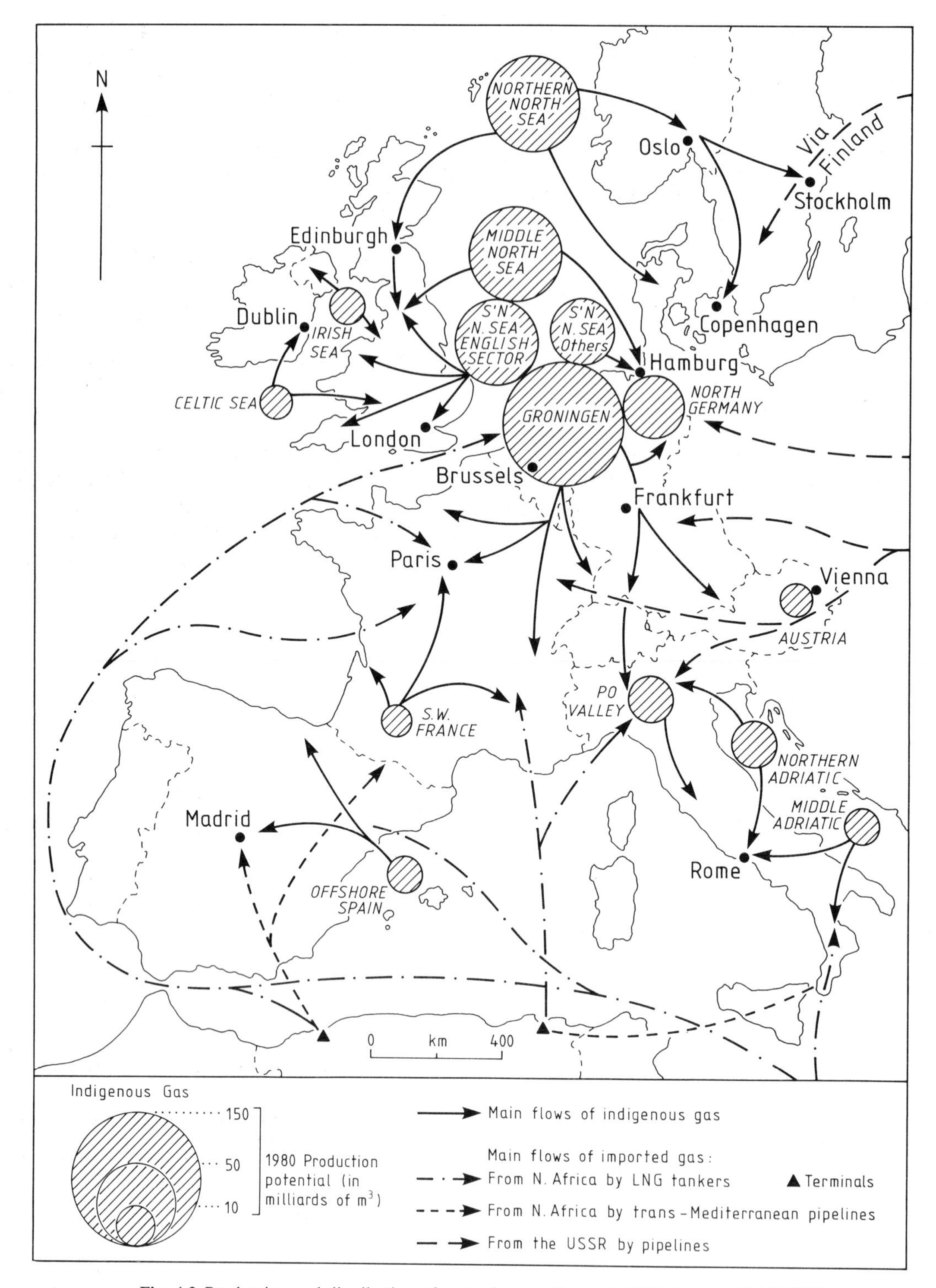

Fig. 4.3 Production and distribution of natural gas in the early 1980s (*source:* Odell 1981)

proven reserve in Morecambe Bay off the Lancashire coast, natural gas could supply up to 20 per cent of the west European energy market by 1990. Despite this, gas imports into western Europe are increasing. These come first from North Africa, via the trans-Mediterranean pipeline (Dean 1982) and secondly, from Iran via Russia to Switzerland, Austria, West Germany, and Italy (Odell 1981).

One of the most important aspects of the energy revolution has been the transformation of primary fuels into secondary source electrical energy. Electricity is an increasingly important source of energy in western Europe and its generation depends upon the use of coal, oil, and natural gas in thermal power stations, together with nuclear power and hydro-electric power stations (Minshull 1978). The demand for electric power in western Europe has been increasing by approximately 7.5 per cent per year over the past 25 years, leading to a doubling in demand every 9.6 years. According to Alting Von Geusau (1975), its increased use is due to four factors:

1. It is the cleanest form of energy both in use and production.
2. It is flexible and can be transmitted and distributed by simple means. Electricity is the most suitable energy for automated industrial processes.
3. In western Europe, electricity will be produced in the future almost entirely from nuclear sources, which are available in large quantities, at competitive prices.
4. There has been an increase in the number of new uses of electrical energy, in place of other energy sources.

The production of electrical energy in western Europe increased by 45 per cent between 1971 and 1983 and the largest producers are West Germany, France, the United Kingdom, and Italy (Table 4.3). This increase has been accompa-

Table 4.3 Production of electrical energy, 1971–83

Country	Net production (GWh)										
	Hydro			Nuclear			Conventional thermal			Total production	
	1971	1983	% total production	1971	1983	% total production	1971	1983	% total production	1971	1983
West Germany	13 809	19 346	5.6	5 470	60 087	17.4	233 601	265 503	77.0	242 880	344 936
France	48 726	71 045	26.7	8 743	103 008	38.7	91 226	92 529	34.6	148 998	266 339
Italy*	39 807	43 809	24.9	3 189	6 587	3.7	74 285	122 946	69.9	119 764	175 968
Netherlands	—	—	—	383	3 674	6.4	42 364	53 903	93.6	42 747	57 577
Belgium	156	1036	2.2	1	14 752	30.8	31 442	32 148	67.0	31 597	47 936
Luxembourg	1056	472	52.3	—	—	—	1213	430	47.7	2 269	902
United Kingdom	4294	5597	2.2	24 013	38 721	15.2	210 292	210 182	82.6	238 599	254 500
Ireland	460	1189	11.4	—	—	—	5618	9250	88.6	6 078	10 439
Denmark	24	26	0.1	—	—	—	17 518	22 386	99.9	17 542	22 412
Greece	2646	3551	16.3	—	—	—	8397	18 212	83.7	11 043	21 763
The Ten	110 978	146 071	12.1	41 799	226 889	18.9	706 259	827 186	68.8	861 499	1 202 772
Spain	32 208	27 042	24.9	2414	8 300	7.6	25 046	73 312	67.5	59 668	108 654
Portugal	6061	6748	46.0	—	—	—	1 596	7 935	54.0	7 657	14 683
Norway	62 965	91 740	99.5	—	—	—	300	415	0.5	63 265	92 155
Sweden	51 507	54 962	56.5	85	37 295	38.3	13 566	5 084	5.2	65 158	97 341
Switzerland	27 563	37 035	70.8	1817	14 276	27.3	2 207	974	1.9	31 587	52 285
Austria	16 602	30 508	73.1	—	—	—	11 104	11 250	26.9	27 706	41 758
Finland	10 470	12 983	31.4	—	15 820	38.3	10 455	12 515	30.3	20 925	41 318
Western Europe	318 354	407 089	24.7	46 115	302 580	18.3	770 533	938 671	56.9	1 137 465	1 650 966

* Italy produced 2626 GWh of geothermal electricity.

Source: Basic Statistics of the Community.

4.3 Uranium research centre at Karlsruhe (Gesellschaft für Kernforschung)

nied by greater efficiency in both production and transfer since the 1960s. Minshull (1978) has shown that the same amount of primary fuel as used in 1958 was, in the late 1970s, generating 45 per cent more electricity, which can be distributed, via the grid, over large areas with a minimum loss of energy.

Table 4.3 indicates that 57 per cent of the electrical energy produced in western Europe comes from conventional thermal sources (70 per cent in 1976) and these are the most important in a large majority of the countries. Only in seven countries—Norway, Austria, Luxembourg, Sweden, Switzerland, Finland, and France—does a larger percentage share come from non-thermal sources. In the first three, hydro-electric power is dominant, whereas in the next three hydro-electric power and nuclear power together provide more electricity than thermal sources; in France thermal sources are second to nuclear energy, which, as can be seen, has developed rapidly in that country. Indeed, nuclear power is becoming more important as a source of electrical energy, increasing its share of total production from 4 per cent in 1971, to 8 per cent in 1976, and over 18 per cent in 1983. It now accounts for over 30 per cent of the electrical energy produced in Switzerland, Finland, Belgium, and France.

Industry (56 per cent) and households (24 per cent) are the most important consumers of electricity and the mobility of this energy source has helped to dismantle the pattern of heavy industrial regions near coalfields, in favour of a more flexible distribution of industry (Chapter six). The movement of electricity within western Europe has been aided by the formation of international organizations, designed to ease the exchange of electricity between member countries. Lucas (1977) points to two such organizations, both of which are essentially private endeavours:

1. Union for the Coordination of Production

and Transport of Electricity (UCPTE) which has eight members, comprising Switzerland, Austria and the original Six.

2. The International Union of Producers and Distributors of Electricity (UNIPEDE) of which all European power-supplying countries are members.

Nuclear power is set to become the major energy source of the future in western Europe (Greenhalgh 1975; Pierantoni 1975; Ilbery 1981a and b). The structure of the nuclear energy market is continuing to develop and the contribution of nuclear energy to the total production of electrical energy is increasing rapidly, having quadrupled between 1971 and 1983. Its rapid rise in importance during the 1970s was related to the availability, efficiency, and cost of fossil fuels, and the availability of uranium. Reserves of uranium, nuclear power's 'raw material', are sufficient for large-scale developments in the production of nuclear power, especially with the development of fast-breeder reactors, which extract 50–60 times more energy from the same amount of uranium as present reactors. In addition, uranium is required in relatively small quantities and is thus easily transportable.

The potential of nuclear power was realized in the 1950s, as exemplified by the formation of Euratom (European Atomic Energy Community) in 1957, but availability of alternative energy sources in the 1960s meant that it was disregarded. However, the increasing cost of oil during and since the 1973/4 oil crisis and the general concern over world energy shortages has encouraged an active programme of nuclear research. Research into the production of nuclear power by fission and fusion processes has been progressing, and the Joint European Torus (JET), for example, which has its headquarters in Oxfordshire, England, is conducting research into the possibility of establishing nuclear fusion as a useful source of energy. If successful, then in the course of the next century, fusion power plants, which should be 10–100 times 'cleaner' than today's nuclear fission power stations, may be built to provide electricity.

Getting power out of fusion means domesticating the hydrogen bomb, and it is this potentially dangerous situation of splitting (fission) or joining (fusion) atomic nuclei that has led to much debate on the environmental and health risks of developing nuclear power. It has been more difficult than anticipated to develop sufficiently safe technology to cope with problems such as the disposal of nuclear waste. The by-product of processing uranium is plutonium and this is produced in greater quantities than can be used. Plutonium remains radio-active for up to 200 000 years and must, therefore, be 'dumped' in safe places. There is much public concern in western Europe over the alleged dangers of such problems as the disposal of radio-active waste and this concern has led to the abandonment of plants in certain countries and the revision of projected developments for the 1980s and beyond. Such was the strength of public feeling in Sweden, for example, that plans to increase the number of nuclear units from 5 to 12 by 1985 caused a change of government in 1976.

Geographically, the production of nuclear power requires large supplies of water and plenty of space. For these reasons, and the dangers of radiation, early nuclear power plants were located away from the major centres of population in western Europe and the overall pattern of distribution was more dispersed than with other sources of primary electricity (Fig. 4.4). Therefore, nuclear stations are often found in peripheral areas, although the proliferation of plants (Table 4.4) means that this is unlikely to remain the case for the foreseeable future. Indeed, Openshaw (1982) has demonstrated how plants are often no longer remote from major urban centres.

The United Kingdom led the rest of western Europe in the development of nuclear power and by 1973 had 28 units, at 14 power stations, in operation. The majority of stations were located in coastal areas and in 1978 they contributed approximately 13 per cent to the national electricity grid. Other west European countries quickly followed suit and their combined nuclear capacity stood at 22 223 megawatts (MW) in 1977. A large increase, to 50 000 MW, was projected for the mid-1980s, supplying one-quarter of the total electricity output. In fact, Alting Von Geusau (1975) felt that nuclear power would cover over 50 per cent of the electricity demand in western Europe by 1990 and 65–75 per cent by 2000; as with the figures for coal, these will prove to be over-generous estimates.

Within western Europe, possibly the most

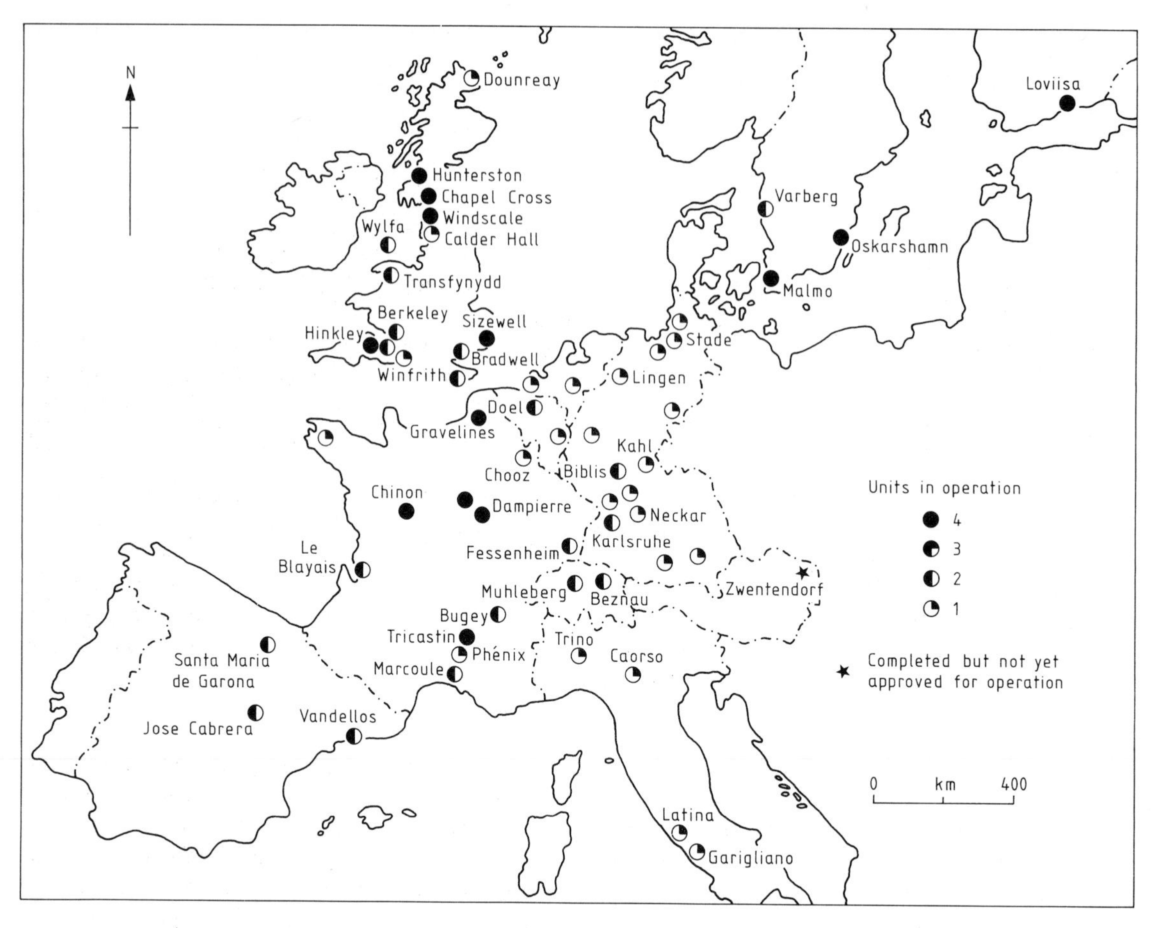

Fig. 4.4 Nuclear power plants in commercial operation in western Europe, 1982

Table 4.4 Nuclear power in western Europe, 1973–85

	No. of units in commercial operation								Expected
	1973	1974	1975	1976	1977	1978	1979	1982	1990
West Germany	8	8	9	11	12	16	16	16	27
France	10	10	10	10	12	13	18	36	61
Italy	3	3	3	3	3	4	4	4	6
Netherlands	2	2	2	2	2	2	2	2	2
Belgium	1	2	4	4	4	4	4	6	8
United Kingdom	28	28	28	31	33	33	33	35	42
Spain	3	3	3	3	3	3	4	6	15
Sweden	1	2	4	5	6	6	7	10	12
Finland	—	—	—	—	1	1	2	4	4
Austria	—	—	—	—	—	—	—	—	1
Switzerland	3	3	3	3	3	3	4	4	6

Sources: Derived from *Nuclear News*, February 1979; *Annual report of the International Atomic Energy Agency, 1983.*

4.4 The 200 MW Latina nuclear power station (Nuclear Power Group)

ambitious programmes of nuclear development are to be found in West Germany and France (Table 4.5). By the end of 1978, West Germany had become the largest single producer of nuclear power and had 16 units in operation, generating over 9500 MW of electricity. Production increased rapidly, as demonstrated by the 49 per cent increase between 1976 and 1977, and this was partly due to the performance of the Stade unit,

Table 4.5 Nuclear power in western Europe 1983–95 (GWh)

	1983	1990	1995
Belgium	3.5	5.5	5.5
Finland	2.2	3.2	3.2
France	26.9	56.0	61.2
West Germany	11.1	25.0	27.6
Ireland	—	0.6	0.6
Italy	1.4	5.4	5.4
Netherlands	0.5	0.5	0.5
Spain	3.8	12.7	15.2
Sweden	7.4	9.4	9.4
Switzerland	1.9	2.8	2.8
United Kingdom	8.3	12.3	14.3
Western Europe,	67.0	133.4	145.7
of which France, %	40.2	42.0	42.0

Sources: International Energy Agency, 1982; International Atomic Energy Agency, 1983.

the most efficient nuclear power station in the world. However, in common with many west Euorpean countries, West Germany has been engaged in a prolonged public debate about the environmental consequences of nuclear technology. Indeed, three authorized plants—Wyhl, Grohnde, and Brokdorf—had their construction licences revoked by the courts, and further plants are still awaiting construction licences. The result was that by 1984 only 16 of the 23 units planned were in operation, producing 11 110 MW of electricity. In an attempt to overcome the problems of waste disposal, the Government has long-term plans to build a fuel-cycle centre at Gorleben, in Lower Saxony, which will consist of a reprocessing plant, a waste vitrification plant, and a waste depository in the local salt domes. The Germans have also shown a strong interest in the fast breeder reactor and a prototype is being constructed at Kalkar in North Rhine-Westphalia, in partnership with the Belgians and Dutch, and should be commissioned in the mid-1980s.

France's nuclear power programme has also developed in an impressive fashion. In 1978, the installed capacity of nuclear power was 6673 MW and by 1984 the 36 units in operation were producing nearly 27 000 MW of electricity, more than twice that of West Germany, the second producer (Table 4.5). By 1990, it was envisaged that nuclear power stations would be supplying approximately 30 per cent of France's total energy needs. One of the world's first prototype fast breeder reactors went into operation at Phénix in 1973 and on the basis of its success, a decision to build the 1200 MW Super Phénix at Creys-Malville was made in 1976. This is a joint project with the Germans, Dutch, Belgians, and Italians, and work on the fast breeder reactor has been under way for some time and the unit is due to open in late 1985. However, as a result of the oil crisis and economic recession, the forecasts of the 1970s have proved to be over optimistic and, although France has met its nuclear targets, the expected doubling of demand has not materialized and the nuclear industry has been faced with a fall in demand for nuclear electricity (Bodanis 1984). The consequence is that the foreign debt for the nuclear programme stood at £5000 million in 1985, with many reactors switched off for long periods.

Similar programmes of nuclear development have been proposed and then revised in other west European countries. In Belgium, 30 per cent of the electrical requirements are met from nuclear power, the fourth highest percentage in western Europe; however, this is a lot less than the 50 per cent predicted for 1985. A lack of indigenous energy sources encouraged Italy to embark on a programme of nuclear development. The commissioning of a nuclear station at Caorso in 1978 brought the nuclear capacity to 1250 MW and a further 1999 MW are under construction. Once again, the initial projection of 14 640 MW by 1990 has been revised down to 5400 MW. Spain has 6 units in operation and a further 9 are under construction, at Almarez (3), Lemoniz (3), Asco (2), and Confrentes (1). Finland began its nuclear programme in 1978 and 1979 with two units at Loviisa and by 1984 four units were producing over 2000 MW of electricity, equivalent to 38 per cent of the country's net production of electrical energy (Table 4.3). Similarly, Sweden and Switzerland are now heavily dependent on nuclear power for their electricity, but public opinion has possibly been stronger here than elsewhere in western Europe and development programmes have been restricted on safety and environmental grounds (Ilbery 1981a). Despite this, referendums held in both countries on the nuclear issue in the late 1970s led to narrow defeats for the anti-nuclear lobby.

Therefore, whilst there is increased European co-operation on the development of fast breeder reactors, there has been a slowdown in the planned projections for nuclear power plants. This relates to a range of factors, including a fall in the demand for electricity, financial and technical problems, safety and environmental concerns, and consent difficulties. According to Odell (1983) this is not surprising as nuclear power is a poor option and came about for three reasons:

1. The influence of the only effective institution in the energy field—Euratom.
2. The remoteness of Brussels from the widespread public opposition to developments in nuclear power.
3. The lack of knowledge of Europe's oil and gas reserves.

However, with North Sea oil and gas, large reserves of coal, and the generation of electricity and nuclear power, the energy scene in western Europe looks prosperous for the rest of the century. This is of prime importance for industrial development and Odell (1973) argued that western Europe offers a highly favourable location for the development of North Sea energy, for four main reasons:

1. Western Europe offers of powerful geography of demand, from the highly industrialized surrounding countries.
2. European funds and skill are available, although American intervention in North Sea development is important.
3. Natural gas offers the advantages of being clean and pollution-free.
4. West European supplies of oil and natural gas may reduce the dependence on supplies from politically unstable parts of the world.

Efficient energy provision is vital for economic development and with increasing demands for energy in western Europe, notably from the Ten, some kind of energy policy is necessary. As with agriculture, the European Community has attempted to develop such a policy, but subsequent discussion shows that to date this has met with little success.

4.3 Towards a Community energy policy

During the Paris Summit, held on 19 and 20 October 1972, the heads of State and heads of Government deemed it necessary to invite Community institutions to formulate, as soon as possible, an energy policy guaranteeing certain and lasting supplies, under satisfactory conditions. However, little progress was made between the member countries of the Nine, and in 1976 two sharp blows were delivered to an energy policy when a proposed meeting of the Council of Research Ministers never took place and when the results of a meeting of the Council of Energy Ministers were inconclusive. In addition, there had been a failure to find alternative energy supplies and little collective effort to conserve energy.

One of the main problems was the divergent interests of member states in regard to the supply of energy. Before the Second World War, western Europe was mainly self-sufficient in energy, thanks to coal, but after 1950 it became increasingly dependent on the outside world for its energy and in particular for oil. By 1959, the energy market of the European Community was dominated by abundant supplies of oil and the difficulty of selling coal increased. The original Six had not prepared for this and, as a result, the various national Governments proposed separate measures to meet their individual demands. Belgium and France were left with labour-intensive coal industries to protect; Italy and the Netherlands were concerned only with securing access to energy at the lowest possible price; and West Germany opted for an open energy economy. This division resulted in the virtual exclusion of the energy sector from inter-governmental moves towards economic integration (Odell 1976). Energy policy at this stage was of minor importance, concerned only with technical matters and tariff equalization on trade in energy products; indeed, energy was not mentioned in the Treaty of Rome.

Wide gaps had emerged in oil prices between member states, and the formation of an executive working party to investigate this problem, represented the first signs of Community action. In 1962, it was proposed by the Executive of the Community that there should be:

1. Free movement of crude oil and petroleum products between member states.
2. Free import of crude oil and petroleum products from non-member countries.
3. Community quotas for imports from east European countries.

Unfortunately, this did little to resolve the differences of interest which existed over energy policy within the European Community. Further committees were established, in 1963 and 1964, to deal with this problem, but they did little more than express the view that a common energy market must ultimately be established. As Odell (1976) clearly showed, integration in the energy market during the late 1950s and the 1960s was achieved more by the activities of international oil companies than by governmental efforts. The oil companies treated western Europe as a unified market for the rapidly increasing production potential of the oilfields of the Middle East. London emerged as the capital city of the European oil system, because most companies established their headquarters there, and only Paris and The Hague offered any competition.

The varied responsibilities of the ECSC, Euratom, and the EC (for oil, gas, and electricity), were brought within a single framework in 1967, and in 1968 a working party of the Commission proposed the guidelines for a common energy policy. The policy was to be based on the principles of fair competition, low-cost supply, security and freedom of choice for the consumer, and stability of supply in respect of the cost and quantity, and all member states adopted minimum stock levels for oil. However, the document was not discussed until the end of 1969 and the proposals were not accepted until 1972, when it was agreed that member states should pool information on oil imports and investment projects. There was simply a lack of interest in energy policy at this time (Lucas 1977).

However, the 1973 oil crisis, when the price of crude oil quadrupled, provided the incentive for the Community to work on an energy policy. The following year witnessed two developments. The first was in April 1974 when the Commission announced new proposals, which the Council accepted in May. The Commission argued that nuclear power was the only acceptable alternative to oil open to the Nine and on the basis of an annual economic growth rate of 3.5 per cent, the Community should by 1985:

1. Use energy more efficiently, leading to a reduction in consumption of 15 per cent of the 1973 forecast for 1985.
2. Limit imports to 40 per cent of their needs (1973 figure was 63 per cent).
3. Limit outside supplies to 75 per cent of their needs (1973 figure was 98 per cent).
4. Increase natural gas's share in energy supply to 25 per cent (1973 figure was 2 per cent).
5. Increase electricity's role in energy consumption from 25 to 35 per cent, and produce 50 per cent of electricity supplies from nuclear power.
6. Maintain existing coal output levels and increase coal imports.

Thus, the aims were to become more self-sufficient, to encourage the conservation of existing energy resources, and to search for and encourage the development of indigenous sources. However, even if all the proposals were met by 1985 the EC would still not be 'energy independent'; in fact it would remain the world's largest importer of energy. The proposals were not readily turned into action and the Community, characteristically, could not agree on the arrangements for establishing the machinery to harmonize the energy market. In September 1974, the Council did commit itself to the idea of cuts in energy consumption, the development of nuclear energy, the use of coal, and an effort in the research field (Bailey 1976; Hafele and Wolfgang 1978).

The second development came in November 1974 and was more concerned with energy policy in the whole of western Europe. The International Energy Agency (IEA) was established, within the framework of the OECD, to implement an energy programme amongst its members. All west European countries, except France, Finland, Iceland, and Portugal, joined the IEA, which had four basic aims:

1. Co-operation between members to reduce the excessive dependence on oil, through energy conservation, the development of alternative energy sources, and energy research and development.
2. An information system on the international oil market, as well as consultation with oil companies.
3. Co-operation with oil-producing and other oil-consuming countries, with a view to developing a stable international energy trade, as well as the rational management and use of world energy resources, in the interest of all countries.
4. Plan to prepare participating countries against the risk of a major disruption of oil supplies and to share available oil in the event of an emergency.

Unfortunately, the IEA was a rather fragile organization and did not lead to rapid integration of the energy economy.

By 1975, the situation had improved, but only just. Disagreements, notably over North Sea oil and gas, continued to prevent co-operation. The United Kingdom and the Netherlands resisted general EC access to North Sea oil and gas fields, even those developed with the assistance of Community funds. There was still further disagreement in 1976: Italy wanted more action on alternative energy sources such as solar, geothermal, and nuclear; France wanted action on Euratom loans, uranium prospecting and research; and Britain thought the most important objective was to secure a minimum safeguard price for oil. All the time, the consumption of energy was rising and the cost of implementing energy policy proposals between 1976 and 1985 was estimated at 204–15 million units of account.

Attempts to formulate a common energy policy had thus been slow and the objectives of the '1985' policy lay in shambles. Few signs of effective effort could be discerned and the Council itself accepted the Commission's view that progress in energy policy had been dismal. However, the second oil crisis, in 1979, saw oil prices double and this led to real concern over the availability and security of energy supplies. All Community members were vulnerable and in 1979 the Council of Energy Ministers re-emphasized their belief in the desirability of common action. In 1980 the Council adopted targets for 1990 and these revolved around three objectives:

1. To save and use energy more efficiently. Oil consumption was to be reduced by 30 per cent and the use of coal, gas, and electricity was to be increased. In terms of efficiency, the ratio between the growth in energy consumption and the growth

of gross domestic product had to be cut from 1.0 to 0.7, a 30 per cent saving.

2. To research novel energy sources, such as the liquefaction of coal, and geo and solar energy, and in particular to encourage the use of renewable sources.

3. To establish international agreements to guarantee energy imports for the rest of the century.

A fall in energy consumption during the early 1980s, due to the economic recession, and a growth in gross domestic product meant that the energy efficiency ratio was soon achieved, suggesting that the 1980 objectives were not strenuous enough. However, it was important for the community not to become complacent about the energy problem, for its was still dependent on imported energy, and oil in particular, and energy consumption began to rise again in 1984 and is likely to increase into the 1990s. With the main oil markets being insecure and European coal uncompetitive, nuclear power was actively encouraged, in the hope of its accounting for 40 per cent of electricity production by 1990. Indeed, policy objectives for 1995 need now to be formulated and the European Commission (1984) has stated that these can only be achieved by the coordinated efforts of Member States. Action is to be concentrated on five priority areas: investment; prices and taxation; research, development and demonstration projects; safety measures against market disturbances; and external relations.

Energy, in all its forms, is a vital factor in industry, agriculture, commerce, domestic comfort, and leisure activities. Therefore, the establishment of a Common Energy Policy for the EC is vital if Europe is going to be lifted from the economic recession. Whilst the development of such a policy was too gradual throughout the 1960s and 1970s, a greater sense of urgency has been apparent in the 1980s.

References

Alting von Geusau, F. A. M. (Ed.) (1975). *Energy in the European Communities*. Sijthoff, Leyden.

Bailey, R. (1976). 'Headings for an EEC Common Energy Policy'. *Energy Policy*, **4**, 308–21.

Bodanis, D. (1984). 'The great French nuclear disaster'. *The Guardian*. 18 January.

Clout, H. D. (Ed.) (1975). *Regional development in western Europe*. Wiley, London.

Clout, H.; Blacksell, M.; King, R., and Pinder, D. (1985) *Western Europe: geographical perspectives*. Longman, London.

Commission of the European Communities (1984). *A European energy strategy*. Brussels.

Dean, C. J. (1982). 'The trans-Mediterranean gas pipeline'. *Geography*, **67**, 258–60.

Greenhalgh, G. H. (1975). 'The nuclear power industry in Europe'. *Energia Nucleare*, **22**, 320–5.

Hafele, W. and Wolfgang, S. (1978). 'Energy options and strategies for western Europe'. *Science*, **200**, 164–7.

Ilbery, B. W. (1981a). 'Nuclear power in Western Europe'. *Tijdschrift voor Economische en Sociale Geografie*, **72**, 242–51.

Ilbery, B. W. (1981b). 'The diffusion of nuclear power in Europe'. *Geography*, **66**, 297–300.

Jensen, W. G. (1967). *Energy in Europe, 1945–80*. Foulis, London.

Lucas, N. J. D. (1977). *Energy and the European Communities*. Europa Publications, London.

Manners, G. (1980). 'Prospects and problems for an increased resort to coal in western Europe, 1980–2000'. In: Landsberg, H. (Ed.), *Selected essays on energy*. Ballinger, London.

Mellor, R. E. H. and Smith, E. A. (1979). *Europe: a geographical survey of the continent*. Macmillan, London.

Minshull, G. N. (1978). *The new Europe: an economic geography of the EEC*. Hodder and Stoughton, London.

Odell, P. R. (1973). 'Indigenous oil and gas developments and western Europe's energy policy options'. *Energy Policy*, **1**, 47–64.

Odell, P. R. (1976). 'The EEC energy market: structure and integration'. In: Lee, R. and Ogden, P. E. (Eds), *Economy and society in the EEC*. Saxon House, Farnborough.

Odell, P. R. (1981). 'The energy economy of western Europe: a return to the use of indigenous resources'. *Geography*, **66**, 1–14.

Odell, P. R. (1983). *Oil and world power*, 7th edn. Penguin, Harmondsworth.

Openshaw, S. (1982). 'The geography of reactor siting policies in the UK'. *Transactions of the Institute of British Geographers*, **7**, 150–62.

Parker, G. (1981). *The logic of unity*. Longman, London.

Pierantoni, F. (1975). 'Some implications of nuclear energy policies in European countries'. *Journal of the British Nuclear Energy Society*, **14**, 303–9.

Sill, M. (1984). 'Coal in western Europe, 1970–81'. *Geography*, **69**, 66–9.

5

Transport

5.1 Modes of transportation in western Europe

A good transportation system is a necessary prerequisite for trade, regional development, and industrial location. Pacione (1974), for example, has shown that since 1958 over 600 new industrial plants have arisen along the Autostrada del Sole motorway in Italy, and Thomson (1976) suggests that the core–periphery contrasts in economic development in western Europe are caused directly or indirectly by the transport system. Western Europe has a well-developed and unique transportation network, which is the result of four main factors (Jordan 1973):

1. The high population density and high level of industrialization, producing a network unequalled in density of routes.
2. The comparatively small size of west European countries, aiding this dense network of routes.
3. The nationalization or semi-nationalization of transport facilities, which means fewer competing lines.
4. The regional variety and density of transport systems, producing small-scale national systems rather than an internationally integrated system.

Therefore, each country's transportation system has been developed, with the aid of government policies, to satisfy particular national needs. With the formation of trading blocs in Europe, it has become necessary to reorientate the various modes of transport and to create an integrated international network which allows the movement of goods and people between, as well as within, different countries. The various attempts to develop an integrated transportation system will be analysed in Section 5.2.

An important factor affecting the density and type of transportation system in any area is transport costs. Total transport costs, comprised of terminal and line-haul costs, tend to 'taper-off' as distance increases. However, the degree of tapering varies greatly from one mode of transportation to another, depending upon their respective terminal and line-haul costs. Figure 5.1 illustrates the theoretical differences in movement costs between three modes of transportation—highway, railway, and waterway—and it can be seen that each mode offers advantages over different lengths of haul. For short distances, up to 80 km, road transport is cheapest. For

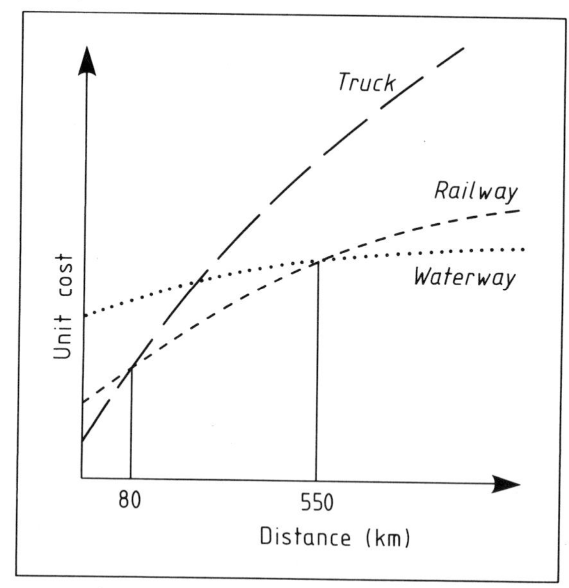

Fig. 5.1 Idealized transport cost curves for three modes of transport

medium-length hauls, between 80 and approximately 560 km, the railway is least costly, whereas over long distances, waterways and air transport are cheapest. In reality, such competition is probably restricted to certain type of freight only, but the principles help one to understand the emphasis placed on particular modes of transport in certain parts of western Europe; for example, the declining importance of waterways in Great Britain can be partially attributed to a lack of haulage distances over 560 km.

Since the 1950s, western Europe has witnessed something of a transport revolution and motorways, air transport, and pipelines now compete with railways and waterways for the long-distance movement of goods and people. Therefore, developments in the various modes of transportation will first be discussed.

1. Water transport: a declining mode of transportation

Inland waterways have played an important role in western Europe since the eighteenth century. The Rhine, navigable from Basle to Rotterdam, overwhelmingly dominates western Europe's system of inland waterways by its length and capacity, as well as providing the focus for numerous tributary systems. Its dominance is clearly indicated in Fig. 5.2 and in the peak year of 1975 it transported 270 million tonnes of goods, compared to just 33 million for the Amsterdam–Rhine canal, 16 million for the Main, and 12 million for the Moselle.

The North European plain is drained by a series of navigable rivers, apart from the Rhine, and consequently it is not surprising to find a

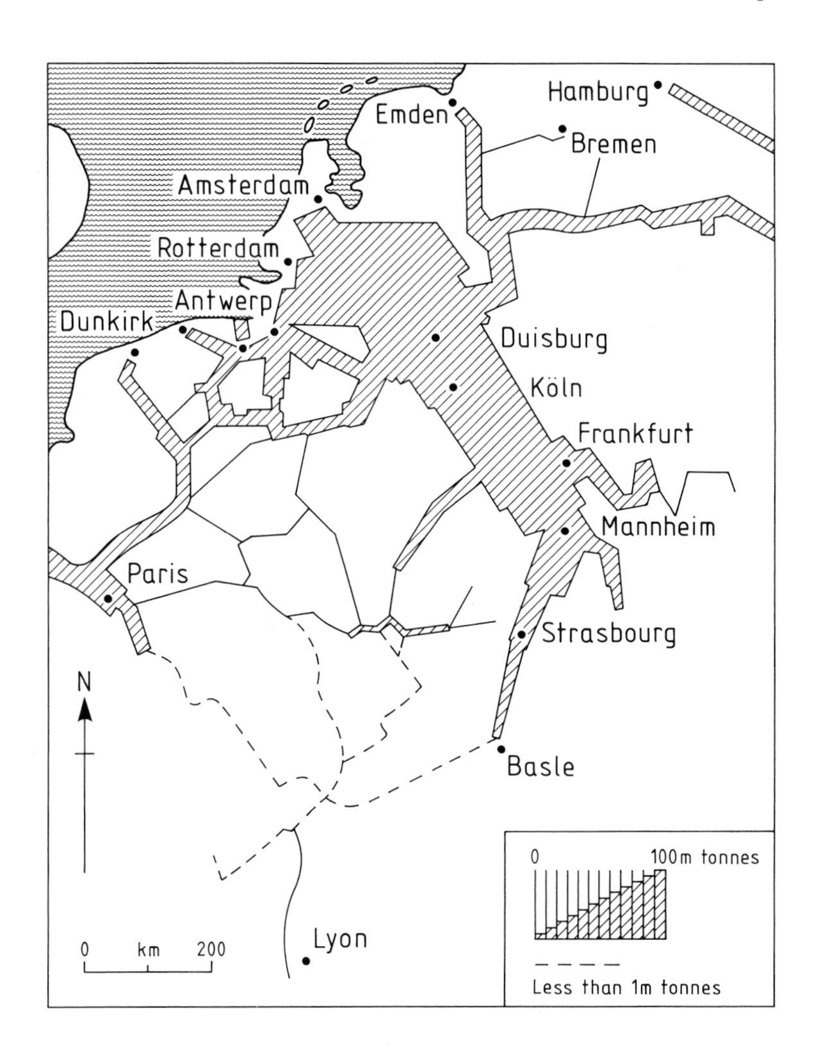

Fig. 5.2 Traffic flows on the Rhine and other major inland waterways (*source:* Tuppen 1975)

concentration of waterways in this part of western Europe. Belgium, the Netherlands, West Germany, and France depend, to varying extents, upon inland waterways. In 1982, these four countries accounted for 94 per cent of freight traffic carried by, and 61 per cent of the total length of, waterways in western Europe. In the Netherlands, waterways play a significant part in the carriage of freight and in 1982, 34.1 per cent of all goods handled went by water (Table 5.1). The country has over 4800 km of waterways in use, more than for the whole of West Germany (4447 km), and they carry over 200 million tonnes of goods per year. Belgium is the second most important country in terms of freight traffic carried by inland waterways (18.3 per cent), and it is the northern part of the country that is best served by waterways, with the Albert canal being dominant. Considering the size of France and its 6496 km of waterways, the minor role played by this mode of transport is surprising. In 1982 only five per cent of inland freight traffic was carried on inland waterways although it has over 20 per cent of the total length of waterways in western Europe. This reflects a high degree of under-utilization, an unequal spatial distribution of the network, and strong competition from the railways and roads for bulk hauls (Tuppen 1975).

The remainder of western Europe is not well served by inland waterways for the purpose of transporting freight. In Denmark and Ireland, there are no commercial waterways of significance, and in Sweden and Finland, the latter with a considerable length of waterways, the problems of winter freezing and road competition have led to large-scale abandonment of waterways for freight traffic. The Gota canal in Sweden, for example, is used more for recreation and tourism. In Italy, only the river Po and its tributaries are of any value for freight traffic, and in the United Kingdom the waterway network has been severely contracted. Waterways in the United Kingdom were built before the railways and have declined in importance because of the absence of European gauge waterways[1] and the rather limited distances covered, which meant that railways, once built, could easily compete for freight traffic (Fig. 5.1). However, the length of waterways in use in the United Kingdom has increased over the past decade, from 569 km in 1975 to 1631 km in 1982. Although this mode of transport accounts for less than one per cent of total freight carried, the amount more than doubled between 1975 and 1982, from 4.2 to 8.7 million tonnes (Table 5.2).

Table 5.1 Percentage of inland freight traffic carried by the various modes of transportation in western Europe, 1982

	Rail	Road	Inland Waterways	Pipeline
France	12.5	77.0	5.1	5.4
West Germany	9.7	80.2	7.8	2.3
Netherlands	2.8	57.7	34.1	5.4
Norway	8.3	86.1	—	5.6
Spain	9.7	87.4	—	2.9
Sweden	10.1	89.9	—	—
Switzerland	11.7	82.6	2.3	3.4
United Kingdom	10.5	86.1	0.3	3.1
Austria	50.2	13.5	6.9	29.4
Belgium	13.7	61.8	18.3	6.2
Finland	5.3	94.7	—	—
Italy	11.3	82.0	0.1	6.6
Luxembourg	21.1	73.7	5.2	—

Source: Annual Bulletin of Transport Statistics for Europe

With increasing competition from railways, and later roads, the waterways began to carry less freight and, in post-war years, new construction has been essentially restricted to certain key routes and greater emphasis has been placed on the improvement and rationalization of the overall network (Pilkington 1966; Martin 1974; Tuppen 1975; Scargill 1976; Mellor 1983). This has led to a decline in the number of waterways in use but, as modernization has occurred, the length of waterway and canalized river of European gauge has increased, notably in the Netherlands but also in Italy and France. Despite this, waterways of large capacity represent a small proportion of the total network. The waterways which have been upgraded to provide access for barges of 1350 tonnes include the Main, the Neckar, the Moselle (Schofield 1965; Martin 1974), the Meuse, the Albert, the Rhine–Rhône (Scargill 1976), and the Dortmund–Ems. Other schemes are of a more international nature and will be discussed in Section 5.2.

The upgrading of waterways led to an expansion in freight traffic during the 1960s, at a time when the railways, their major competitor, car-

[1] Waterways with a gauge of 1350 tonnes, the critical minimum limit agreed in 1953 by the European Conference of Ministers

Table 5.2 Freight transported by waterways in western Europe, 1963–82

	Volume of goods (million tonnes)				Percentage change		
	1963	1971	1975	1982	1963–71	1971–75	1975–82
Belgium	65	95	77	87	+46	−19	+13
France	77	107	87	70	+39	−19	−19
West Germany	167	230	215	210	+38	−6	−2
Italy	3	4	4	3	+33	0	−25
Netherlands	151	245	211	208	+62	−14	−1
United Kingdom	9	5	4	9	−44	−20	+125

Source: Tuppen (1975) and *United Nations Statistical Yearbook*

ried less goods. Between 1963 and 1971, the volume of goods transported by waterways, in selected countries, increased by over one-third, reaching an increase high of 62 per cent in the Netherlands (Table 5.2). However, when the figures are projected to 1975, a dramatic reversal can be seen, with all countries witnessing a decline in traffic between 1971 and 1975. This reflected the continuing emergence of motorways, air transport, and pipelines as major competitors for long-distance hauls of goods, a trend continued into the 1980s, except in Belgium and the United Kingdom where the volume of goods carried on the waterways actually increased.

2. *Railways: a change in emphasis*

Europe has approximately one-third of the total rail mileage in the world but, whilst the railway remains a major means of transport in eastern Europe, its importance in western Europe declined throughout the 1960s and 1970s. The basic rail network in western Europe was completed by about 1900 and on the mainland, unlike the situation in Britain, railways preceded major industrialization. Railways became a major means of transport because they could carry large volumes of heavy goods relatively quickly and cheaply. For distances of over 80 km they were more economical than roads (Fig. 5.1), and the problem of competing lines was largely avoided because of state ownership. However, the motorway revolution has led to a decline in the importance of railways and necessitated a change in emphasis.

In terms of freight traffic, length of line, and density of network, marked national variations exist in the importance of railways. All west European countries, with the possible exception of Denmark, have experienced a decline in the amount of freight carried on the railways since the late 1960s. France and West Germany account for over 50 per cent of all freight carried on railways in western Europe, yet Table 5.1 indicates that in these two countries railways are a rather insignificant mode of transportation compared to roads. The situation in France is interesting in that, up to the mid-1960s, the transport system was firmly based on the railways, which carried as much freight as roads and waterways combined. Today, roads carry over 75 per cent of all freight (Table 5.1). However, the railways remain significant and, although experiencing a large relative decline in comparison with other modes, have experienced only a small absolute decline, a situation which is characteristic of most west European countries. Railways remain the most important carrier of freight in Austria, and in Luxembourg they still account for over 20 per cent of total freight carried.

The length of rail line in operation has fallen in most west European countries, with the United Kingdom experiencing one of the highest rates of decline, especially during the Beeching era when the network was severely pruned. Little new building has taken place, except in northern Norway and Finland, where missing links in the network have been completed. A better and more comparable indicator of the importance of railways is the actual density of network in the individual countries. Fig. 5.3a shows that the concentration of lines is highest in north-west Europe, reaching a peak of 129 km per 1000 km^2 in Belgium. The density of network appears to decline in a fairly regular manner away from a

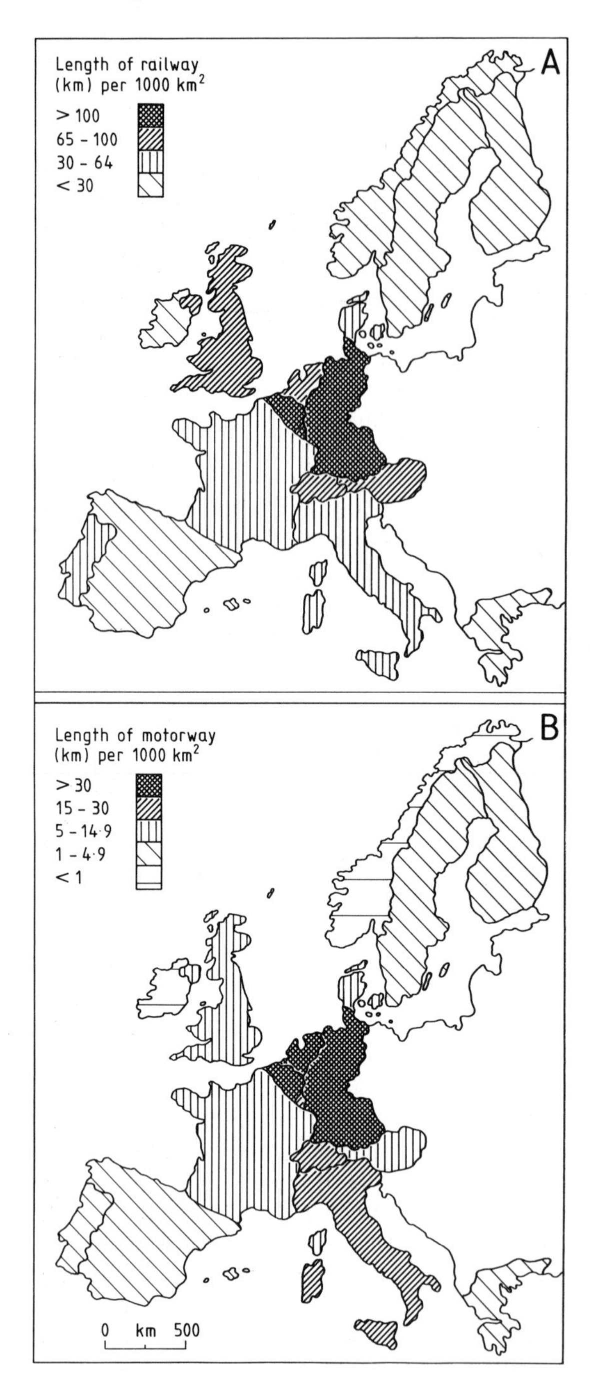

Fig. 5.3 Density of (a) rail and (b) motorway networks in western Europe, 1981

core area based on Belgium, Luxembourg (103), and West Germany (114).

In response to a declining situation, the railway system has been continuously improved and modernized, especially on selected lines. Double-tracking and electrification of lines have taken place on international routes such as Paris to Cologne, and West Germany to Italy, via the Brenner and St. Gotthard passes. Apart from the improvement of certain lines, the railways have changed their emphasis, away from freight traffic and towards passenger traffic, and the transporting of businessmen in particular. This has meant a concentration on fast inter-city links, a system which has been implemented in many countries. The German inter-city was inaugurated in 1971 (Clout 1975), and 12 lines link 73 major cities. The emphasis is on speed and many inter-city links record speeds of 120–140 km per hour. A similar system is in existence in Italy, France, and the United Kingdom and fast inter-city links include London to Cardiff, London to Edinburgh and Glasgow, Paris to Bordeaux, and Hamburg to Dortmund.

Reform on the railways also led to the establishment of the Trans-Europ Express service (Chapman 1968). To stem the flow of potential passengers to the roads and airways, the European railways had to improve the speed, comfort, and attractiveness of their services and, in order to remain viable, long distances had to be covered and international boundaries crossed. The Trans-Europ Express, created in 1957 and enlarged in 1964, served the original Six, plus Switzerland and Austria, and is one of the earliest examples of transport integration in Europe. The main aim of the express service was to transport businessmen between the major centres of industry and commerce, and among its major attractions were schedules to fit the needs of businessmen, an average speed of 140 km per hour, plenty of space per passenger, and border crossings free of delay. A modification was introduced in 1964, when it was decided to accept into the system trains connecting points within one country only, as long as they fulfilled the requirements of international trains. The famous Paris to Nice train was one such link to be added to the network. Further integration of the west European railways has taken place since the Trans-Europ Express through extensive modernization, and as will be seen in Section 5.2, the most notable achievement has been the establishment of links between Italy and the core areas of western Europe.

5.1 Canalization of the Moselle (ECC)

5.2 One of the early Trans-Europ luxury expresses

3. Roads: the motorway revolution

The road network in western Europe is characterized by a modern system of motorways, superimposed upon a system inherited from the past. Early routes were influenced by the physical environment, but more recent 'Euroroutes' have managed to overcome physical obstacles and Italy, a country that had been previously isolated from the core areas of western Europe, has benefited considerably. Fast road links between the European countries have enabled this mode of transport to become the principal means of transporting inland freight, even over very long distances. In most west European countries, roads account for over 75 per cent of the total freight carried, reaching a high of 94.7 per cent in Finland (Table 5.1). Austria is the only country where more freight is carried by the railways, and waterways compete with roads only in the Netherlands.

The importance of road transport has greatly increased with the explosion in car ownership. In 1976, there were 85 million private cars on the roads of western Europe, of which nearly 71 million were registered in the European Community, and this represented an increase of approximately 250 per cent on the 1960 figure. By 1982 the total had reached 112 million, 82 per cent of which were concentrated in the Ten. West Germany leads the way in the total number of automobiles, with 24 million, followed by Italy (19.6), France (19.3), and the United Kingdom (17.5). In terms of the number of private cars per 1000 people, significant national differences exist, ranging in 1982 from 102 in Greece to nearly 400 in West Germany (Fig. 5.4). In Sweden there are

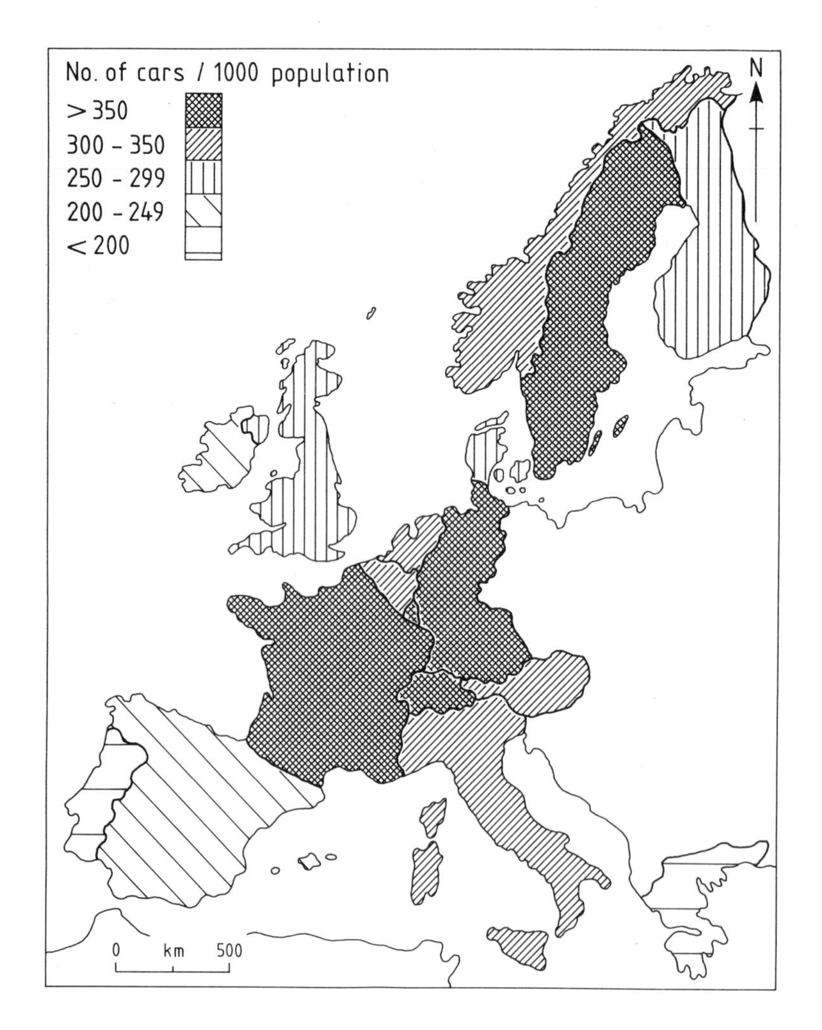

Fig. 5.4 Car ownership in western Europe, 1982

382 private cars per 1000 people (compared to 42 per 1000 in 1950), and this is one of five countries to record a value of over 350. A core–periphery contrast is again in evidence, even if the periphery is small in extent and comprised of Greece, Portugal, Spain, and Ireland.

The rapid diffusion of the automobile in western Europe has led to inevitable problems. Jordan (1973) felt that the motor-car was adopted before the burden of creating a first-class highway system was accepted and Clout (1975) mentions the problems of congestion, pollution, and accidents. In 1975, 53 000 people were killed and 1 530 000 injured on the roads of the Nine alone. The number of casualties has been falling ever since, but in 1982 there were still 45 000 deaths and 1 514 900 injuries on the roads of the Ten. These problems enhanced the need to improve the major road links and to construct motorways. Motorways are a relatively new form of transport and, with the exception of those in West Germany, have been mainly built since the Second World War. Motorway construction has been essentially restricted to the EC countries, which have increased their length of motorway from 1500 km in 1952 to 26 400 km in 1982. West Germany, France, and Italy account for 72 per cent of this total figure, with the United Kingdom responsible for a further 11 per cent. High densities of motorway serve Belgium, the Netherlands, and the Rhinelands of West Germany, the core areas of economic activity in western Europe (Fig. 5.3b). Away from the core area, motorways tend to radiate out from the capital cities, as in the cases of London, Paris, and Madrid, or from the ports and main areas of economic activity.

Two of the most advanced motorway systems are the West German autobahn system and the Italian autostrada system. The autobahns were introduced in the 1930s and have been continually extended. Consequently, many were built for a unified Germany, linking Berlin to the major industrial areas in an east-west direction. Since 1945, these have declined in economic importance, compared to the more recently constructed north–south links. There are two main north–south lines of movement (Minshull 1978): the Rhineland, linking the Ruhr with Bonn, the Saar, Frankfurt, Mannheim, and Stuttgart; and a line running from the North Sea ports of Hamburg and Bremen south to Hanover, Brunswick, and Kassel, and on to Nuremburg and Munich. By 1982, West Germany had 7900 km of motorway and a total of 13 000 km is planned for 1990. The construction of the autostrada system is a post-war phenomenon and a critical element in the Italian economy. Between 1961 and 1971, the motorway network increased in length from 480 to 4000 km (Pacione 1974) and by 1983 it stood at nearly 6000 km. The most famous motorway is the Autostrada del Sole, which links Milan to Naples over 750 km and helps to counteract the extremely long north–south distances in Italy, but probably the most important development is the construction of motorways and tunnels through the Alps to link Italy with the economic heart of western Europe.

Unfortunately, most of the motorways built in the 1960s were designed to serve national rather than international needs. Only since the 1970s have they regularly crossed international boundaries, upon EC recommendation and funds, to provide a truly European motorway system: international developments will be discussed in Section 5.2.

4. Air transport and pipelines: new modes of transportation

Air transport is a modern-day competitor to the more firmly established modes of transportation. In the twentieth century western Europe has been characterized by remarkable growth in airport traffic and is covered by a fairly dense network of air routes; it has become the focal point in the world air-traffic pattern. A number of major airlines operate international services and seven can boast a revenue of over £100 million: in order of importance these are British Airways, Lufthansa (West Germany), SAS (Denmark, Sweden, and Norway), KLM Royal Dutch Airlines, Swissair, Alitalia, and UTA (France). Civil aviation is dominated by passenger traffic, on national, international, and continental scales. The United Kingdom, and London (Heathrow) in particular, handled over 30 per cent more passengers in 1980 than the second most important country for passenger traffic, France. The largest non-EC carrier of passengers is Spain but collectively, the countries outside the Community account for less than one-third of passenger traffic in western Europe. Air freight traffic is rather less important, being restricted by limited capacity and high freight charges. However, security and speed of delivery mean that air is an efficient form of transport for high-value goods over long distances and during the 1970s freight traffic increased remarkably, especially between the major airports of London (Heathrow), Paris, and Frankfurt.

The distribution of major airports (Fig. 5.5) reflects their greater dependence on passenger traffic. Inland sites, near capital cities, dominate and location is related to passenger, rather than industrial, demands and major routeways. With the exception of West Germany and Switzerland, the largest airports of each country, in terms of passenger traffic, are those of its capital city. The attractions of the capital are obvious and Parker (1981) notes two in particular: first, as a result of political, financial, and business concentration, it has the largest number of potential passengers; and secondly, it is normally a major centre of the

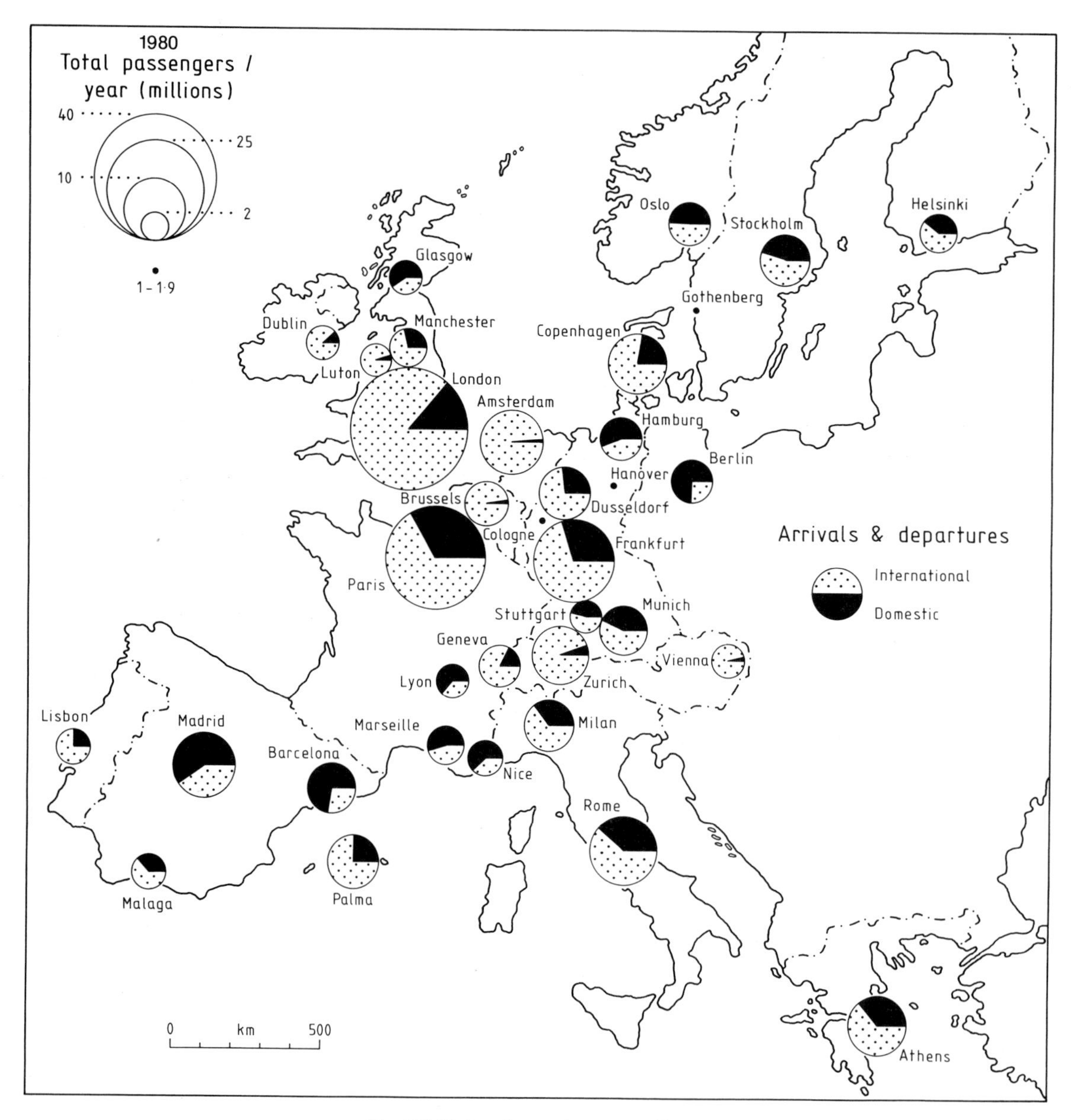

Fig. 5.5 Major airports in western Europe

national transportation system and hence relatively accessible to many people. Frankfurt, rather than Bonn, is the principal airport in West Germany and has grown in importance for a number of reasons:

1. The city is situated at the junction of the Rhine and Main, a natural focus of routes.
2. Frankfurt has a central location in West Germany.
3. Bonn, the capital, is rather small and of a non-industrial nature.
4. The Federal Republic has a policy of decentralization away from the capital city.

Figure 5.5 highlights the dominance of London, Paris, and Frankfurt in the air-traffic system of western Europe. London, comprising Heathrow, Gatwick, and Stansted airports, registered over 37 million arrivals and departures in 1980, 43 per cent more than the Orly, Le Bourget, and Charles de Gaulle airports of Paris. The international nature of the major airports is clearly demonstrated and in 1980, only 23 per cent of all arrivals and departures in London, Paris, and Frankfurt were of a domestic nature. In addition to the major airports, a number of important regional airports exist in western Europe. Airports such as those at Milan, Marseille, Manchester, and Munich supplement the major airports and many have grown in importance with the rapidly increasing number of charter flights in western Europe. The Mediterranean lands are the major destination of charter flights, which account for nearly 40 per cent of air traffic inside Europe.

Pipelines represent a revolutionary long-distance form of transportation that has grown in importance since the formation of custom unions in Europe. Before the European Community came into existence, distances within individual countries were too short to make the transportation of oil, petrochemicals, and natural gas by pipelines a viable proposition. Pipelines primarily carry liquids and the disadvantages of inflexibility and fixed carrying capacities are offset by the very low unit costs, the lack of physical barriers, and the ability to cover very large distances. Where volume and market demand are sufficiently great and steady, pipelines are more economical than other forms of transportation. International co-operation has therefore been necessary for the construction of pipelines in western Europe. France has the longest length of pipeline which, at over 5300 km, is significantly higher than the United Kingdom (3166), Italy (3069), and West Germany (2086). Outside the Ten, pipelines are most important in the land-locked countries of Austria and Switzerland, although the greatest length is found in Spain (1535 km). The growth of large inland markets for oil in such places as Cologne and Vienna has led to a considerable mileage of crude-oil pipelines across western Europe. In fact, pipelines are the second most important mode of transportation in Austria, after railways (Table 5.1).

In 1958 the role of pipelines in western Europe's transport system was significant. In the 1980s they are the main means of transporting oil, the most important source of energy, and provide increasing competition for the more ubiquitous forms of transport (Blacksell 1977). The major crude-oil pipelines have a south–north orientation and lines go from the Rhône delta, Genoa, and Trieste northwards into the interior of western Europe, converging on the middle Rhinelands. These are supplemented by another set of lines leading into southern Germany from Rotterdam and Emden. Oil and chemical products are transported direct to centres like Munich, Stuttgart, and Ingolstadt, and the southern *Landër* of West Germany have been transformed into a fast-growing industrial region (Chapter 6). North Sea oil is also leading to a concentrated network of pipelines which serve neighbouring countries.

Important gas pipelines also radiate out from the North Sea to the consuming regions of central England, northern Germany, and the lower Rhinelands. Another series of lines links the producing areas of south-west France and the Po valley to Paris and Milan. One gas pipeline requiring even greater international co-operation was that built under the Six Nation Gas Project. This project, concluded in 1975, was concerned with the export of gas from Iran to West Germany, France, and Austria, but via the Soviet Union and Czechoslovakia. The two Comecon countries built their respective sections of the pipeline, in return for specified amounts of free gas.

In summary, two major aspects have emerged in pipeline development in western Europe (Mellor and Smith 1979): first, the orientation is

5.3 Lufthansa aircraft at Frankfurt airport passenger terminal (ECC)

essentially from oil terminal ports to inland refineries; and secondly, there are internal systems from deposits to consumers.

5. Transfer points: the growth of Euroports

Western Europe has long taken advantage of its seas for transport; the Mediterranean, the North and Baltic Seas, and the Atlantic Ocean have been extensively used for trading. There are 65 sizeable seaports in western Europe, of which 50 are concentrated within the European Community (Bird 1967; Bird and Pollock 1978). The EC ports are mainly located along three stretches of coastline (Parker 1981):

1. The Channel and North Sea coasts, which contain the major ports of Rotterdam, Antwerp, Le Havre, Hamburg, London, Bremen, and Amsterdam. Some ports in this group have benefited from the growth in importance of ferries across the Channel.
2. The Mediterranean coast, dominated by the ports of Genoa and Marseille and including Cartagena, Venice, Naples, and Trieste.
3. The Atlantic coast, dominated by the British ports of Liverpool, Glasgow, and Severnside but including the more specialized ports of Milford Haven, Bordeaux, and Nantes.

Since the late 1950s, a number of 'Euroports' have emerged in western Europe, handling a larger tonnage and serving a much wider area than the remaining ports, which serve the needs of local areas. The Euroports, which compete with 'regional' ports whilst serving the needs of the Community, are dominated by Rotterdam, which handles more cargo than any other port in the world (North 1968). Rotterdam witnessed a six-fold increase in cargo between 1938 and 1975 and handles approximately three times the amount of cargo than either Marseille or Antwerp. The port has coped with containerization by expanding westwards onto the island of Rozenburg, where two projects have been undertaken: first, the Botlek scheme in east Rozenburg, which added 1417 hectares to the port area and is used for oil refining, chemicals, and shipbuilding yards; and secondly, the Europort plan in west Rozenburg, which began in 1958 and handles the huge supertankers. Other major Euroports include Marseille, Antwerp, London, Genoa, Le Havre, and Hamburg, with Amsterdam showing signs of marked growth (Mills 1978).

Indeed, Antwerp has taken a leading role in the field of modernization and is a good example of adaptation to modern trends of rational cargo-handling techniques. Modern methods are employed for the handling of containers, iron and steel, forest products, and roll-on/roll-off traffic. There are 14 container gantries in the various container terminals and all have been designed for multi-purpose use. Similarly, there are 18 special berths for roll-on/roll-off operations, some of which permit several vessels to be handled simultaneously. The growth in container traffic and roll-on/roll-off techniques reflects the large volume of general cargo which passes through Antwerp (32.2 million tonnes in 1983); indeed 'rolling cargo' surpassed two million tonnes for the first time in 1982. Therefore, the adoption of modern technology and its favourable geographical location have enabled Antwerp to develop into a very flexible port and compete as Europe's second major Europort.

These Euroports dominate total port traffic in western Europe. A glance at Table 5.3 reflects this dominance and the largest non-EC port, Gothenberg, handles only one-tenth the cargo of Rotterdam. Aggregating the total tonnage handled at the ports listed in Table 5.3, one finds that the

Table 5.3 Total traffic handled at the major port of individual west European countries, 1982

Country	Tonnage at major port (million tonnes)	Major port
Belgium	84.5	Antwerp
Denmark	10.5	Copenhagen
France	85.2	Marseille
Germany	62.5	Hamburg
Italy	49.7	Genoa
Netherlands	246.5	Rotterdam
United Kingdom	62.6	London
Finland	11.5	Skoldvik
Greece	14.8	Eleusis
Norway	20.2	Norvik
Portugal	10.9	Lisbon
Spain	17.4	Bilbao
Sweden	24.9	Gothenberg

Source: Lloyd's Register of Shipping

5.4 Car assembly plants of GMC at the port of Antwerp (ECC)

eight EC ports listed handle over seven times as much cargo as the remaining five ports.

Brief mention should be given to inland ports, some of which have increased in importance since the rationalization and modernization of European waterways. In France, Paris and Strasbourg are important, but not as important as Duisburg, Frankfurt, and Mannheim on the Rhine, a waterway that contains some of the world's largest inland ports. Nuremburg has also developed into an inland port, with the completion of the Main canal (Burley 1970).

5.2 An integrated transportation network?

One of the objectives of the EC since its inauguration has been the development of an internationally integrated transportation network, replacing the separate national networks and serving the Community as a whole. Attempts to integrate the various individual transport systems have been hampered by such factors as the lack of standardization between member countries as regards the relative importance of transportation modes, the changing importance of transportation modes, and the introduction of new modes of transportation like pipelines, air transport, and motorways. One of the main tasks is to complete the missing links in the Community transport network, like the new proposals for the Channel Tunnel, as this will assist the flow of goods between members and strengthen economic integration (Clout 1975). Integration is certainly not easy, but as Parker (1981, p. 24) states 'it is made easier by the fact that many methods of transport have grown up almost entirely in the conditions of the Community and so have been adapted from the start, particularly in the original Six, to a continental rather than national scale of operations'. Pipelines, for example, have been developed on a continental scale from the outset.

Economic integration in Europe is highly unlikely without the development of a Common Transport Policy (CTP), the aim of which is the formation of a single, efficient, and economically viable transport unit. The first memorandum on the proposed CTP was published in 1961 and stated that the coordination of transport was essential. Five principles were laid down, aptly summarized by Thomson (1976, p. 278–9):

1. Equality of treatment: the same rules should be observed for different operators within a particular mode of transport and for different modes; equally there should be no discrimination between transport users.
2. Financial autonomy: given equality of treatment, all operators should be financially self-supporting.
3. Free enterprise: in pursuit of financial autonomy, operators should be left free to conduct their business according to their own judgement.
4. Freedom of consumers' choice: the use of private transport—for either goods or passengers—should, as far as possible, be placed on a comparable footing with that of public transport.
5. Coordination of investment: the Commission emphasized the need for coordination between investment in infrastructure and investment in vehicles, and between one mode and another, and it mentioned the need for studies in this field, but it did not make any specific recommendations.

However, the idea of replacing existing national policies with a new CTP met with expected opposition and little progress towards implementation was made. The major problem was that each country's transport system was unique and emphasized different modes of transport; in France and Luxembourg, for example, railways were dominant and in the Netherlands there was a complex waterways system. There was also at this stage no attempt to develop a common policy for air transport, pipelines, or ports.

In 1965, the EC recommended that attention be given to filling the gaps in the motorway system, to facilitate access between, as well as within, member countries and in 1966 a system of consultation was instituted, which obliged governments to consult one another before developing new communication links or expanding existing ones. However, little was achieved before the merging of the three Communities (ECSC, EEC and Euratom) in 1967. In that year, a firm programme was agreed, but few subsequent developments have occurred. The situation was not helped by the changing relative importance of the various modes of transportation (Despicht 1969), or by the formation of the Nine in 1973 (Goergen 1972). The addition of two islands to the Euro-

5.5 Site of the Europoort scheme, west Rozenburg, Rotterdam (ECC)

pean Community increased the difficulty of achieving an integrated transport system. It was in 1973 that the Commission renewed efforts to develop a common policy. Unfortunately, this new initiative had little impact and was concerned more with minor details than with real network policy issues (Thomson 1976).

The most important steps towards an integrated transportation network within western Europe have been taken by organizations with a larger area of interest than that of the Community. Thomson (1976) has shown that the Trans-Europ Express system was built up and coordinated by the International Union of Railways, whilst the designation and improvement of international E roads have been undertaken by the United Nations Economic Commission for Europe in Geneva. Similarly, both the coordination and control of air traffic have been supervised by Eurocontrol, a body set up by several governments (Collester and Burnham 1975).

The need for co-operation between the west European countries was recognized in the early 1950s and resulted in the formation of the European Conference of Ministers of Transport (ECMT) in 1953. This permanent body, comprising all west European countries except Iceland and Finland, had two main objectives:

1. To take whatever measures may be necessary to achieve the maximum use and rational development of an international inland transport system in western Europe.
2. To coordinate and promote the activities of international organizations concerned with inland transport, taking into account the work of the supranational authorities in this field.

A major stumbling block to integration was the fact that the ECMT did not have the power or the resources to achieve its objectives and the pattern of progress has been as slow as that for the development of a CTP. Agreement on the aims and principles was not reached until 1968 and subsequent work has produced nothing more than a series of academic studies.

Despite the lack of progress made towards integration by the ECMT and the CTP, the late 1960s and 1970s witnessed certain developments and, broadly speaking, fusion was attempted in two ways:

5.6 The petroleum port and BP refinery at Lavera, west of Marseille (Port Autonôme de Marseille)

1. By encouraging modernization and diversification; and
2. By increasing the scale of developments, particularly routeways linking member countries.

Notable achievements have been made. Motorways, originally built to serve national interests, developed a number of cross-border links in the 1970s, many with the aid of funds from the European Investment Bank (Pinder 1978). Projects such as the Brussels to Paris motorway, the Riviera coastal motorway linking France and Italy, the Antwerp–Liège–Aachen motorway, the Metz to Saar motorway, and the motorways via the St. Bernard and St. Gotthard passes and the Mont Blanc tunnel linking the Italian autostrada system with the German, Swiss, Austrian, and French motorways have helped to increase fast international communications and to overcome the relative isolation of Italy. The motorway system has been designed to link together the major urban and industrial regions of the EC, as well as acting as a stimulus to economic activity; examples of the latter include the Autostrada del Sole in southern Italy and the M4 corridor in southern England. Extensive modernization of the railways is necessary for integration, but electrification has taken place on many routes and priority has been given to international links. Those to be established include the trans-alpine routes between Munich and Verona (via the Brenner pass), Basle and Milan (St. Gotthard and Simplon links), and Turin and Chambéry; Narbonne to Barcelona; Paris to Cologne, and Paris to Metz and Mannheim. Regarding waterways, a number of major links of European gauge have been established. The 'Europa' canal, connecting the Rhine and the Danube; and the canal linking the North Sea to the Mediterranean, via Antwerp, the Canal de l'Est, and the Rhône–Saône Canal, are the two major internationally planned waterways. The Rhône corridor is also benefiting from the extension of the Rhône–Rhine canal (Scargill 1976). Mellor (1983) has commented upon the progress of the Rhine–Main–Danube waterway and the political controversy surrounding the proposed completion of the final 55 km between Hilpoltstein and Kelheim. The oil crisis of 1973 and subsequent economic difficulties slowed the planning process, but the target date for completing the final leg of the Europa canal is 1989.

If the individual networks created by the various modes of transportation are superimposed, a concentration of routeways can be seen to exist in western Europe. Parker (1981) has identified a number of 'Euroroutes', each characterized by modernized methods of transport (Fig. 5.6):

1. The Rhine Euroroute, followed by road, rail, and pipelines besides the waterway itself, is easily the most important. This route has important additional transport axes following such tributaries of the Rhine as the Moselle, Main, and Neckar.
2. The Rhône–Saône corridor, a major connection between the Mediterranean and the Rhinelands.
3. The trans-alpine Euroroute, between Germany and Italy and through Austria and Switzerland.
4. The British Euroroute, from the south-east to the north-west, which carries mororways, electrified rail, pipelines, and transmission line connections.

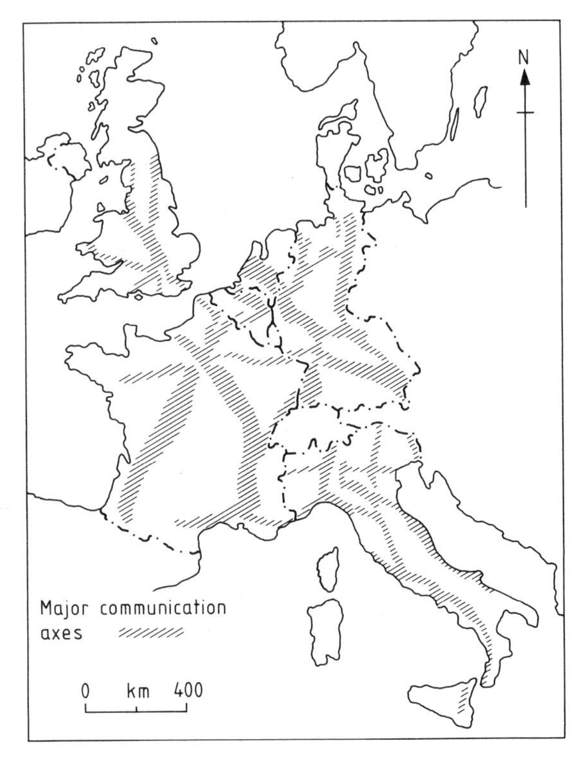

Fig. 5.6 Euroroutes within western Europe (*Source:* Parker 1981)

Further agreements in the Community have been reached on how to regulate the transport market and establish fair competition. The intention is to achieve a regulated transport market by controlling freight rates, by maintenance and stabilization of the railway system, and by a common system of containerized freight to integrate the road, rail, inland waterway and sea transport network (Minshull 1978). With regards to fair competition, it is necessary to implement standardized working hours and conditions and a uniform system of transport taxation, with a system of licences which give road-haulage operators unlimited rights to make journeys throughout the Community.

Even accounting for the various 'international' developments, Blacksell (1977) points out that the members of the Ten continue to disagree on how to implement a Common Transport Policy. Indeed the CTP is still fragmented and as the same author states (p. 172), 'transportation remains the preserve of national governments and private enterprise'. The situation is adequately summarized in the following statement from Thomson (1976, p. 286): 'the transport system in Europe has grown up as a number of separate, but connected, systems and, despite considerable achievements by various international bodies, it is still a long way from operating as a single integrated system'.

Indeed, the entry of Greece into the EC in 1981 and the accession of Spain and Portugal in 1986 has meant that the sphere of influence will continue to be pushed southwards, leading to changes in the pattern of trade. There is a shortage of modern road networks in these countries and with transport interests spread over a wider geographical area, current differences between members will be accentuated and it will be less easy to reach common decisions. This was emphasized in May 1985 when the European Court of Justice censured EC governments for failing to introduce a common transport policy—27 years after they promised to do so! In response, a master plan for transport, from the current Italian EC president, was discussed in late June 1985. However, the plan has already been criticized for being vague and there is continued disagreement over such items as working hours for commercial drivers, a more liberal aviation system, and the right of road hauliers to operate outside their own country. It is clear that many problems remain to be solved before an integrated system of transportation is a real possibility.

References

Bird, J. H. (1967). 'Seaports and the European Economic Community'. *Geographical Journal*, **133,** 302–27.

Bird, J. H, and Pollock, E. E. (1978). 'The future of seaports in the European Communities'. *Geographical Journal*, **144,** 23–48.

Blacksell, M. (1977). *Post-war Europe: a political geography*. Dawson, Folkestone.

Burley, T. M. (1970). 'Nürnberg to become a port'. *Geographical Magazine*, **42,** 536.

Chapman, A. S. (1968). 'Trans Europ Express: overall travel time in competition for passengers'. *Economic Geography*, **44,** 283–95.

Clout, H. D. (ed.) (1975). *Regional development in western Europe*. Wiley, London.

Collester, J. B. and Burnham, H. (1975). 'Eurocontrol: a reappraisal of functional integration'. *Journal of Common Market Studies*, **13,** 345–67.

Dean, C. J. (1982). 'The trans-Mediterranean gas pipeline'. *Geography*, **67,** 258–60.

Despicht, N. S. (1969). *The Common Transport Policy of the European Communities*. PEP, London.

Goergen, R. (1972). 'Transport policy in the European Community after the British entry'. *The Chartered Institute of Transport Journal*, **34,** 343–53.

Jordan, T. G. (1973). *The European culture area*. Harper and Row, London.

Martin, J. E. (1974). 'Some effects of the canalisation of the Moselle'. *Geography*, **59,** 298–308.

Mellor, R. E. H. (1983). 'A future for the Rhine–Main–Danube canal'. *Geography*, **68,** 338–40.

Mellor, R. E. H. and Smith, E. A. (1979). *Europe: a geographical survey of the continent*. Macmillan, London.

Mills, D. G. (1978). 'Changes in the port of Amsterdam'. *Geography*, **63,** 209–13.

Minshull, G. N. (1978). *The new Europe: an economic geography of the EEC*. Hodder and Stoughton, London.

North, G. (1968). 'Gargantuan Rotterdam'. *Geographical Magazine*, **40,** 1464–78.

Pacione, M. (1974). 'Italian motorways'. *Geography*, **59,** 35–41.

Parker, G. (1981). *The logic of unit*. Longmans, London.

Pilkington, R. (1966). 'Joining the Rhine and the Rhône'. *Geographical Magazine*, **39,** 214–28.

Pinder, D. A. (1978). 'Guilding economic development in the EEC: the approach of the European Investment Bank'. *Geography*, **63,** 88–98.

Scargill, D. J. (1976). 'New hope for the Rhône–Rhine waterways'. *Geography*, **61,** 160–3.

Schofield, G. (1965). 'The canalisation of the Moselle'. *Geography*, **50,** 161–3.

Thomson, J. M. (1976). 'Towards a European transport strategy. In Lee, R. and Ogden. P. E. (eds), *Economy and society in the EEC*. Saxon House, Farnborough.

Tuppen, J. (1975). 'Canals and waterways in the EEC'. *European Studies*, **21,** 1–4.

6

Industry

6.1 Industrial development in western Europe

A study of the industrial geography of western Europe must necessarily concern itself with three main items:

1. The areal aspect, or distribution, of industry.
2. The factors which have shaped the pattern of industrial development.
3. Temporal changes in industrial development.

The distribution of industry in western Europe

The industrial revolution had its origins in Great Britain in the early eighteenth century. Gradually the ideas diffused to other parts of western Europe, notably those in close proximity to sizeable coal deposits. This diffusion process is still continuing and has yet to reach some of the more peripheral areas. As with agricultural production and population density, western Europe has an industrial core area and a distinct core–periphery relationship is in evidence (Fig. 6.1). This core area is relatively small although, after being restricted to the main coalfield areas in the nineteenth century, it expanded in the present century into southern and central Sweden, Switzerland, northern Italy, and the Barcelona district of Spain. Areas such as southern Italy, southern Iberia, Ireland, Iceland, and the Scottish Highlands form the outer periphery, which is much less industrialized than the core.

Within the industrial core area one major zone of industrial activity is prominent. This axial belt of development stretches from the English midlands, through Rotterdam and Antwerp, and inwards along the Rhine, Meuse, and Schelde rivers into the Ruhr, Saar–Lorraine, Sambre–Meuse, and Swiss Plateau industrial districts of western Europe. The so-called Lotharingian axis has been expanding southwards and could well now include the industrial district of northern Italy as well as the Rhône-Saône valley in France. The heavy industrial triangle (Fig. 6.1), which completely dominated the pattern of heavy industry in the nineteenth century and still contains over 30 per cent of western Europe's steel capacity, is located on the Lotharingian axis and is bounded at its apices by the Nord coalfield of France, the Ruhr coalfield of West Germany, and the Lorraine iron-ore field.

Although western Europe remains one of the world's most highly industrialized areas, many of its more advanced economies have experienced a process of decentralization, which has quickened since the economic recession of the mid-1970s. A reduction in the workforce has not only characterized such traditional industries as textiles, shipbuilding, and metal manufacture, especially in north-western Europe, but also more recent growth industries like motor vehicles and electrical engineering (Knox 1984). Indeed, emphasis in recent years has shifted away from manufacturing and towards the service and quaternary sectors, as demonstrated by the decline in unskilled and manual jobs and growth in white-collar, professional, and technical work. Many 'core' areas of western Europe can now confidently be classified as post-industrial.

Factors affecting the location of industry

A combination of factors, both historical and contemporary, has produced the pattern of industrialization in western Europe. The most important factors at work include:

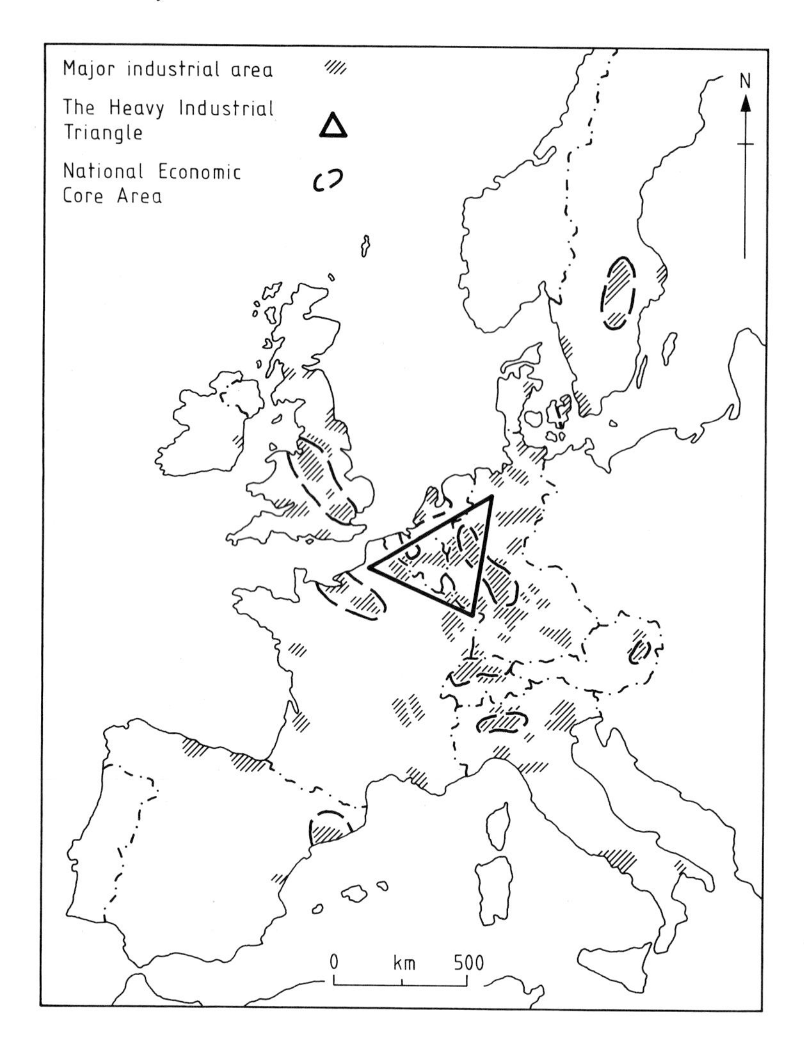

Fig. 6.1 The distribution of industry in western Europe

1. *Raw materials*. Indigenous raw materials acted as a strong locating factor during the industrial revolution and industry was attracted to coalfield and iron-ore sites. However, the twentieth century dispersal of industry in western Europe, characterized by a movement to coastal locations, reflects the weakening ties of local raw materials. For certain types of industry, location near raw materials is still important and Jordan (1973) noted four examples:

(a) Industries that locate near the source of raw materials out of physical necessity, such as mining and lumbering industries.

(b) Industries that lose considerable weight in processing, such as in the pulp and paper industries.

(c) Industries which use perishable goods as their raw materials, such as in the canning of fruit and vegetables.

(d) Industries which consume large quantities of fuel and especially coal.

The emergence of oil, natural gas, and most notably electricity, as alternative energy sources to coal, has helped to reduce the dominance of coalfield locations, especially as these new sources of energy can be transported to most parts of western Europe. Despite this greater flexibility in industrial location, there is a tendency for industry to continue to concentrate in the older industrial districts, even though local raw materials are either exhausted or no longer used. This is known as geographical inertia, a process whereby industry remains in an area to benefit from such factors as a skilled labour supply, the existence of linkages with other industries, the fixed capital invested in plant, and the large urban consumer markets. Good examples are the heavy industrial triangle and the textile industry in Lancashire and the West Riding of Yorkshire.

2. *Transport.* According to Weber (1909), most industries attempt to minimize transport costs by locating at either the source of raw materials or the market, depending on the type of industry. Very rarely would industry locate in between the two, at the so-called breaking point, as this would mean a doubling of transport costs. As well as cost, the ease of access was an important locating factor and the Ruhr industrial district, for example, was dependent on the Rhine waterway for its development and survival, especially when foreign raw materials began to be imported into western Europe. However, transport costs are no longer a strong locating factor as the network of communications, pipelines, and grids, for the movement of raw materials and finished goods, is rapidly improving. In addition, break-of-bulk points, and hence ports, have become efficient locations for industry in western Europe as they are ideally sited for the importation of raw materials.

3. *Factors of production.* The availability of land, labour, and capital has played an important role in industrial development in western Europe. Supplies of merchant capital helped to make possible early industrial enterprises and the availability of investment capital, either foreign or indigenous, is just as important today. Capital is more readily available in the major cities and this helps to explain why centres like London, Paris, and Brussels have become important industrial areas. In contrast, most of the peripheral areas of western Europe lack the necessary capital to promote industrial development. An adequate labour supply is another important factor, although the increased mobility of the European population has lessened its importance. Many industries, especially those where machines do most of the work, require little skill and are consequently attracted to large concentrations of people and hence to cities. However, a skilled labour force is different and can greatly influence the location of industry, as with the watchmaking industry in the Jura district of Switzerland and market gardening in the Vale of Evesham.

4. *The market.* The purpose of industry is the production of goods for sale and there is little point in producing goods unless there is a market for them. The market is dependent upon the number of potential purchasers and their standards of living and it follows, therefore, that the areas of greatest market potential are the areas of greatest population density (Harris 1954). In western Europe these coincide with the major industrial districts, suggesting that the market is a very important locating factor and one of increasing significance. However, one should note that the areas of dense population were often the result rather than the cause of industrial development. Certain types of industries have little choice but to locate at the market and these include weight-gain industries, such as bottling, and industries where the finished product is either more fragile (glass), or more perishable (bread), than the raw material. The concentration of industry near the market enhances the possibility of agglomeration economies, linkage, and increases in the size of production; these factors can lead to 'regional swarming' of industry, as in the West Midlands of England.

5. *Government intervention.* Government interference, direction, or control may be responsible for industrial location and development. The role played by the governments of western Europe in the location of industry is increasing in importance and state intervention has reached its greatest extent in the United Kingdom, although it is very important in France and Italy. Government intervention has helped to guide the necessary adjustments in west European industry and it has three basic aims (Minshull, 1978):

(a) To control areas of production which are basic to the economy.

(b) To support declining industrial areas.

(c) To support high technology and research industries.

Support for areas of severe decline in western Europe has been at the forefront since the 1950s and numerous projects have been undertaken; one good example is the Fund for the South, created in 1950 to ease the problems of the Mezzogiorno region of Italy. Regional problems and planning will be discussed in more detail in Chapter 10. An example of Government interference in the location of industry in Great Britain is the establishment of trading estates, which are part of a complex organization to arrest industrial decline (Bale 1977). Industrial location has also been profoundly affected by the political reorganization of western Europe. The division of Germany and the creation of the European Community and the European Free Trade Association have changed the shape and nature of the political units within which industrialists are operating (Parker 1981).

6. *Multinational corporations.* The rise of giant business organizations, both indigenous and foreign, helped to transform western Europe's industrial system, especially in the 1960s. Spatially, they had the effect of intensifying the core–periphery contrast in industrial development. This was because US investment was biased towards 'leader' regions (Hamilton 1976), and indigenous companies only developed 'branch plants' in peripheral areas whilst maintaining their headquarters in the central core. An international division of labour emerged (Chapter 10), modifying the industrial geography of western Europe (see below).

Temporal changes in the pattern of industrialization

The pattern of industrial development in western Europe has witnessed two distinct trends:

(a) A nineteenth century trend towards the concentration of industry.
(b) A twentieth century trend towards the dispersal of industry.

The major characteristics of nineteenth century industrialization were: first, the overwhelming reliance upon heavy and staple industries; and secondly, the exceptionally strong correlation between the distribution of industry and the major west European coalfields. Early industrial areas were largely, but not wholly, confined to West Germany and the United Kingdom and four in particular stood out above the others:

1. *The Ruhr industrial district of Westphalia*, which had developed into the most important industrial area in western Europe by the end of the nineteenth century. With the diffusion of ideas from the United Kingdom, this district changed from a pre-1850 situation of being characterized by small-scale industrial activity to being a major coal and iron and steel district by 1900. Annual coal output between 1800 and 1850 had increased by a modest seven times, but during the 1850–1900 period it had increased 33 times (Jordan 1973). Local supplies of iron-ore were soon exhausted and the area became dependent upon imported ores from Sweden and Spain. The development of the Ruhr was thus dependent on the highly productive coalfield, which provided excellent coking coal, and the Rhine river, which provided the major routeway for the transportation of goods and raw materials. Industrial development was accompanied by a population explosion, which in turn created a large market demand for industrial goods. Today, the Ruhr is essentially a steel and heavy metallurgical district, although specialized sub-regions of clothing, textiles, chemicals, and engineering have emerged. The district remains the industrial heart of western Europe, even though its importance has faded in relative terms.

2. *The Saar–Lorraine industrial district*, which was affected by much political rivalry and changing national boundaries. This situation arose from the fact that the German Saar contained the coal whilst the French Lorraine contained the iron-ore. Consequently, the Germans attempted to seize the iron-ore deposits and the French tried to detach the Saarland from Germany. The district was mainly developed by the Germans but it never rivalled the Ruhr in importance. Development did not take place until the 1850s and the area was further hampered by having poor transport links with other areas and coal which was not suited to coking. Therefore, coal was imported from the Ruhr in exchange for supplies of iron-ore, a movement which was later to benefit from the canalization of the Moselle river. The Lorraine area also contains natural salt deposits which form the basis of important modern chemical industries.

3. *The Sambre–Meuse industrial district of Belgium*, which was unparalleled up to 1850. The district surpassed the Ruhr until the 1860s and Riley and Ashworth (1975) noted that the first Belgian coke-fired blast furnace was erected in 1823, whereas the Ruhr had to wait until 1849.

This coalfield area, stretching from the Borinage to Liège, had attracting advantages similar to those of the Ruhr, including good supplies of coking coal, local iron-ore, and good communications in the form of the navigable Meuse river. Ultimately, the district became dependent on imported raw materials and, in line with other industrial areas, it entered the twentieth century with a narrow economic base. Although iron and steel remains the most important industry in the 1980s, the district has managed to diversify, and both chemicals and textiles are important.

4. *The United Kingdom*, which led the way in the industrial development of western Europe. During the nineteenth century industry became concentrated in two main areas. In the early part of the century, development was based on the coalfields, notably those in South Wales, the north-east, and Scotland. Although experiencing a relative decline, these areas are still of considerable importance, specializing in heavy and semi-finished products such as tin plate (South Wales), and heavy engineering (the north-east). To help diversify the industrial base, light industry has been introduced into these areas, as in the trading estates of South Wales (Bale 1973). In the latter part of the century, an industrial district developed in central England, based around the southern Pennines and composed of the Black Country, the Potteries, south Lancashire, Sheffield, west Yorkshire, and a few smaller areas. This district specialized in metallurgy, pottery, precision steelmaking, and textiles and it was in close proximity to local coal supplies. After the 1960s, the district was revitalized by such growth industries as electrical engineering, transport equipment, and chemicals.

The main non-coalfield areas to experience industrial growth in the nineteenth century were the larger urban centres such as Berlin, Paris, Hamburg, and London. The main attracting factors for industry were the market and a large supply of labour; this was especially true of London which, as well as being the major centre of commerce and trade in the country, had the added advantage of readily available supplies of capital from the merchant class (Jordan 1973).

In the twentieth century, different types of industries assumed importance and the traditional heavy industries were challenged by a group of industries which included motor vehicles, aircraft, electrical equipment, electronic and 'sunrise' industries, man-made fibres, and synthetics. These modern growth industries did not use coal directly and their energy was provided by electricity, oil, and gas. Consequently, their choice of location was made more flexible, resulting in a general trend towards the dispersal of industry into new areas of western Europe. As Parker stated (1981, p. 94) 'the twin keynotes of the process of change have been a dispersal from existing centres accompanied by a tendency to concentrate into certain new ones'. Factors instrumental in this process of dispersal are five-fold:

(a) The availablity of new sources of power, such as oil, natural gas, and nuclear energy.

(b) Improved means of transport, resulting from the development of motor vehicles and better roads.

(c) Improvements in technology, as in the iron and steel industry where successively smaller amounts of coal have been required over the years.

(d) The ever-increasing need to rely on imported raw materials.

(e) The rise of multinational companies and the international spatial division of labour.

A major characteristic of twentieth century dispersal has been the movement towards coastal locations, as ports became the cheapest assembly points for industry. Indeed, a number of new industrial districts, not all coastal and not dependent on coalfields, have emerged and Jordan (1973) mentioned four in particular:

1. *Central and southern Sweden*. Iron was produced in central Sweden as early as 1500, based on local charcoal and iron-ore reserves at Bergslagen. However, Sweden was handicapped by a lack of coal and the southern and central parts did not fully develop into an industrial district until the twentieth century, when hydro-electric power became the major source of power. In comparison to West Germany and the United Kingdom, the quantity of steel produced is small, but industrial potential is based primarily on the quality of Swedish manpower and technique. The area specializes in certain industrial activities for which it is best endowed—forest products, machinery,

instruments, and ships. Although steel production remains concentrated in the Bergslagen area, engineering industries are more widely distributed, with centres like Göteborg and Malmö increasing in importance.

2. *Northern Spain*. This part of Spain benefits from local supplies of raw materials and hydro-electric power and has developed into an area of industrialization. Industry is concentrated in the Barcelona district on the Mediterranean coast and around Bilbao and San Sebastian on the Basque Biscay coast. In the latter area, high grade iron-ore mines and coalfields near Oviedo have resulted in iron and steel industries, which supply important shipbuilding industries along the well-watered Atlantic coast. In the Barcelona district, streams have been harnessed in the hills to yield substantial amounts of hydro-electric power, which supply textile and chemical industries in the valleys and in the suburbs of Barcelona. The city itself has specialized in cotton trade and manufactures and the larger Catalonian metropolitan area is the centre of diverse manufacturing.

3. *The Swiss Plateau–Jura district*. This area contains the two most urbanized parts of Switzerland: the northern triangle, Zurich–Bern–Basel; and the south-western corner of the shores of Lake Geneva. As with Sweden, a lack of indigenous raw materials meant a heavy reliance on skilled labour and the production of high quality goods, although hydro-electric power from the Swiss Alps has helped to widen the industrial base since the 1950s. Today, the Swiss Plateau–Jura district has a diversified industrial character but retains its high quality. St. Gallen specializes in linen and laces; Basel is the centre of the Swiss chemical industry, especially dyes and pharmaceutical products; and Geneva is the capital of the famous watchmaking industry as well as being an international centre for industrial and financial concerns.

4. *The Po valley industrial district*. Lacking coal and iron-ore, this district was revived in the twentieth century by hydro-electric power from the Alps. Industry is concentrated in Piedmont and Lombardy, in an industrial triangle which is bounded at its apices by Turin, Milan, and Genoa. This is the part of Italy which is closest to the core area of north-west Europe and it had good access to the port of Genoa in the south. Since World War II, the Po valley has witnessed remarkable industrial growth and, after the Ruhr, it is the most important industrial district in western Europe. As a result of its rather late growth, the area has successfully managed to develop a diversified industrial base, which includes iron and steel industries based on imported raw materials, motor vehicles in Turin, chemicals and textiles in Milan, and shipbuilding in Genoa.

Although at the macro-scale industrial dispersal is a clearly recognizable feature, the process has been complicated by two superimposed trends: first, the rise of multinational firms during the major phase of post-war industrial growth; and secondly, the relative growth of manufacturing employment in rural areas, generally known as the 'urban–rural shift'.

Multi-national corporations, especially from North America but also from the Far East and Europe itself, came to dominate industrial decision-making in western Europe during the favourable economic climate of the post-war period. By the early 1970s, US investment in western Europe was four times as great as European investment in the USA and approximately one-third of the USA's annual foreign investment was channelled into west European economies (Hamilton 1976; Clout *et al.* 1985). The most popular initial target was the UK (Young and Hood 1976), where US investment reached $127 per capita by 1970, before attention was concentrated in or near the golden triangle, comprising Belgium, south Netherlands, northern France, and West Germany (Hamilton 1976). Eventually the process diffused and US firms became interested in both the subsidies and cheaper labour available in the peripheral areas of Italy, Iberia, and Scandinavia.

By the time of the oil crisis, foreign-controlled firms accounted for approximately 20 to 30 per cent of industrial production in all the leading European economies, compared with 10 per cent or lower in the periphery (Clout *et al.* 1985). One exception to this pattern was Ireland, which actively encouraged foreign investment and industry (O'Farrell 1978; Breathnach 1985). Hamilton (1976) was able to demonstrate how foreign multinationals were spatially more concentrated than their European counterparts, preferring such major cities, ports, and regional centres as London, Paris, Ghent, Antwerp, Randstad, the Ruhr, Hamburg, and Milan. In contrast, the headquarters of indigenous firms were often located in second and third order provincial centres, like ICI in Billingham, Philips in Eindhoven, and Michelin in Clermont-Ferrand.

Foreign firms established plants in western Europe, especially during the 1960s, for numerous reasons, including:

(a) Production costs were approximately 15 per cent lower than in the USA.

(b) Western Europe was an affluent and rapidly growing market.

(c) There was abundant and relatively cheap labour, with good skills.

(d) There was a need to overcome the tariff wall of the European Community and to manufacture within its market.

(e) American companies wanted to avoid the perceived threat of competition from European manufacturers.

Multinational developments in western Europe did not have wholly positive effects. Whilst the supply of capital for growth and development was increased and the technology gap between western Europe and the USA reduced by new skills and know-how (Hamilton 1976), foreign firms were parasitic in so far as profits often returned to the home country, just as materials and components came from outside the local area. The economic recession of the mid-1970s changed the situation and as surplus capacity occurred a process of rationalization began. Once again the multinational companies emphasized their flexibility as well as their parasitic nature by transferring production to cheap labour areas in the Third World.

During the 1960s and 1970s a process of industrial decentralization, away from major metropolitan areas, also occurred in many west European countries (Fothergill and Gudgin 1982; Knox 1984). However, the detailed pattern of this urban–rural shift was not established until the survey of EC members by Keeble *et al.* (1983). These authors analysed the 'shift' in 107 administrative regions of the EC, in terms of three indicators: gross domestic product, industrial output, and manufacturing employment. The regions were classified into four groups—highly urbanized, urbanized, less urbanized, and rural—and in the 1970s there was striking evidence of an urban–rural shift in all three indicators (Table 6.1). Heavy relative losses occurred in the most urbanized centres like Hamburg, Dusseldorf, Paris, Manchester, and Rome, whereas major relative gains characterized the smaller towns and rural settlements. This situation was true of small, large, southern, and northern regions of the EC, leading Keeble *et al.* (1983, p. 417) to conclude that 'the Community's urbanised regions are declining despite possession of relatively favourable manufacturing structures, while rural areas are growing, at least relatively, despite very unfavourable structures biased towards industries which are declining fastest at the EC level'.

Three possible explanations for this urban–rural shift were advocated:

(a) *Production cost theory*, involving the higher operating costs in urban centres.

(b) *Constrained location theory*, relating to factory floorspace constraints in cities, as machines replace much manual labour.

(c) *Capital restructuring theory*, where the urban–rural shift is seen to be one outcome of the process of uneven spatial development under

Table 6.1 The urban–rural shift in the European Community

	Change in % share:		
	Gross domestic product, 1970–7	Industrial output, 1970–7	Manufacturing employment, 1973–9
1. Highly-urbanized regions (22)	−1.7	−2.9	−6.0
2. Urbanized regions (22)	−0.3	+0.2	−6.9
3. Less-urbanized regions (32)	+0.9	+1.1	−1.6
4. Rural regions (30)	+1.1	+1.6	−2.9

Source: Keeble *et al.* (1983, p. 410–11).

the capitalist mode of production, especially as production shifts to areas characterized by exploitable, unskilled, and cheap labour.

Evidence to support aspects of all three theories was forwarded by Keeble *et al.*, although the constrained location theory was thought to be the most appropriate.

Overall, the process of industrial dispersion has meant that the traditional core areas of western Europe are attracting a disproportionate amount of new industrial investment. Nevertheless, industry remains concentrated in a number of selected areas and it has been argued, by Friedmann and Alonso (1975), that certain institutional factors prevent decentralization and have the effect of increasing the pattern of concentration. These are:

(a) The insufficient availability of social capital in the poorer, peripheral regions of western Europe for items such as transport.

(b) The reduction in regional wage differentials as a result of trade union activity and the introduction of social considerations in wage policies. Industry was formerly attracted to developing areas because they offered plenty of cheap labour, which helped to offset the many other disadvantages. This incentive no longer acts as a locating factor.

(c) The differences in local taxation within countries such as the United Kingdom, Sweden, and West Germany. Low levels of income in the peripheral areas meant higher tax rates, to meet local expenses, and this had an adverse effect on the location of industry.

(d) The imperfections in the credit market, whereby centralized banking systems supplied credit to large and well-established firms only. Small and medium-sized firms in poorer areas were obliged to pay more to secure funds.

(e) The principle of tapering rates in transport costs over long distances, which theoretically releases industry from a raw material location and encourages dispersal. However, it led to increased centralization because it reduced the advantage of processing products on the spot rather than in the industrialized areas. In other words, it encouraged industry to concentrate near the market.

(f) The increasing use of electricity in western Europe, in place of coal, could facilitate decentralization, but there are examples where regional systems of power rates discouraged this. The cost of electricity in the Greek provinces, for example, was three times higher than in Athens in the early 1970s.

Collectively, these and other factors already mentioned help to militate against industrial development in the peripheral regions of western Europe. Since the 1960s government policy has attempted to overcome these problems by encouraging the general dispersal of industry (Chapter 10).

6.2 Regions of relative growth and decline

With the twentieth century trend towards the dispersal of industry in western Europe, industrial areas based on the coalfields have suffered a relative and sometimes absolute decline. In contrast, industrial areas removed from the coalfields have witnessed relative and absolute growth. Some of the main areas of relative growth and decline are listed in Table 6.2 and in preference to analysing each area, attention will be focused on two countries, West Germany and Belgium, which contain both types of area.

West Germany: the Rhine–Ruhr and Bavaria

Since 1945, the Ruhr industrial district has de-

Table 6.2 Regions of relative growth and decline in western Europe

Growth	Decline
1. Southern Germany (Bavaria)	1. The Rhine–Ruhr
2. Northern Italy (Po Valley)	2. The German Saarland
3. The Swiss Plateau	3. Southern Belgium (Sambre–Meuse)
4. Central and southern Sweden	4. South Wales
5. Brussels–Antwerp axial belt	
6. Southern France (lower Rhône)	

clined in relative terms compared to Bavaria in southern Germany. Bayern is the largest *Land* in West Germany and the second largest in terms of total population. However, the density of population (155/sq.km), is well below the average for the whole country (248/sq.km). Before 1945, industry in Bavaria (Bayern and Baden-Württemberg) was small-scale, remote, and located near the occasional nucleated settlement. Certain pockets of industrial development could be discerned:

1. Timber, pulp, paper, and rayon industries based on the Spessart forests.
2. Agricultural industries based on the better soils of Franconia.
3. Cement, brewing, and canning industries in the Main valley, based on local raw materials and good communications.
4. Skilled handicrafts in Nuremberg, a famous toy centre.
5. The chemical triangle, at the eastern end of the Bavarian Alpine Foreland, based on the river Inn and obtaining coal and limestone from the Ruhr.

Before world war II, 28 per cent of the working population of Bavaria was employed in agriculture and there was a strong conservative attachment to the land. After the 1950s and especially from 1960 onwards, the situation changed dramatically and employment in agriculture in 1980 was approaching the national average figure of just 5.2 per cent in Baden-Württemberg, although it was still relatively high in Bayern (9.8 per cent). Regional development has taken place in Bavaria, greatly aided by the designation of development areas in the 1950s, and the movement of people to the area and away from the Ruhr and East Germany resulted in 16 per cent of the population in 1968 being comprised of refugees. This remarkable change-about can be attributed to a number of interrelated factors:

1. Although presented with many post-war difficulties, Bavaria did not have a large coal-mining area like the Ruhr to renovate.
2. An influx of refugees meant an initial burden on an already weak economy, but many were skilled and set up traditional textile, glass, and jewellery industries in former military buildings.
3. The breaking of links with East Germany, along with Federal support for industrial development, encouraged firms from the east to locate in the small towns of Bavaria where labour was plentiful. This gave industry a scattered spatial structure, even though the area contained two of West Germany's larger urban agglomerations, Munich and Nuremberg.
4. Bavaria was originally remote from the main axis of West German development, a situation made worse by the formation of the European Community and restrictions on trade with eastern Europe. However, the area has been greatly aided by developments in transport, such as the electrification of rail routes to the Rhineland and the north German ports, motorway construction, oil pipelines, and the canalization of the Main river.
5. The change in energy sources in western Europe, from coal to oil, natural gas, and electrical energy benefited southern Germany. Before 1960, there was no oil refining in Bavaria but, with industry demanding oil, a building programme was undertaken by the oil companies themselves. As a result, oil prices fell by one-third and by 1965 there were 10 oil refineries in southern Germany, representing over 30 per cent of West German refining capacity. By 1969, Bavaria alone accounted for 18.1 per cent of West German production, compared to zero production in 1961, and was second only to North Rhine-Westphalia, which had experienced a decline from 60.1 to 34.4 per cent over the same time period.
6. Oil refining in Bavaria led to a number of associated growth industries and acted as a major stimulus to the economy. The outdated iron and steel industry had the opportunity to modernize itself and take full advantage of improved production techniques. During the 1960s, there was rapid growth in electrical engineering, especially in the labour-intensive electro-technical industries. Bayern and neighbouring Baden-Württemberg boasted more than 1200 factories between them and by 1969 they accounted for approximately half of the production, by value, of all West German electro-technical industries (Table 6.3). These two southern *Länder* increased their percentage share of employment and production, whereas the formerly important *Länder* of West Berlin and North Rhine-Westphalia, in northern Germany, lost ground (Burtenshaw 1974).
7. Bavaria benefits from internal tourism, both in summer and winter. The Alpine area, and Munich in particular, is the major attraction and

Table 6.3 Production and employment in electrical industries, 1969

Land	Value of production in DM million	% of production	Workforce '000	% of total
Bayern	9010	23.3	248	24.9
Baden-Württemberg	8845	22.9	226	22.7
North Rhine-Westphalia	8409	21.8	192	19.2
West Berlin	3737	9.7	92	9.2

Source: Burtenshaw (1974, p. 122)

between 1970 and 1980, 25 per cent of all holidays taken by West Germans were in Bavaria. Increasingly, Bavaria has become a *Land* of second homes, with figures approaching 10 000, and these too are concentrated in the Alpine foreland. Tourism has stimulated the building and agricultural industries and encouraged the development of a better infrastructure.

8. Agriculturally, Bavaria has made important steps forward and the consolidation of farm holdings and the resettlement of farmers have helped to reduce the numbers employed in agriculture. However, there is still a long way to go and the majority of agricultural units remain too small and severely fragmented.

Therefore, Bavaria's achievements are remarkable, especially as it is peripheral to the main centre of west European economic activity. The situation in the Rhine–Ruhr is quite different, even though North Rhine-Westphalia remains the leading industrial and commercial *Land* in West Germany. Its role in the European Community is quite clear: it is responsible for one third of West Germany's and one-eighth of the EC's total exports; it produces over one-quarter of the Community's coal and steel; and 15 per cent of the crude oil used in the Ten is refined here. In 1982, it had a population of 17 million, 6.3 per cent of the EC total, yet it covers just two per cent of the surface area of the Community. Despite this impressive list, the *Land* has declined, relative to growth in the country as a whole, and since the 1950s it has been confronted with problems associated with coal-mining and related industries (Spelt 1969; Barr 1970; Hellen 1974; Thomas and Tuppen 1977; Hull and Kenny 1983).

A series of crises has affected the Ruhr since the war. The output of coal declined from a post-war peak of 124.6 million tonnes to just 90 million tonnes in 1967 and 70 million in 1981. This was accompanied by a corresponding decrease in the number of pits in operation, from 161 in 1939 to 97 in 1964 and just 31 in 1980. Employment declined from 470 000 in 1957 to 140 536 in 1980. The decrease in production was thus less dramatic than the decrease in employment, indicating an increase in output per worker, and the Ruhr maintained its position of having the highest output per worker in western Europe. Bochum typifies the demise of coal mining in the Ruhr. In 1955 the town produced 9.5 million tonnes of coal and employed 65 000 in the industry; by the late 1970s all thirty mines had been closed and less than 200 people were left in the industry (Thomas and Tuppen 1977). Similarly, the Ruhr has been affected by falling steel production since the oil crisis, from 32.2 million tonnes in 1974 to 22.3 million in 1981. A total of 13 000 jobs was lost between 1977 and 1980 alone (Hull and Kenny 1983). Between 1964 and 1968, the area lost population for the first time, mainly from the major cities, and the unemployment rate rose to 3.6 per cent, almost twice the national average. (In 1983 unemployment was over 13 per cent!)

A process of general decay had established itself and the cause was not a depletion of resources but a combination of the following factors:

1. Improved fuel efficiency in the iron and steel industry, which demanded less and less coal.
2. A movement away from the use of coal by the railways.
3. Competition from oil and natural gas for energy production and heating.
4. The use of oil and gas as raw materials in the chemical industries.
5. An outmoded and inadequate infrastructure.

6. Lack of government awareness of the problems; a national energy policy, for example, did not emerge until 1967.

7. The unfavourable image of the Ruhr, as an area of coal-mining and heavy iron and steel industries, inhibited the attraction of new industries.

8. The economic recession of the mid-1970s.

The response to these problems was numerous planning measures from the 1960s onward, which have been well documented by Hall (1966 and 1977). Planning in the Ruhr has been the responsibility of a unique regional planning body, the Ruhr Planning Authority (known as SVR). The SVR produced a planning atlas in 1960 and regional plans in 1964 and 1966. The development plan of 1964 divided the Ruhr into a core area and a development zone (Fig. 6.2a). The core area contained the most intense problems of the Ruhr—congestion, pollution, lack of open space, and an unplanned mixture of land uses—and its industrial based needed diversifying. To the north, the development zone contained much open space and gave the opportunity for positive planning. Within this zone, four growth points were selected to become the foci of a balanced industrial structure: moving in a west to east direction these were Dinslaken–Wesel, Dorsten, Marl, and Datteln–Waltrop. The main elements of the 1966 structure plan are shown in Fig. 6.2b and include a series of north–south belts of urbanization within the core area. Each belt contains a zone of industrial development, surrounded by a regional green belt system. A buffer zone separates the core from the northern development zone. In addition, the SVR planned a series of new transport links and area parks as well as reducing the levels of pollution.

The SVR achieved a creditable record of environmental improvement. Unemployment and pollution levels were reduced and new science-based industries moved in. However, the SVR became too small to deal with the problems and in 1976 ultimate control of planning strategies for the Ruhr was transferred to the planning authorities of North Rhine-Westphalia. In 1979 the *Land* government announced an extra 5 billion DM for the Ruhr over a five year period. There is little doubt that environmentally the Ruhr has improved remarkably in recent years. Nearly 60 per cent of the region is classified as 'green areas' (Hull and Kenny 1983), and much recreational space has been provided. However, the area still provides 80 per cent of West Germany's coal and 65 per cent of its steel; with nearly 40 per cent of industrial employment in these two industries, it has not really diversified sufficiently to solve the long-term employment difficulties.

Belgium: the Sambre–Meuse trough and the Brussels–Antwerp axial belt

A similar north–south contrast exists in Belgium but, unlike the Ruhr, the area of relative decline occurs in the south. There have been two phases in the economic development of Belgium:

1. A nineteenth century phase, concentrated in the south.

2. A modern phase, in the coastal area to the north.

This contrast between north and south has been exaggerated by the two linguistic groups which exist in Belgium: the Walloons in the south (Wallonia) and the Flemings in the north (Flanders).

The Sambre–Meuse trough in the south experienced similar problems to those of the Ruhr (Elkins 1956; Davin 1969; European Studies 1971; Gay 1981), and, with the collapse of coal-mining, people began to migrate northwards at the rate of two to three per cent per year. Before the First World War, this area accounted for the entire output of Belgian coal, which stood at approximately 24 million tonnes per year. After the wars, output in the south levelled out and the increased production came from the Flemish Campine mines in the north. Coal production in Belgium reached a peak of 30 million tonnes in 1952, with only 68 per cent coming from the Sambre–Meuse area. In 1963, the south produced just 11 million tonnes of coal and by 1970, output stood at less than 10 million tonnes per year. This represented a major collapse in just 20 years and, although all coalfield areas were affected by the coal crisis of the 1960s, output in southern Belgium had decreased by 56 per cent compared to just 14 per cent in northern Belgium. The decline in output was accompanied by a fall in employment and the numbers employed in coal-mining in the Sambre–Meuse trough fell from 119 000 in

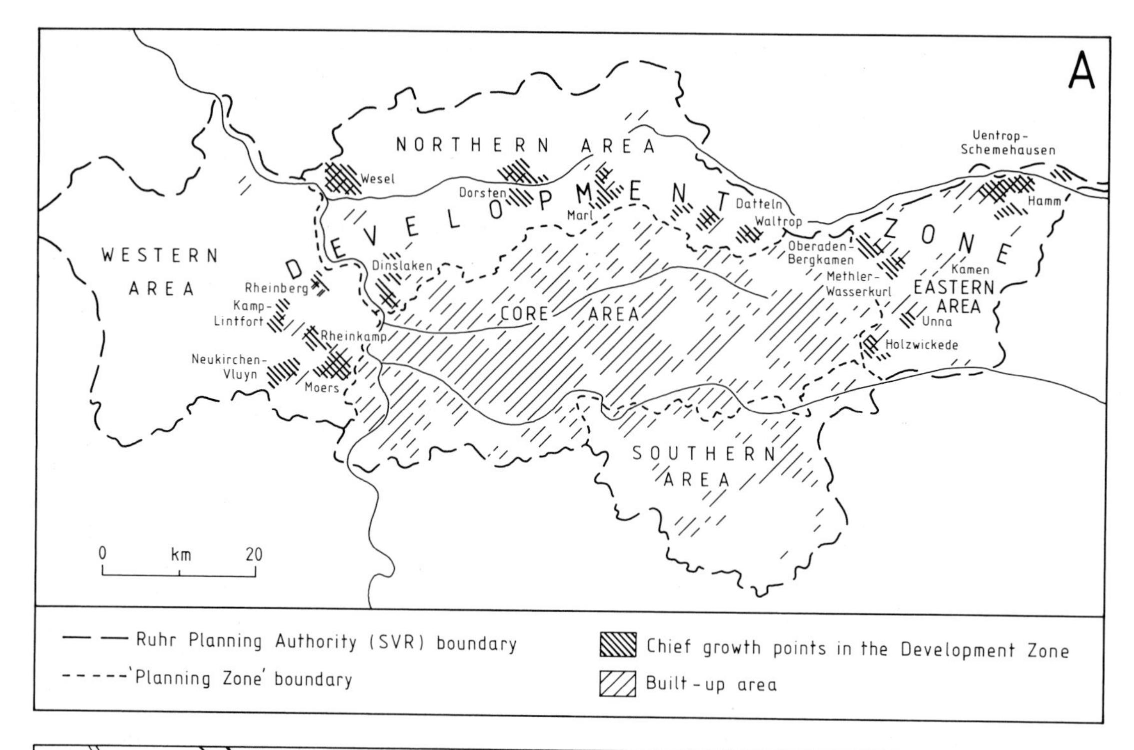

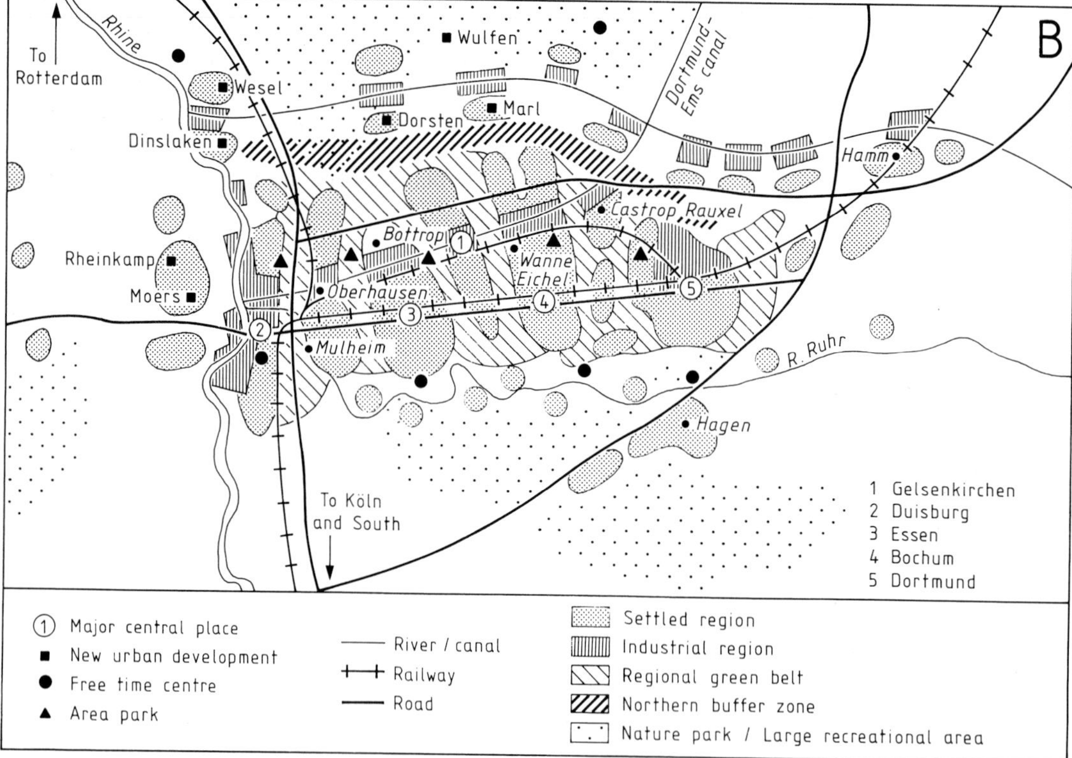

Fig. 6.2 Planning in the Ruhr: (a) the 1964 development plan (*source:* Hall 1966, p. 136); (b) the 1966 structure plan (*source:* Burtenshaw 1974, p. 214)

1952 to 44 000 in 1965 and 20 000 in 1970. Three main elements helped to cause this decline:

1. *A technical element*, concerning the nature of the coal deposits. The coalfields suffered from faulting and overthrusting and inaccessible seams became uneconomic to mine.

2. *A competitive element*, from alternative energy sources and in particular from oil and natural gas. The consumption of oil doubled in Belgium between 1957 and 1965, but the south again suffered because the major oil refineries were located near the coast in the north and it was too expensive to send petrol and fuel to the south. Competition from oil meant that coal prices had to be reduced, but with increasing costs of production this was impossible.

3. *A structural element*, in the industry itself. The Wallonian coal-mining industry was fragmented and characterized by too many small firms. In 1938, there were 163 firms involved in coal-mining; by 1952, the figure was still as high as 136 and it remained impossible to obtain economies of large-scale production. Rationalization did not occur until after the coal crisis, but by 1969 only 22 firms remained.

The coal-mining industry continued to decline throughout the 1970s, with total Belgian output levelling out at around 6 million tonnes per year. Less than 5 million tonnes now come from Wallonia and the numbers employed in coal-mining have been more than halved since 1970. The index of industrial production has continued to fall in Wallonia since 1974, in comparison to a steady rise in Flanders. Associated with the demise of coal-mining, the Sambre–Meuse trough has experienced demographic stagnation and there are no more people in the area than in 1930, unlike Flanders which has seen its population increase by 1.3 millions since that date (Gay 1981). Wallonia's population, which is distributed at a density of less than 200 per km^2 (Flanders is over 400), has an unbalanced structure and is one of the most aged in the world, with 17.4 per cent so classified in 1981. Unemployment is particularly high in the coal basins, where industrial redevelopment is difficult. This problem is compounded by a declining infrastructure, weak unions, and lower incomes than further north.

As with the Ruhr industrial district, these problems have been tackled by political measures since the 1960s. Regional development legislation has been passed, notably in 1959, 1966, and 1976 (Fig. 6.3). In 1959, 15 development areas were designated in Belgium and these were to receive special help, especially with structural unemployment. However, early policy favoured the north and not Wallonia in the south, which continued to decline. Of the 42 billion francs invested in Belgium between 1959 and 1965, only 10 per cent went to Wallonia. Regional policies were not successful in the south and Davin (1969) felt this was because the authorities confused 'development' and 'criti-

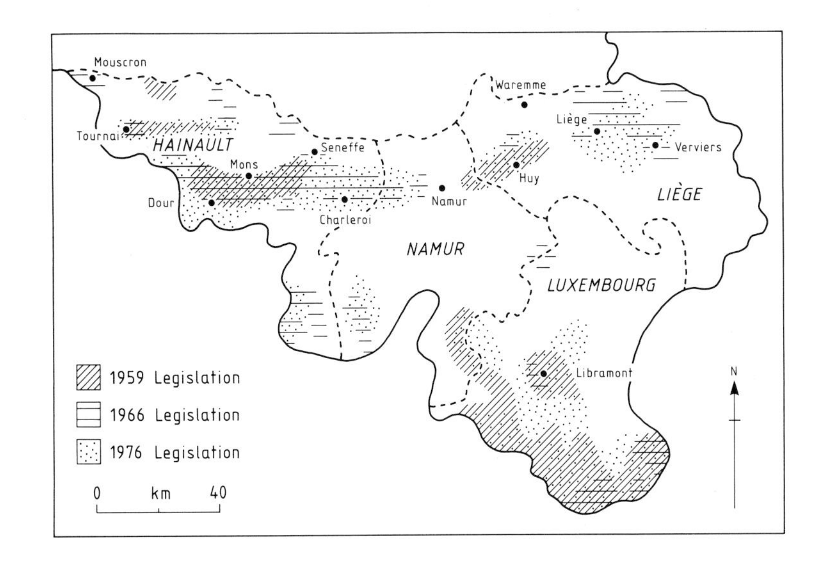

Fig. 6.3 Regional development in the Sambre–Meuse trough

cal' areas; the Lower Meuse region, for example, was retained as a development area when according to Davin it was a depressed area which required a different type of policy. Similarly, Gay (1981) advocated that early policies focused too sharply on employment criteria rather than on management problems and alternatives. The 1966 legislation set up six development areas, covering 678 communes (380 in the Flemish area and 298 in the Wallonian area), and for the first time assistance in Wallonia was greater than that assigned to the Flemish area. Also, special measures were introduced to revitalize the local economies of coal-mining and textile areas. After 1970, additional criteria—low living standards and slow rates of economic growth—were used in the delimitation of development zones. The expanded zones covered one-third of the country and 35 per cent of the population, but contravened EC rules. Consequently, new legislation in 1976 led to the redesignation of zones (Fig. 6.3). As Gay (1981) suggested, there has been a blow-by-blow approach to regional problems in Belgium rather than a coherent policy. The situation has been made worse by the two main linguistic groups, with 'any major investment in one socio-cultural region having to be compensated by an equivalent measure in the other' (Gay 1981, p. 189).

Regional policy, in the form of financial incentives and the provision of industrial estates, has been successful in attracting new industry to the south. As a result of their combined actions, 145 new plants were established in the coalfield area between 1959 and 1970 alone (Riley and Ashworth 1975). More important than the actual number of plants, was the type of industry attracted and it was encouraging to see modern growth industries among the new entrants. As Riley and Ashworth note, out of the 27 largest firms to enter the area, five were in oil-based chemicals, four in electronics, four in transport equipment and components and three in aluminium fabrication. However, whilst over 268 000 jobs were created on the 300 industrial estates established in Belgium between 1959 and 1975, only 18.6 per cent were in Wallonia. This is because industrial estates were opened earlier in the north, where planning authorities actively encouraged foreign investment. Therefore, Wallonia still has a long way to go, but improvements in the basic infrastructure, notably in transport and the E41 motorway in particular (which most of the new industries depend on), have helped in the revitalization of the Sambre–Meuse trough.

The Brussels–Antwerp axial belt has witnessed rapid post-war growth, primarily based on the increased prosperity of the two named cities. Brussels is the capital city of Belgium and has become the centre of the European Commission and Antwerp has developed into a major international port. With a decline in the coal-mining industry, especially in southern Belgium, it was not surprising to find northern Belgium increasingly attractive to migrants. Suburbanization has caused Brussels and Antwerp virtually to merge and together they form a belt of rapid development, which according to Riley and Ashworth (1975) is beginning to move south towards Liège.

It is not difficult to appreciate why Antwerp and Brussels have grown so rapidly. Antwerp, with a population of 800 000, has developed into a major port for three main reasons: first, the importance of foreign trade to Belgium; secondly, its location in relation to the major industrial areas of the European Community; and thirdly, the availability of large tracts of suitable land for industrial development and the construction of docks. Brussels (1.3 million) has attracted manufacturing and tertiary activities partially because of its role as capital city. By 1968, 165 major American companies had selected Brussels for their European headquarters and the city contained over half of all company head offices in Belgium (Clout *et al.* 1985). As the centre of Belgium's political activity and the European Commission, Brussels generates a high demand for goods and services. Two-thirds of Belgium's banking and insurance companies are located here and there was an upsurge in office building in the early 1970s, especially after the expansion of the European Community in 1973; the Berlaymont Building (European Commission) is one good example. Riley and Ashworth (1975) estimated that 473 790 m^2 of offices would be constructed in Belgium between 1972 and 1976 and a major phase of modernization began to transform Brussels, with the development of an underground railway system and a network of inner ring roads. This process was halted by the financial crisis of the mid-1970s and part of the ring road system, for example, is still to be completed.

The city was, to many people, being destroyed and a major conservationist backlash occurred (Clout *et al.* 1985). Nevertheless, per capita income in Brussels is the highest in Belgium and the city is the single most prosperous region in the Ten.

Good communications, by rail, road, and water, have aided the expansion of manufacturing in the axial belt. Characteristically, the belt contains many growth industries, including modern technology and engineering, chemicals, food processing, and printing and publishing. The completion of the E10 and E40 autoroutes should encourage the southward movement of the axial belt.

6.3 Case studies of west European industries

1. Traditional industries: Iron and steel

The iron and steel industry in post-war Europe has gone through two distinct phases: first, a period of rapid expansion up to 1974; followed secondly, by a serious crisis and declining output. In the initial part of the first phase, the whole framework of production was altered by the formation of the ECSC in 1952. Not only did this organization lead to the removal of frontier tariffs between the six member countries, but it had other far-reaching effects: freight rates were lowered; a single market, many times the size of the largest national market, was created; conditions of equal competition and a regular supply of raw materials were ensured; the industry was rationalized and modernized; and assistance was given, in the form of investment grants, to new steelwork development projects. The conditions of production were thus changed, both geographically and economically, and western Europe became one of the world's major steel producers (Sinclair 1969; Warren 1975).

Between 1952 and 1972, steel output in western Europe increased 2.5 times, from 62.5 to 161.8 million tonnes. The bulk of this increase came from the EC, although its share of total west European output decreased from 93.3 per cent in 1952 to 86 per cent in 1972. Steel output continued to rise, reaching a peak of 185 million tonnes in 1974. However, figures for 1982 (Table 6.4) show a marked reversal, with a total of 111.4 million tonnes for the European Community and 137.3 for western Europe; in relative terms the Community's share of total output has further declined, to 81.1 per cent. One possible explanation for this changing pattern of production is

Table 6.4 Crude steel output in western Europe, 1952–82 (million tonnes)

	Output 1952	%	Output 1972	%	Output 1982	%
West Germany	18.6	29.8	43.7	27.0	35.9	26.2
France	10.9	17.4	24.1	14.9	18.4	13.4
Italy	3.6	5.8	19.8	12.2	24.0	17.5
Netherlands	0.7	1.1	5.6	3.5	4.4	3.2
Belgium	5.1	8.2	14.5	9.0	10.0	7.3
Luxembourg	3.0	4.8	5.6	3.5	3.5	2.6
United Kingdom	16.4	26.2	25.4	15.7	13.7	10.0
Ireland	—	—	0.1	0.1	0.1	0.1
Denmark	—	—	0.5	0.3	0.6	0.4
Greece	—	—	0.6	0.4	0.9	0.7
Norway	0.1	0.2	0.9	0.6	0.8	0.6
Sweden	1.7	2.7	5.2	3.2	3.9	2.8
Switzerland	0.2	0.3	0.5	0.3	0.8	0.6
Austria	1.1	1.8	4.0	2.5	4.3	3.1
Portugal	—	—	0.4	0.2	0.5	0.4
Finland	0.2	0.3	1.5	0.9	2.4	1.8
Spain	0.9	1.4	9.5	5.9	13.2	9.6
EC (the Ten)	58.3	93.3	139.8	86.4	111.4	81.1
Western Europe	62.5	100.0	161.8	100.0	137.3	100.0

Source: United Nations Statistical Yearbook

that the main nineteenth century producers have tended to decline, relative to post-war growth in non-EC members such as Spain, Sweden, and Austria. The post-war pattern of change is well demonstrated in Table 6.4 and three main groups of countries can be identified:

1. Those countries in which the share of west European steel production has decreased. The country experiencing the largest relative decline is the United Kingdom, whose share of total production was reduced from 26 per cent in 1952 to 10.0 per cent in 1982. Whilst output managed to increase by over 50 per cent between 1952 and 1974, it has since tumbled by 33 per cent. Other countries in this category, but where the decline has been less dramatic, include West Germany (the largest steel producer), France, Belgium, and Luxembourg. The recession has affected the last two countries in particular as both were heavily dependent on the steel industry and led the world in the production of steel per inhabitant (1670 kg) (Evans 1980).
2. Those countries in which the share of total steel production has increased. Dramatic increases in both output and the percentage share of west European output have occurred in certain countries away from the traditional coalfield areas. The two most notable examples are Spain and Italy; in the latter, 80 per cent of total steel capacity is less than 40 years old. Output in these countries increased by over 700 per cent between 1952 and 1982, and Italy's contribution to total west European output increased from just 5.8 per cent in 1952 to 17.5 per cent in 1982, just as Spain's increased from 1.4 to 9.6 per cent.
3. Those countries which make little contribution to overall steel production, but which showed signs of increase before the economic recession. These are led by Austria, Sweden, and the Netherlands, who together produced 14.8 million tonnes of steel in 1972 (12.6 million in 1982).

Therefore, the centre of gravity of the steel industry was pulled away from the main nineteenth century industrial areas and one of the most dramatic locational changes has been the shift from coalfield to coastal locations (Fleming 1967; Warren 1967; Parker 1981). In 1938, western Europe did not possess a single coastal integrated iron and steel works. By 1966, there were eight plants on tidewater locations, each with a steel capacity of over one million tonnes; the figure had risen to 14 by 1972, and stood at 17 in 1984. In 1955, coastal steelworks accounted for just 5.5 per cent of the ECSC's steel, but this had increased to 14 per cent in 1965, and nearly 30 per cent in 1981. Coastal steelworks are located in two main areas (Fig. 6.4):

1. Along the North Sea coast, from Dunkirk to Hamburg and including the British coastal steel works. This group contains some very large plants, such as Ijmuiden and Zelzate, with capacities of four and six million tonnes respectively. Two outlying plants, the Mo-i-Rana plant in Norway, developed in the 1950s and based on the Syd-Varanger iron ore mines, and the Government-managed plant at Luleå in Sweden, can be added to this group.
2. Along the Mediterranean coast, from Valencia in Spain to Taranto in Italy. This group is dominated by the four Italian plants, which have benefited from state participation and control and which demonstrate the characteristic growth and dramatic locational change of the steel industry over the past twenty years.

This coastal movement was aided by new technology and the increasing dependence on imported raw materials, brought in on huge bulk carriers. The carriers revolutionized the concept of economic distance for water-transported cargo and between 1952 and 1974, the peak period, there was over a 500 per cent increase in the importation of iron-ore. The switch from heavy steel lines to lighter ones also encouraged a coastal movement; old plants needed to be reconstructed and this represented an ideal opportunity to build new plants on the coast and in the process eliminate the cost of transporting imported raw materials into the interior.

Location of iron and steel prior to the ECSC. The west European iron and steel industry was traditionally located in close proximity to local raw materials, with the presence of coal being the most important single factor. Coal was found in a central coal belt, which stretched from the Ruhr to Calais and into England and South Wales, and the carboniferous strata also contained iron-ore and limestone, other important ingredients for iron and steel production. As iron and steel was a weight-loss industry, it was advantageous to

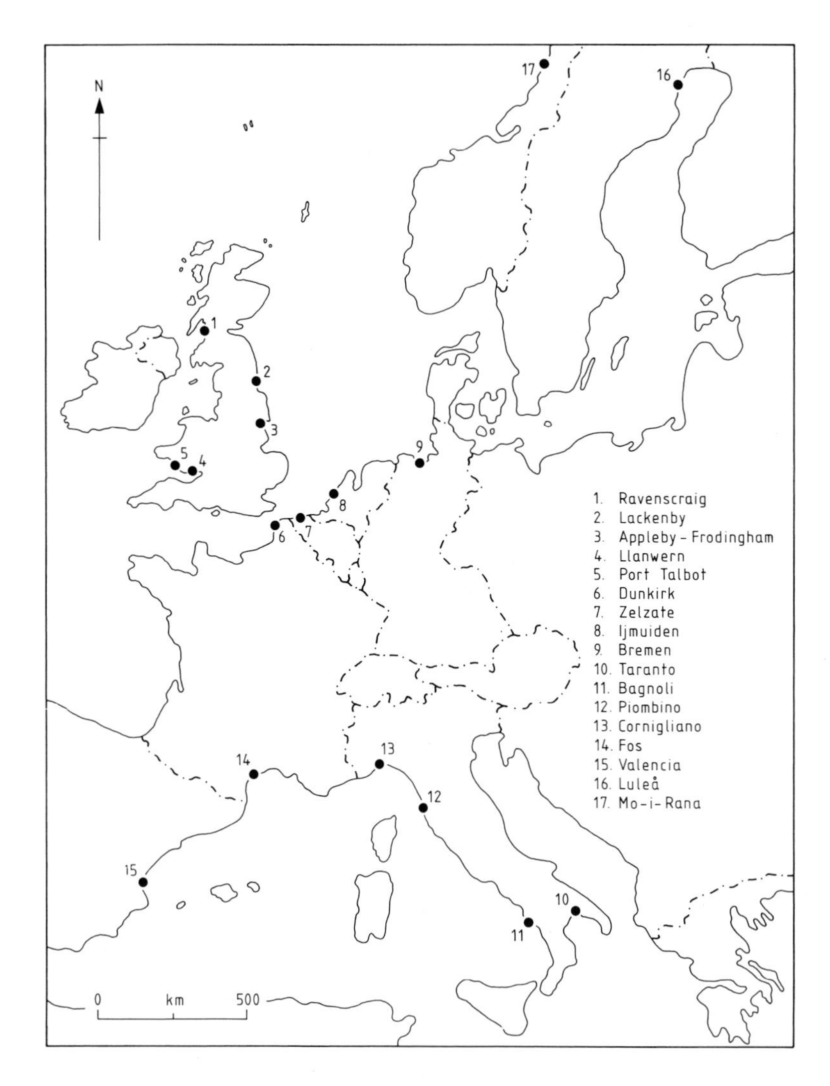

Fig. 6.4 Coastal steelworks in western Europe

locate as near as possible to these raw materials and consequently the central coal belt provided an unrivalled site for the industry. Iron smelting was pioneered in Britain, which became the principal centre, and it later spread to the continent, where the Ruhr provided ideal conditions, especially the large quantities of high-grade coking coal. Even when local ore supplies were exhausted, the Ruhr was well served by the Rhine waterway which permitted the importation of foreign ore. Apart from the Ruhr, the Nord–Pas de Calais region of northern France and the Sambre–Meuse trough in southern Belgium developed substantial iron and steel industries, despite inferior quality coal and difficult mining conditions.

Iron-ore reserves provided an alternative location for the iron and steel industry (Riley 1973). Apart from reserves in the Bergslagen district of Sweden, the main deposits of iron-ore were found in an area which stretched from Lorraine to Luxembourg and southern Belgium. However, these ores had a high phosphorous content and were not used until the introduction of the basic process in the 1880s. After this date, the industry flourished as it was easy to import coal from the Ruhr and Franco-Belgium coalfields.

Therefore, with the exception of the United

Kingdom, the iron and steel industry became heavily concentrated in a relatively small area, known as the heavy industrial triangle (Fig. 6.1), which was bounded by the Nord, the Ruhr, and Lorraine (Parker 1981). In 1952, the triangle accounted for 86 per cent of the steel and 90 per cent of the pig iron produced in the ECSC.

Location of iron and steel since the formation of the ECSC. The spatial pattern of steel making showed signs of change in the late 1950s and early 1960s and this was closely related to the declining importance of coalfield locations. The causes of this change are numerous and following Parker (1981) include:

1. Improved production techniques, whereby smaller quantities of coal were required in the smelting process. With the increasing use of large integrated plants, which made noticeable savings on heat, less coal was required and with the development of oxygen and electric converters, coal was not needed at all.
2. Improved transport facilities, which enabled coal and iron-ore to be moved more easily and cheaply than in the past.
3. The increasingly uncompetitive nature of European coal in world markets. Both Italy and the Netherlands lacked indigenous coal resources but were able to import quantities of American coal more cheaply than those from the Ruhr.
4. The inadequate nature of local iron-ore sites. Indigenous ore was usually poor in quality and quantity and it became cheaper to import high-grade, bulk-carried ores.
5. The ECSC itself, whereby the removal of quotas and tariffs meant that the iron and steel industry was serving a new and different market structure. Optimum conditions for plant location changed and the market replaced raw materials as a dominant locating factor.
6. Regional development policy, whereby the governments were encouraging the iron and steel industry to move to areas which experienced high rates of unemployment and were unattractive to new industry. The coastal steelworks at Taranto was the result of government policy, in this case to act as a growth point in southern Italy.

Collectively, these factors led to changes in the iron and steel industry, both outside and within the heavy industrial triangle. Allowing for the unmistakable movement to coastal locations, the triangle remains the centre of western Europe's, and more specifically the Community's, steel making. However, this area has experienced a relative and absolute decline and, although the output of steel rose from 61.4 million tonnes in 1965 to 70 million tonnes in the early 1970s, it had fallen back to just 50 million tonnes by 1982. Consequently, the triangle's share of total EC production had been reduced from 70 to 45 per cent in just 17 years. Reasons for this include the growth of steelmaking in Italy, the entry into the European Community of the United Kingdom, and the general drift to coastal locations. It may be surprising that the triangle retains its importance, but it still possesses certain advantages. The area has been modernized and new plants have been built, mainly as a result of government investment. Parker (1981) analysed the pattern of investment in the steel industry and between 1955 and 1965 the majority of investment in France and West Germany, for example, went to areas within the triangle; it was only after 1965 that more and more investment went to areas outside the triangle. Other factors aiding this area include good communications, large reserves of coal, a huge urban market, ready supplies of labour, and geographical inertia. Indeed, the triangle is ideally placed within the European Community and there is no reason why it should not continue to remain the heart of iron and steel making in western Europe.

Developments outside the triangle have been dominated by the movement to the coasts. There was some coastal steel production before the formation of the ECSC, at Ijmuiden in the Netherlands and Piombino in Italy, but output was small. Since 1952 however, coastal production has escalated, compared to that in the triangle (Fig. 6.5), and in 1972/3 the 14 large coastal plants in the European Community produced 41 million tonnes of steel. The movement to the coasts is one of the major trends in the British iron and steel industry (Warren 1976), and is well-marked in south Wales and the North-East. This reflects the increasing dependence on imported ores and the growing use of oxygen and electric processes for steel making. In 1975, 47.5 per cent of total steel output in the United Kingdom was produced by oxygen processes; by 1982, 73 per cent of the entire EC output was produced in this way.

Fig. 6.5 The growth in importance of coastal steelworks, 1952–82 (*source:* Minshull 1980, p. 72 and updated)

The non-coastal iron and steel centres outside the triangle have had varied success. Central England declined in importance with the decline in coal and it has only been maintained by the specialist markets of the Black Country and Sheffield, which now depend more on electric and oxygen processes. In northern Italy, the industry had to rely on scrap and hydro-electric power as it lacked coal and iron-ore. Unlike England, this area has maintained its importance, through heavy investment and the large markets of Turin, Brescia, and Bergamo. Specialization of production, as a response to mallocation, is thus common in western Europe and many inland producers tend to concentrate on special steels or finishing branches of the iron and steel industry. Sulzbach in Bavaria and Ebbw Vale in south Wales are good examples of such inland producers.

Therefore, as with the axial belt of industrial development, the iron and steel industry of western Europe has been dispersed from its original core area and the heavy industrial triangle is being extended in both north and south directions, towards the North Sea and the Italian coast. However, it must be stressed that in the past decade the steel industry has been characterized by a 'manifest crisis' (Evans 1980; Commission of the European Communities 1985). Steel production within the EC fell by 13 per cent between 1975 and 1983, before showing signs of recovery, and 316 000 jobs were lost. Over-investment in the industry caused capacity to reach 200 million tonnes in 1980 (output was 128). The

6.1 Houses for Dutch miners built with funds from the ECSC (ECC)

crisis is related to the world recession, but also to a decline in steel consumption in western Europe and competition from newly industrialized countries. Greater Community involvement was needed and came with the Davignon Plan in 1978, a package of subsidization and modernization. Over 4000 million ECUs were loaned to help soften the social and regional impact of the steel crisis, and since 1980 a mandatory restriction has been imposed on steel production and deliveries in the EC. This is enforced in the form of quarterly quotas and covers over 80 per cent of steel output. The system expires at the end of 1985, but negotiations to extend state aid until 1990 are well advanced. In the meantime, another 100 000 jobs will have to be lost, with retraining and the creation of new jobs in steel areas becoming more vital. This must be achieved, for as the Commission of the European Communities (1985, pp. 10–11) states 'the successful modernisation of this industry must be one of the Community's top priorities'.

2. Growth industries: chemicals and electrical engineering

The major twentieth century growth industries include motor vehicles, aircraft, electrical and electronic equipment, machine tools, and chemicals. Unlike traditional industries, most growth industries depend on electricity for their energy and this has released them, from the beginning, from the ties of the coalfields. Therefore, growth industries have a wide choice of possible locations, but as Parker (1981) notes they are not footloose because:

1. They use a variety of raw materials and semi-finished materials and so need to be well placed to obtain them without incurring excessive transport costs.
2. They must have good communications with their component and raw material suppliers and their market.
3. They must be ensured of an adequate labour supply.

The initial effect was for these industries to be attracted to the main centres of economic activity within western Europe and to the heavy industrial districts, for steel and other metals, and the large cities, for labour, component supplies, and communications, in particular. This pattern was subsequently modified by economic and technical developments and some decentralization was achieved through government regional policies.

The chemical industry was one of the major growth industries in western Europe before the economic recession and grew almost twice as fast as the combined rate for all industry. Between 1970 and 1976, chemical production witnessed an average annual growth rate of 5.6 per cent, which compared with an annual increase of 2.7 per cent for industrial production. This growth was not characteristic of all divisions of the chemical industry and, whilst petrochemicals, pharmaceuticals, plastics, and man-made fibres experienced boom conditions, fertilizers and carbo-chemical industries lagged behind. One important reason for the success of the chemical industry was the key role it played in the industrial economy. According to Hudson (1983), it was a central element in the growth of the capitalist economy, with the EC being the location for much of this growth. It continued to supply an increasing number of raw materials to such industries as textiles (bleaches, dyes, synthetic fibres), and footwear (synthetic resins, rubber, plastic, tanning materials). West Germany became and remains the most important country for chemical production, with a labour force, at its peak, of over 550 000 employees, followed by the United Kingdom (465 000), Italy (340 000), and France (325 000).

The capital-intensive chemical industry was influenced by a number of locating factors, including raw materials and energy sources, transport, water supplies, labour, markets, and geographical inertia. Minshull (1978) has shown that there are three main types of location for the chemical industry in western Europe:

1. near a raw material and energy source;
2. at the point of importation or trans-shipment of bulky raw materials;
3. near the market for the product.

Early developments in the chemical industry were based on raw materials like coal, salt, and potash. This led to a concentration of activity near the major coalfields and in the heavy industrial triangle in particular. The Ruhr, Saar, Sambre–Meuse, and Kempenland coalfields became centres of production, and Duisburg, Recklinghausen, and Marl were important for carbo-chemicals, whilst Liège and Charleroi specialized in heavy chemicals such as fertilizers. The Rhine river was an excellent highway for the importation of raw materials and the shipment of finished goods, and Leverkusen and Ludwigshaven, both on the Rhine, became important chemical centres.

Outliers of chemical production also occurred near deposits of potash and salt in such areas as Lorraine, Mulhouse, Merseyside, and Lower Saxony and the so-called chemical triangle, of upper Bavaria and the Inn valley, arose as a result of local hydro-electric power, limestone, and salt, and imported coal from the Ruhr. Hydro-electric power was also important for the production of nitrate fertilizers at Terni and Crotone in Italy, just as gas reserves led to chemical production in the Po valley and sulphur extraction at Lacq and St. Marcet in the Pyrenees.

Changing energy sources, notably from coal to oil, and new technology led to a rapid diffusion of petrochemicals in western Europe. Pre-war chemical sites retained their importance, especially if they received oil via pipelines, but oil refineries, and hence ports, became one of the main locating factors for petrochemicals. The growing influence of oil companies was possibly the most important development in the chemical industry after the Second World War (Chapman 1974), and by the late 1950s, Shell, BP, and Esso were in direct competition with the multinational ICI chemical giant. These companies had the necessary capital to develop petrochemical industries as well as controlling the raw material sources. Fig. 6.6 shows that the group chemical plants in western Europe belonging to BP, Shell, and Esso in the early 1970s had an essentially coastal location and were dominated by a concentrated pattern in the Rhine–Maas delta and along the Rhine axis. Shell's main strength lies in the Netherlands, the United Kingdom, and France, although it does manufacture petrochemicals, independently or in joint ventures with other companies, at Antwerp, Rho, and Tarragona. Esso is second only to Shell in its international involvement in petrochemicals; the company's major investments are in France (Marseille), West Germany (Hamburg and Cologne), Sweden (Stenungsund), and Denmark (Kalundberg), although Fawley, Salonika, Zaragoza, and Castello also have important Esso petrochemical plants. As its name suggests, BP is dominated by the Grangemouth, Baglan Bay, and Humberside complexes in the United Kingdom, but the company has joint ventures in France (Marseille), and West Germany (Vohburg, Dinslaken, and Cologne).

Petrochemicals are also important in inland centres along rivers and again West Germany leads the way in western Europe. Centres along the Rhine, which provides an excellent means of transport and plentiful supplies of water, include Duisburg, Dusseldorf, Cologne, Leverkusen, Ludwigshaven, and Mannheim. Leverkusen is the site of the Bayer chemical works, which employs over 20 000 people and is supplied by the Dinslaken refinery of BP. Cologne is important for cosmetics, pharmaceuticals, and artificial fibres, and Dusseldorf has the largest detergent works in West Germany. But probably the most important development is at Ludwigshaven, where the huge BASF works, employing 45 000 people, is sited. BASF has over 280 associated companies in West Germany (Mellow 1978), and is the fourth largest manufacturer of chemicals in western Europe (Table 6.5). Ludwigshaven is an ideal centre because raw materials, energy, and transport are readily available: limestone and rock salts are found locally; coal is imported from the Saar; crude oil comes from the Mannheim refinery; natural gas comes from the Rhine valley and via pipeline from the Netherlands; refinery products come via pipeline from Karlsruhe; and BASF has its own electricity generators. The result is a wide range of chemical products, from heavy chemicals and pharmaceuticals to plastics and organic acids.

Petrochemical plants in West Germany do not possess their own large oil refineries and, in order to expand production, it was necessary to establish links with suppliers. Therefore, inter-plant linkages became very important and in the Rhine–Ruhr, for example, 21 plants were linked by over 650 km of pipelines, transporting more than 20 products (Burtenshaw 1974). Once the links had been established, new sites near to refineries were found; this was exemplified by the joint BASF/Shell plant at Wesseling, which produces one million tonnes of ethylene and is fed by the Godorf refinery.

High value chemical products, such as cosmetics and pharmaceuticals, are found scattered

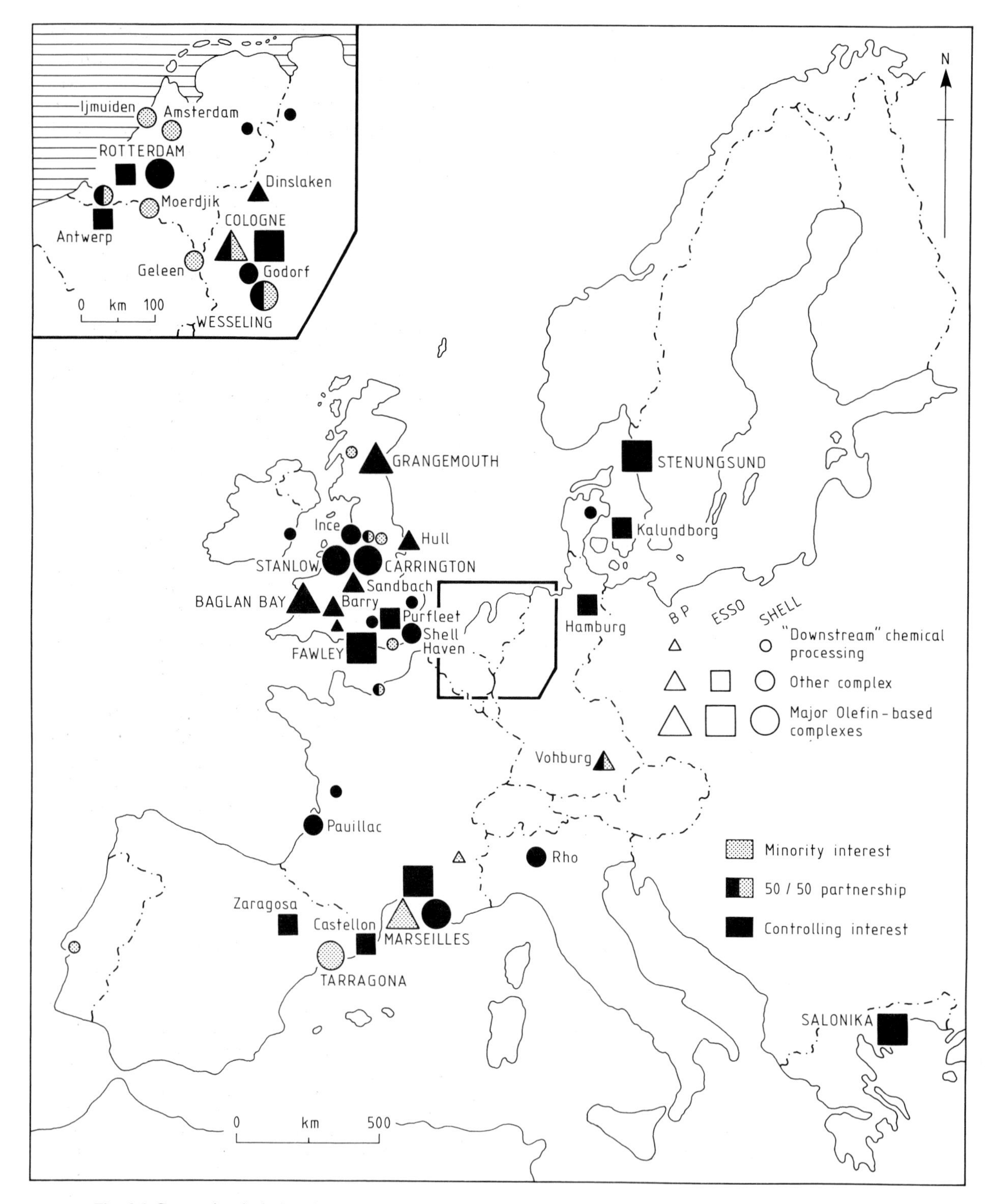

Fig. 6.6 Group chemical plant locations in western Europe of BP, Esso, and Shell (*source:* Chapman 1974, p. 129)

throughout the larger cities of western Europe and, for financial reasons, in the development areas. Cities like Paris, London, Brussels, Antwerp, Milan, and Frankfurt are important centres, and Frankfurt, for example, is the home of the Hoechst ammonia plant and is important for pharmaceuticals and artificial fibres.

Improved linkages in the west European chemical industry led to an increase in the scale of production and during the 1960s and 1970s the industry became organized into very large units owned by a few giant companies (Table 6.5). For example, the West German chemical industry is dominated by Hoechst, BASF, and Bayer; the British by ICI and Unilever; the Dutch by Unilever NV; the Swiss by Ciba-Geigy; and the Italians by the ENI group and Montedison. Indeed, Bayer, BASF, Hoechst, and ICI are among the world's five largest chemical concerns. The industry also became dominated by American investment (Young and Hood 1976), and Minshull (1978) estimated that up to one-quarter of total US investment in western Europe in the mid-1970s was in the chemical industry of the Nine alone. Similarly, Hudson (1983) had shown that US investment in the chemical and allied industries in western Europe had increased from $74 million in 1950 to $1073 million in 1964. Undoubtedly, the chemical industry benefited from the large internal market produced by the European Community but, in order to compete with American investment, more Community mergers became necessary.

However, Burtenshaw (1974) produced an early warning of the twin dangers of increasing the scale of production and over-production. He felt that increases in the scale of production during the early 1970s had only brought marginal decreases in costs and indeed the golden age of the chemical industry, and especially petrochemicals, was brought to an abrupt end by the oil crisis. As with the steel industry, over-capacity, increased competition, and declining profits became characteristic, although the spatial concentration of new fixed-capital investment was continued and enhanced (Hudson 1983). BP lost nearly $1000 million on chemicals between 1980 and 1982 alone and both BASF and DSM (Netherlands) cut their polyethylene capacity by 20 per cent (Clout *et al.* 1985). A process of disinvestment and reorganization had begun and the international division of labour intensified as high value-added products (e.g. pharmaceuticals) became increasingly concentrated in western Europe whilst the mass production of basic chemicals was transferred, by the multinationals, to Third World sites (Hudson 1983). Therefore, a similar cycle of events, which Hall (1981) likened to a Kondratieff cycle, has characterized both the chemical and steel industries, although it has been of shorter duration in the former.

Western Europe has also been characterized by rapid post-war growth in electronics and electrical engineering which, with an average annual growth rate of 10 per cent per year, is one of the fastest growing sectors of the economy. An in-

Table 6.5 Leading chemical and electrical manufacturers in western Europe, 1983

	Chemicals				*Electrical*		
	Company	Country	Turnover 1983 £m		Company	Country	Turnover 1983 £m
1.	ENI Group	Italy	16 093	1.	Siemens AG	West Germany	10 337
2.	Hoechst AG	West Germany	9017	2.	Philips Lamps Holding	Netherlands	9883
3.	Bayer AG	West Germany	8978	3.	Générale d'Eléctricité (Cie)	France	4502
4.	BASF AG	West Germany	8373	4.	General Electric Co	United Kingdom	4190
5.	Unilever NV	Netherlands	7554	5.	Thomson-Brandt	France	4039
6.	ICI	United Kingdom	7358	6.	Bosch (Robert) GmbH	West Germany	3560
7.	Unilever plc	United Kingdom	5447	7.	AEG Telefunken AG	West Germany	3417
8.	Ciba-Geigy AG	Switzerland	4295	8.	Brown Boverie & Cie AG	Switzerland	3017
9.	DSM NV	Netherlands	4178	9.	Electrolux AB	Sweden	2718
10.	Montedison SpA	Italy	3921	10.	Asea AB	Sweden	2213

Source: Times 1000 (1983–4).

crease in per capita incomes and levels of living gave rise to a large demand first, for clothing and housing items and secondly, for sophisticated consumer durables such as refrigerators, televisions, videos, and motor cars. Thus, the private consumer began to dominate the market and, whilst the pre-war branches of electrical engineering expanded, new sectors like electronics emerged.

Consumer durables are 'a footloose congeries of assembly industries' (Mellor and Smith 1979, p. 116) that are often labour-intensive and closely interlinked. Labour represented an important element of cost and became one of the prime locating factors. As supplies of labour diminished in the traditional pre-war centres of electrical engineering, new locations had to be sought. This gave the industry a dispersed distribution pattern, with branches being established initially in the larger cities and subsequently in the smaller market towns where available labour, often female, could be found. Large reserves of labour in southern Italy, for example, meant that Italy became one of the major producers of domestic electrical goods in the 1950s. However, the more advanced west European countries, and especially West Germany, quickly developed electronic and consumer durable industries, in order to satisfy the ever-increasing demand for such goods. Familiar core–periphery contrasts appeared during the 1960s, only to be influenced by rising labour costs in the 1970s. Producers were forced to consider the possibilities of using low-cost labour, which was to be found in the 'peripheral' countries of western Europe and in Iberia in particular. Areas with cheap and plentiful supplies of labour were often designated development areas, which had the added advantage of financial incentives to attract the prospective employer.

In common with other major growth industries, electronics and electrical engineering require a vast expenditure on research. This led to the inevitable invasion of American firms into western Europe and the rise of multinational electrical companies. American companies dominate the field and command about 80 per cent of world sales in electronics, computers, and telecommunications. One such company is IBM which, according to Hamilton (1976), controls over two-thirds of the EC market in electronic computers, with the sole exception of the United Kingdom where domestic firms retain at least 50 per cent of all sales. Indeed, the United Kingdom is the only country in western Europe to possess an individual computer industry outside US control; this is International Computers Limited (ICL) which has about 3 per cent of world sales.

If one excludes American firms, it is not surprising to find that the electronics and electrical engineering industry is also dominated by a few large European groups (Table 6.5). The leading manufacturers, in terms of turnover, are to be found once again in West Germany, with three out of the first seven—Siemens (which heads the list), AEG Telefunken, and Bosch. The Netherlands, France, Switzerland, the United Kingdom, and Sweden are represented in the top ten companies in western Europe. Siemens is a well-studied company (Krumme 1970; Burtenshaw 1974), and is a microcosm of the trends that have characterized the west European electrical engineering industry. The headquarters of Siemens was transferred from Berlin to Munich after the war, the latter offering large supplies of qualified labour and an attractive city with good amenities and research institutions. In the 1950s, plants were established in larger cities like Karlsruhe and Regensburg, for their labour supply, and Siemens introduced various schemes to attract labour, such as flexi-hours, free transport, and cheap accommodation in hostels. During the 1960s, smaller-scale plants were set up in the market towns of Bayern. When local supplies of labour were exhausted in the late 1960s and early 1970s, Siemens hired immigrant labour with Munich being well situated to attract immigrants from Italy, Yugoslavia, and Turkey.

To meet American competition, international mergers within western Europe became necessary. There has been investment between west European companies, and the Swiss company of Brown Boverie, for example, manufactures domestic equipment in Stuttgart and motors and machinery in Mannheim, just as Philips has large plants in Bremen, Aachen, and Hamburg. Siemens has joined forces with the Dutch and French to produce a European computer industry (Burtenshaw 1974), but it is in this sector, and in telecommunications and automation, that western Europe lags considerably behind America.

6.4 Economic potential in western Europe

It has been demonstrated in the preceding sections that industry and people are concentrated in a few core areas within western Europe. This is somewhat surprising, as industry is less tied to either raw materials or markets and has become increasingly footloose. However, the location of industry and poeple is tending towards an 'ever-increasing concentration in a limited number of areas' (Clark *et al.* 1969, p. 197). The core–periphery concept is again relevant and a study of economic potential in western Europe emphasizes this point. Such a study was initially conducted by Clark *et al.* (1969), but their work has been updated and improved by Keeble *et al.* (1982a and b). As core–periphery contrasts represent one of the underlying themes of this book, the results of the latter study will be analysed in a certain amount of detail.

The concept of potential has been used by Stewart (1947), in the form of population potential, and by Harris (1954), in the form of market potential. Clark *et al.* (1969, p. 198) stated that the 'economic potential of any point is defined by summing the regional incomes around it, each regional income having been first divided by the distance costs of reaching it'. Regional incomes and distance costs were thus considered to be the most important variables determining the location of economic activity. However, Keeble *et al.* (1982a) used improved data for the 1965–77 period and based their calculations on regional GDP and the shortest road-distance costs (including shipping and tariff-barrier costs where necessary). For the calculation of distance, a node (most important city economically), was chosen to represent each region in the EC. Ten neighbouring countries, with trading links with the Community (Norway, Sweden, East Germany, Czechoslovakia, Austria, Switzerland, Yugoslavia, Spain, Greece, and Portugal), were included and each was represented by one mass value based on the largest city. Potential values were calculated for different time periods and situations and, for mapping purposes, each region's potential value was expressed as a percentage of the highest value in the EC in that year. The final isopleth maps (Fig. 6.7) were interpolated at 10 per cent intervals up to the maximum, which in each case was recorded at Rheinhessen–Pfalz.

Therefore, the final index measures a region's relative potential for interaction with economic activity, and regions with higher potential values have access to more economic activity within a given distance than those with lower values (they have a comparative advantage for economic development).

1. *1965 to 1973* (Fig. 6.7a and b). During this period there was complete freedom of movement of goods, labour, and capital among the original 'Six', with the remaining countries cut off by a tariff barrier. Between these two dates there was a clear trend towards increasing regional disparities in relative accessibility, as well as a general decline in potential values. This was most marked for the regions of the United Kingdom, although the relative decline in Ireland and Denmark was also considerable. Within the European Community itself, the decline was most noticeable in the peripheral regions of southern Italy and south and west France. Therefore, this period witnessed a relative concentration of economic growth in the more central regions of the original EEC.

2. *1973 enlargement of the EC* (Fig. 6.7c). The removal of the tariff barrier between the Six and three new members—Ireland, Denmark, and the United Kingdom—saw a general rise in economic potential values, especially in the new 'three'. It was the peripheral areas of Ireland, Denmark, and the United Kingdom which benefited most, with an increase in potential values of between 40 and 76 per cent. Similarly, the periphery of the 'Six' gained more than the central regions. However, this reflected the mathematics of percentage calculations on small figures and in reality the centre gained much more than the periphery, as demonstrated by the increase of just 93 million EUA per km in Calabria compared to 365/km in Rheinhessen–Pfalz. Keeble *et al.* (1982a) related this to the greater proximity of the central regions to the three new members.

3. *1973 to 1977* (Fig. 6.7d). The accessibility surface for 1977 again emphasized the widening of core–periphery disparities, although at a slower rate than during the late 1960s and early 1970s. The gain in Rheinhessen–Pfalz was nearly nine times that of Calabria, emphasizing the faster growth of economic activity in the more accessible regions.

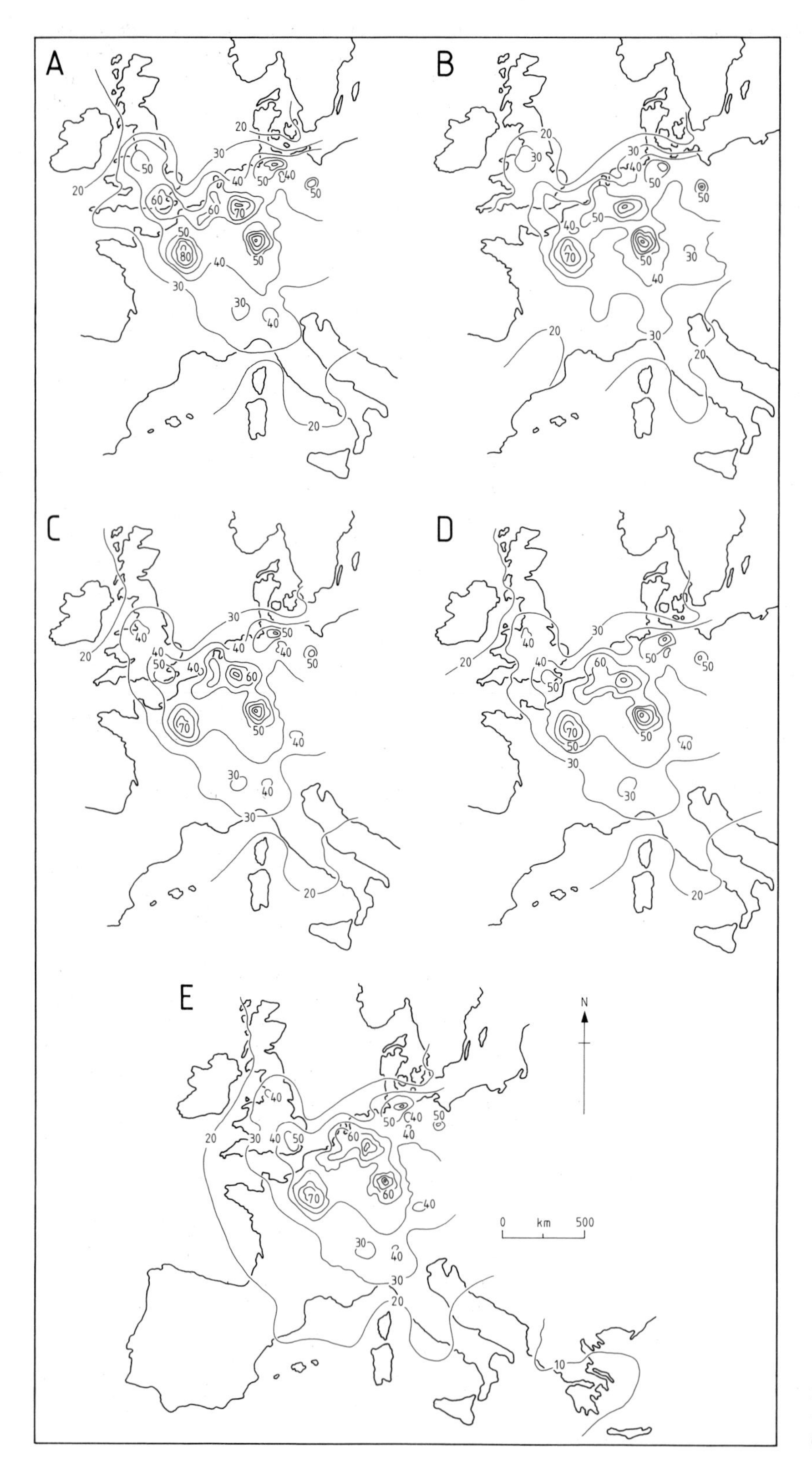

Fig. 6.7 Economic potential in western Europe (*source:* Keeble *et al.* 1982a)

4. *The '12' in the 1980s* (Fig. 6.7e). Using 1977 gross domestic product data for the Nine and regional estimates for Greece, Portugal, and Spain, a further widening of regional accessibility disparities occurred. The lowest potential value, at just 7 per cent of the Euro-12 maximum, was for the Aegean region of Greece; indeed six other Greek regions recorded values of less than 10 per cent, with only Athens rising to 13.4 per cent. In Spain, potential values rose to 22 per cent of the maximum, at Catchuna, and this was higher than 16 other regions in the Nine. Values for Portugal varied between 10.7 and 12.7 per cent of the maximum, with the former recorded for the Sud interior region and the latter for Porto. Regions already in the Nine were little affected by the further removal of tariff barriers, except for the French peripheral regions bordering Spain, and Keeble *et al.* (1982a) attributes this to the preferential treatment accorded to manufacturing goods entering the EC from prospective member countries after 1973.

Therefore, the quantitative index produced by Keeble *et al.* (1982a) shows how regional comparative advantage for economic growth has changed over time. However, it is clear that a triangular plateau of high accessibility exists in the north-east, with outlying peaks around Paris, London, and West Berlin. Similarly, extensive areas of low potential characterize the southern, western, and northern peripheries. This overall pattern reflects the 'historic processes of industrialisation, urbanisation and investment in fixed capital' (p. 430).

References

Bale, J. R. (1973). 'Industrialists' attitudes towards location on industrial estates: the case of south Wales'. *Tijdschrift voor Economische en Sociale Geografie*. **64**, 320–5.

Bale, J. R. (1977). 'Industrial estate development and location in post-war Britain'. *Geography*, **62**, 87–93.

Barr, J. (1970). 'Planning for the Ruhr'. *Geographical Magazine*, **40**, 280–9.

Breathnach, P. (1985). 'The impact of rural industrialization in the west of Ireland'. In Healey, M. J. and Ilbery, B. W. (Eds), *Industrialization of the countryside*. GeoBooks, Norwich.

Burtenshaw, D. (1974). *Economic geography of West Germany*. Macmillan, London.

Chapman, K. (1974). 'Corporate systems in the United Kingdom petrochemical industry'. *Annals* [illegible] *Association of American Geographers*, **64**, 126–[illegible]

Clark, C.; Wilson, F. and Bradley, J. (1969). 'Industri[illegible] location and economic potential in western Europe'. *Regional Studies*, **3**, 197–212.

Clout, H. D.; Blacksell, M.; King, R. and Pinder, D. (1985). *Western Europe: geographical perspectives*. Longman, London.

Commission of the European Communities (1985). *The European steel policy*. Brussels.

Davin, L. E. (1969). 'Structural crisis of a regional economy—a case study: the Walloon area'. In Robinson, E. A. G. (Ed.), *Backward areas in advanced countries*. St. Martin, London.

Elkins, T. H. (1965). 'Liège and the problems of southern Belgium'. *Geography*, **41**, 83–98.

European Studies (1971). 'Economic problems of Belgium's Wallonia'. *European Studies*, **11**, 1–4.

Evans, I. M. (1980). 'Aspects of the steel crisis in Europe, with particular reference to Belgium and Luxembourg'. *Geographical Journal*, **146**, 396–407.

Fleming, D. K. (1967). 'Coastal steelworks in the Common Market countries'. *Geographical Review*, **43**, 48–72.

Fothergill, S. and Gudgin, G. (1982). *Unequal growth: urban and regional employment change in the UK*. Heinemann, London.

Friedmann, J. and Alonso, W. (1975). *Regional policy: readings in theory and application*. MIT Press, Cambridge, Mass.

Gay, F. (1981). 'Benelux'. In Clout, H. D. (Ed.), *Regional development in western Europe*. Wiley, London.

Hall, P. (1966). *The world cities*. Weidenfeld and Nicolson, London. Also revised 2nd edn. (1977).

Hall, P. (1981). 'The geography of the fifth Kondratieff Cycle'. *New Society*, 535–7.

Hamilton, F. E. I. (1976). 'Multinational enterprise and the European Economic Community'. *Tijdschrift voor Economische en Sociale Geografie*, **67**, 258–77.

Harris, C. D. (1954). 'The market as a factor in the location of industry in the United States'. *Annals of the Association of American Geographers*, **44**, 315–48.

Hellen, J. A. (1974). *North Rhine–Westphalia*. Oxford University Press, Oxford.

Hudson, R. (1983). 'Capital accumulation and chemicals production in western Europe in the post-war period'. *Environmental and Planning A*, **15**, 105–22.

Hull, A. and Kenny, S. (1983). 'Ruhr economy in decline'. *Geographical Magazine*, **55**, 516–21.

Jordan, T. G. (1973). *The European culture area*. Harper and Row, London.

Keeble, D.; Owens, P. L. and Thompson, C. (1982a). 'Regional accessibility and economic potential in the European Community'. *Regional Studies*, **16**, 419–31.

…. L. and Thompson, C. (1982b). …l and the Channel Tunnel'. *Area*,

… P. L. and Thompson, C. (1983). … manufacturing shift in the Euro… . *Urban Studies*, **20**, 405–18.

… *The geography of western Europe: a …vey*. Croom Helm, London.

Krumme, G. (…). 'The interregional corporation and the region'. *Tijdschrift voor Economische en Sociale Geografie*, **61**, 318–33.

Mellor, R. E. H. (1978). *The two Germanies: a modern geography*. Harper and Row, London.

Mellor, R. E. H. and Smith, E. A. (1979). *Europe: a geographical survey of the continent*. Macmillan, London.

Minshull, G. N. (1978). *The new Europe: an economic geography of the EEC*. Hodder and Stoughton, London. Also revised 2nd edn (1980).

O'Farrell, P. N. (1978). 'An analysis of new industry location: the Irish case'. *Progress in Planning*, **9**, 129–229.

Parker, G. (1981). *The logic of unity*. Longmans, London.

Riley, R. C. (1973). *Industrial geography*. Chatto and Windus, London.

Riley, R. C. and Ashworth, G. J. (1975). *Benelux: an economic geography of Belgium, the Netherlands and Luxembourg*. Chatto and Windus, London.

Sinclair, D. J. (1969). 'Steel in Europe'. *Geographical Magazine*, **41**, 610–11.

Spelt, J. (1969). 'The Ruhr and its coal industry in the middle sixties'. *Canadian Geographer*, **13**, 3–9.

Stewart, J. Q. (1947). 'Empirical mathematical rules concerning the distribution and equilibrium of population. *Geographical Review*, **37**, 461–85.

Thomas, S. and Tuppen, J. (1977). 'Readjustment in the Ruhr: the case of Bochum'. *Geography*, **62**, 168–75.

Warren, K. (1967). 'The changing steel industry of the European Common Market'. *Economic Geography*, **43**, 314–32.

Warren, K. (1975). *World steel: an economic geography*. David and Charles, Newton Abbot.

Warren, K. (1976). 'British steel: the problems of rebuilding an old industrial structure'. *Geography*, **61**, 1–8.

Weber, A. (1909). *Theory of the location of industries*. University of Chicago Press: Chicago.

Young, S. and Hood, N. (1976). 'The geographical expansion of U.S. firms in western Europe: some survey evidence'. *Journal of Common Market Studies*, **14**, 223–9.

7

Trade

7.1 Trading organizations

One of the major objectives, if not the most important, of developing an integrated transportation network is to encourage the growth of trade between the countries of western Europe. Post-war trade patterns in Europe have been fundamentally changed by the formation of political and trading organizations; indeed, the political evolution of western Europe in recent years has been spectacular (Clout *et al.* 1985). The European Community (EC), and European Free Trade Association (EFTA), dominate intra-European trade and only two west European countries are not full members of one of these organizations: Finland, which became an associate member of EFTA in 1961: and Spain, which is joining the EC in 1986. Denmark and the United Kingdom left EFTA in 1972, to join the EC, and Portugal will do the same in 1986 (Fig. 7.1).

After World War II, the countries of western Europe needed to recover their economies and encourage trade. Revival was largely based on the Organization for European Economic Co-operation (OEEC), which was founded in 1948, at the insistence of the United States, to distribute monies from the Marshall Plan among individual states and guide post-war reconstruction of the west European economy. The organization had 16 original members and was joined in 1949 by West Germany and in 1959 by Spain, thus making every country of western Europe a member. The United States and Canada became associate members in 1950. The main aim of the OEEC was 'the achievement of a sound European economy through the economic co-operation of its members' (OEEC, 1955, p. 237). European prosperity in the 1950s and 1960s was due in part to the creation of a liberalized trading system between members of the OEEC, and the average increase in trade between 1950 and 1960 was 47 per cent, with figures of 182 per cent for West Germany and 104 per cent for Austria (Clout *et al.* 1985).

Certain members of the OEEC wished to go further in the removal of economic barriers and in 1952 the ECSC, a customs union of coal and ferrous products, was formed by Belgium, France, Italy, West Germany, Luxembourg, and the Netherlands. However, by the late 1950s it was not very effective in dealing with the declining market for coal, iron, and steel, especially in traditional heavy industrial regions. A direct outcome of this was the signing of two treaties in 1957 which led to the formation of the European Economic Community (EEC) and the European Atomic Energy Community (Euratom). Other countries in the OEEC became concerned about these developments and in 1960 seven members formed EFTA: Denmark, Norway, Sweden, Portugal, Switzerland, Austria, and the United Kingdom. Finland became an associate member in 1961 and Iceland was admitted to EFTA in 1968.

In 1961, the OEEC was replaced by the Organization for Economic Co-operation and Development (OECD), and four non-European countries also became members: Australia, Canada, Japan, and the United States. The OECD is pledged to promote the general expansion of world trade, by encouraging the economic development of its members and by taking active steps to raise standards in the developing world (Blacksell, 1977). EFTA and the OECD do not have political integration as one of their main objectives, which contrasts with the European Community (EC), formed in 1967 by the amalgamation of the EEC,

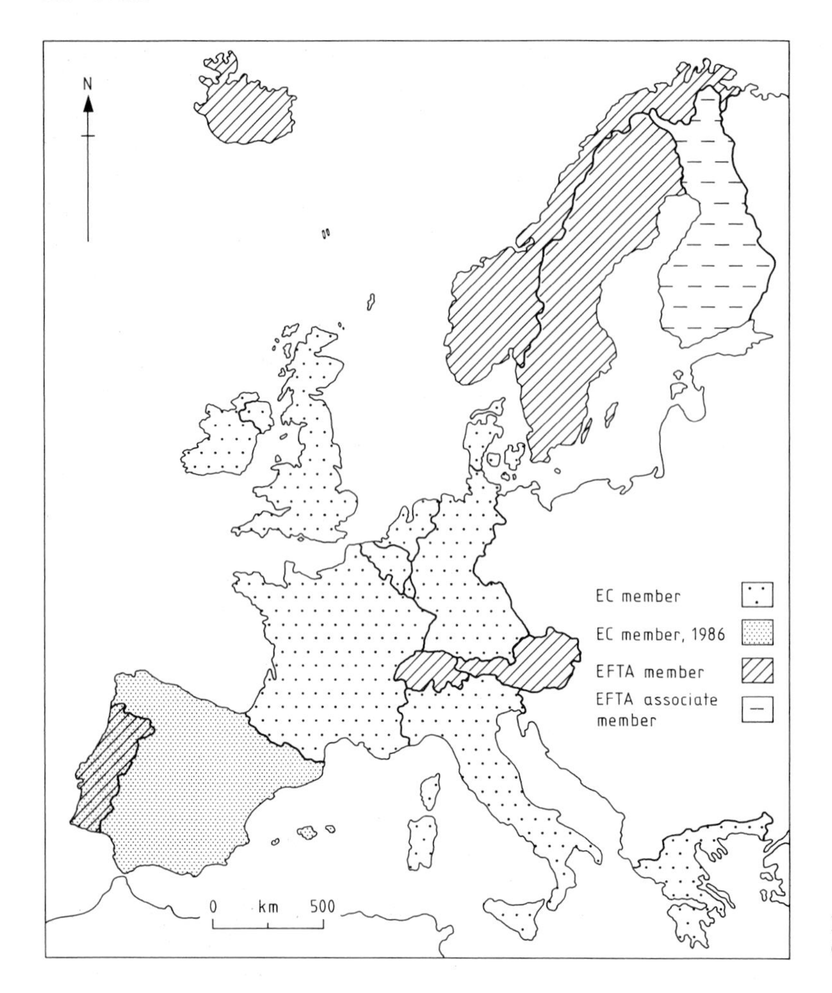

Fig. 7.1 Membership of the European Community and EFTA

ECSC and Euratom, and whose ultimate aim is supranational government.

Therefore, the pattern of trading in western Europe is dominated by EFTA and the EC, with seven and ten members respectively. The seven EFTA members are associate members of the EC and free trade in industrial goods between the 17 countries became possible in 1977. A free trade area, covering all the members of the former OEEC, will thus become a reality when Spain joins the EC in 1986.

7.2 Trade in the European Free Trade Association

EFTA was created in 1960, under the terms of the Stockholm Convention, and had four main objectives:

1. To create free trade in industrial goods between its members, with certain agreements for agriculture and fish. The aim was to ensure that trade between members took place in conditions of fair competition.

2. To assist in the creation of a vast single market which included all the countries of western Europe. This aim is being achieved in successive stages: in January 1972, the United Kingdom and Denmark signed treaties for membership of the EC; in July 1972, five EFTA members (Austria, Switzerland, Sweden, Iceland, and Portugal), signed free trade agreements with the EC; and in 1973, similar agreements with Norway and Finland were made. By 1977, free trade in industrial goods between 16 west European countries, representing a single market of over 300 million people, was a reality. This has since been

extended, with Greece's entry into the EC in 1981, and will be complete with the accession of Spain in 1986.

3. To contribute to the expansion of world trade. Between 1960 and 1975 alone, EFTA's sales to the rest of the world increased by over 400 per cent and total imports increased by approximately the same amount.

4. To promote economic expansion and an increase in living standards, and to minimize spatial disparities between members.

Integration in EFTA is thus through a free trade area and not a customs union. This exempts members from all restrictions on trade in industrial goods, such as import duties and quota restrictions, but allows them to retain their own external tariff towards non-member countries. Complete abolition of all restrictions on trade was to be achieved by 1970, except in the relatively underdeveloped countries of Portugal and Iceland where certain restrictions could be retained until 1980. In reality, restrictions were removed by 1967. The major problem facing EFTA was how to deal with imports from outside the trading area. A common external tariff was needed, to counteract the possibility of unfair trading, and it was necessary to restrict free trade to products manufactured within EFTA. To be eligible for free trade status within EFTA, an exporter had to show either that more than 50 per cent of the value of his product had been added to within EFTA, or that it had undergone a specified number and variety of manufacturing processes so that the original raw materials became a completely new product (Blacksell, 1977, p. 114).

The geographical rationale of EFTA is not too clear. With no supranational ambitions, it attracted a wide spectrum of countries which were widely dispersed in a peripheral ring round the EEC heartland of western Europe. In terms of position, there was a basic unity between members and they stood to gain from any relaxation in the terms of trade, particularly in face of the EEC. However, marked spatial disparities in economic development still exist between the members of the 'outer seven', especially in terms of the standard of living (Table 7.1). Gross domestic product per person in 1982 varied from $15 076 (Switzerland) to just $2114 (Portugal), a sevenfold difference, and in Switzerland it was two-thirds higher than that of its neighbour and fellow member, Austria. In terms of the level of living, wide disparities exist in facilities such as televisions, telephones, and cars and in terms of employment in the main sectors of economic activity, differences between countries are marked. Portugal remains strongly orientated towards primary activities such as agriculture and fishing, and this contrasts with the low level of agricultural employment in Sweden, Norway, and Switzerland and the strong service sector development in the Scandinavian countries. Trade statistics for 1983 also reveal basic differences between members. All countries continue to increase the value of their exports, but the percentage share of total exports going to other EFTA members varies from just 8.5 and 10.7 per cent in the

Table 7.1 *Some key statistics for the EFTA countries*

	Austria	Finland	Iceland	Norway	Portugal	Sweden	Switzerland
Population (millions), 1983	7.5	4.9	0.2	4.1	10.0	8.3	6.5
Population density/sq.km., 1981	90	14	2	13	106	20	156
GDP/person ($US), 1982	9066	10 140	13 014	13 902	2114	11 846	15 076
Telephones/1000 inhabitants, 1980	401	496	477	452	138	796	727
Televisions/1000 inhabitants, 1980	296	322	270	292	140	381	314
Private cars/1000 inhabitants, 1981	308	266	310	312	128	348	370
% labour in primary activities, 1982	10.0	11.2	21.0	8.0	26.8	5.6	7.1
% labour in manufacturing, 1982	40.0	33.4	34.8	29.4	36.5	30.3	38.4
% labour in services, 1982	50	55.4	44.2	62.6	36.7	64.1	54.5
% exports to EFTA, 1983	10.7	18.3	14.3	13.9	10.8	19.4	8.5
% exports to EC, 1983	53.7	36.1	34.9	69.6	59.0	48.7	49.2
% imports from EFTA, 1983	7.8	17.4	23.2	25.6	5.9	13.8	6.6
% imports from EC, 1983	62.7	33.5	44.9	45.3	39.5	41.9	65.6

Sources: United Nations Statistical Yearbook, and *OECD Statistics of Foreign Trade.*

relatively isolated countries of Switzerland and Austria, to 18.3 and 19.4 per cent in Finland and Sweden. Exports to the European Community also vary, from 34.9 per cent of total exports from Iceland to 59 per cent from Portugal. The non-Scandinavian members also import the smallest percentage share of total imports from EFTA, whilst importing the largest share from the Ten (with the exception of Portugal).

The main question to be asked is whether such diversity between members is a weakness or a strength. EFTA has always been aware of the imbalance which exists, as demonstrated by certain concessions for Portugal and Iceland, and there can be no doubting the massive increase in foreign trade since 1960. Excluding Denmark and the United Kingdom, exports from EFTA increased from $7773 million in 1962 (first full year after the completion of the OECD), to $86 910 million in 1983, an increase of over 1000 per cent. Similarly, imports increased by 890 per cent over the same time period. EFTA accounts for approximately seven per cent of world trade and, as they themselves state, 'international trade is more important for EFTA than for any other trading area in the world' (1975, p. 5).

The focus of EFTA trade has changed over time (Table 7.3). Intra-EFTA trade increased between 1962 and 1977, and the percentage share of EFTA imports from member countries increased from 11.4 to 14.3 per cent, just as the share of exports increased from 13.8 to 17.4 per cent. Indeed, intra-EFTA trade increased by an average of 15 per cent per year between 1961 and 1971, and by 34 per cent between 1972 and 1975. However, between 1977 and 1983 imports into EFTA from members began to decline, whilst those from the USA and 'others' increased (Table 7.3). Similarly, the percentage share of exports to EFTA began to fall back towards the 1962 level. Trade with the European Community has declined in relative terms: in 1962, 59.5 per cent of EFTA imports came from the EC and by 1983 this figure had declined to 50.6 per cent; in terms of exports, the share first decreased, from 51.9 to 46.7 per cent between 1962 and 1977, before rising again to 52 per cent by 1983.

All members of EFTA depend on foreign trade

Table 7.2 *Some key statistics for the EC countries*

	West Germany	France	Italy	Netherlands	Belgium	Luxembourg	United Kingdom	Ireland	Denmark	Greece
Population (millions), 1983	61.4	54.7	56.8	14.4	9.9	0.4	56.3	3.5	5.1	9.9
Population density/sq.km., 1981	248	100	190	346	323	141	231	49	119	74
GDP/person ($US), 1982	10 735	10 470	6074	9617	9709	13 698	7887	4857	11 898	3711
Telephones/1000 inhabitants, 1980	464	459	337	509	369	547	477	187	641	289
Televisions/1000 inhabitants, 1980	337	297	234	296	298	247	394	187	362	158
Private cars/1000 inhabitants, 1981	385	349	325	323	325	365	283	225	267	94
% labour in primary activities, 1982	5.5	8.4	12.4	5.0	2.9	5.1	2.7	17.3	8.5	30.7
% labour in manufacturing, 1982	42.7	34.6	37.0	28.7	32.4	37.5	34.7	31.1	26.3	29.0
% labour in services, 1982	51.8	57.0	50.6	66.3	64.7	57.4	62.6	51.6	65.2	40.3
% exports to EFTA, 1983	15.6	8.0	9.1	6.1		6.6	9.7	4.6	22.5	2.7
% exports to EC, 1983	48.1	49.2	46.2	72.2		70.0	43.8	69.1	47.9	52.4
% imports from EFTA, 1983	13.1	6.5	7.7	7.1		6.6	13.6	4.4	24.8	4.9
% imports from EC, 1983	49.3	49.8	42.8	53.4		67.4	45.6	67.3	48.1	48.0

Sources: United Nations Statistical Yearbook, and *OECD Statistics of Foreign Trade*.

Table 7.3 *The pattern of trade in EFTA and the EC, 1962–83*

Trading Group		Imports (%) from:				Exports (%) to:			
		EC	EFTA	USA	Others	EC	EFTA	USA	Others
EFTA	1962	59.5	11.4	8.3	20.8	51.9	13.8	7.4	26.9
	1977	54.1	14.3	6.1	25.5	46.7	17.4	5.1	30.8
	1983	50.6	12.3	7.2	29.9	52.0	13.9	6.6	27.5
EC	1962	40.0	9.7	10.3	40.0	43.1	14.5	7.6	34.8
	1977	49.2	8.5	7.6	34.7	50.7	11.5	5.4	32.4
	1983	50.2	10.1	8.0	31.7	52.4	10.6	7.8	29.2

Source: Derived from the *United Nations Statistical Yearbook*, and *OECD Statistics of Foreign Trade*.

and the changing patterns of imports and exports between 1962 and 1983 are summarized in Figs. 7.2 and 7.3. Figure 7.2 indicates that the percentage share of imports into the individual countries of EFTA, from EFTA itself, increased in five out of seven cases; the two exceptions being Portugal and Finland. The largest increase occurred in Iceland (from 11.0 per cent in 1962 to 23.2 per cent in 1983), and Norway (21.0 to 25.6 per cent). On the other hand, the percentage share of imports into each of the seven EFTA members from the European Community has declined in all but two cases—Austria, where the situation remained unchanged, and Iceland, where imports from the EC accounted for 44.9 per cent of total imports in 1983 compared to 41.5 per cent in 1962. Figure 7.3 shows a different picture for exports. Only three of the seven EFTA members have increased their percentage share of total exports going to EFTA—Portugal, Sweden and Finland. The most notable increase occurred in the case of Finland, where 18.3 per cent of total exports went to EFTA members in 1983, compared to just 12.1 per cent in 1962. Similarly, the percentage share of exports going to the EC countries has decreased in five out of seven cases, the two exceptions being Norway and Portugal, the latter having increased its share of total exports to the European Community from 37.5 to 59 per cent.

Therefore, the success of EFTA in foreign trade is clear. Concentrating on economic matters only, EFTA members have a shared belief in the efficacy of free trade for stimulating economic activity and growth. It is difficult to generalize, because strong national identities have been retained, but belief in the further viability of EFTA derives from the encouragement of regional development programmes, aimed at rectifying regional imbalance within their community. Having said that, it would appear that EFTA's viability is only in the context of the format of the European Community. If the EC had not been formed, it is unlikely that EFTA could have been justified. There appears to be no particular geographical unity or historical allegiance and political homogeneity uniting these countries. Indeed, when Denmark and the United Kingdom left EFTA in 1971, to join the EC, the effect was extreme and remaining members were forced to negotiate formal trade agreements with the EC. Although EFTA has continued its work Clout *et al.* (1985) rightly observe that since 1973 it has become a more peripheral part of the political and economic infrastructure of western Europe.

7.3 Trade in the European Community

The EEC was formed in 1957, under the terms of the Treaty of Rome, and it had as one of its major objectives, the creation of free trade in a common market. Besides this, it sought to establish joint economic policies and to allow the free movement of people, capital and goods. The original six members felt that they had something different to gain from the union and, although the ultimate aim of the Community was the harmonization of legislation, each country was determined to fight for its own interests and aspirations (Blacksell, 1977). Robinson (1973) grouped together the means by which the European Community was to achieve its aims into 11 major areas:

1. The elimination, as between Member States, of customs duties and of quantitative restrictions

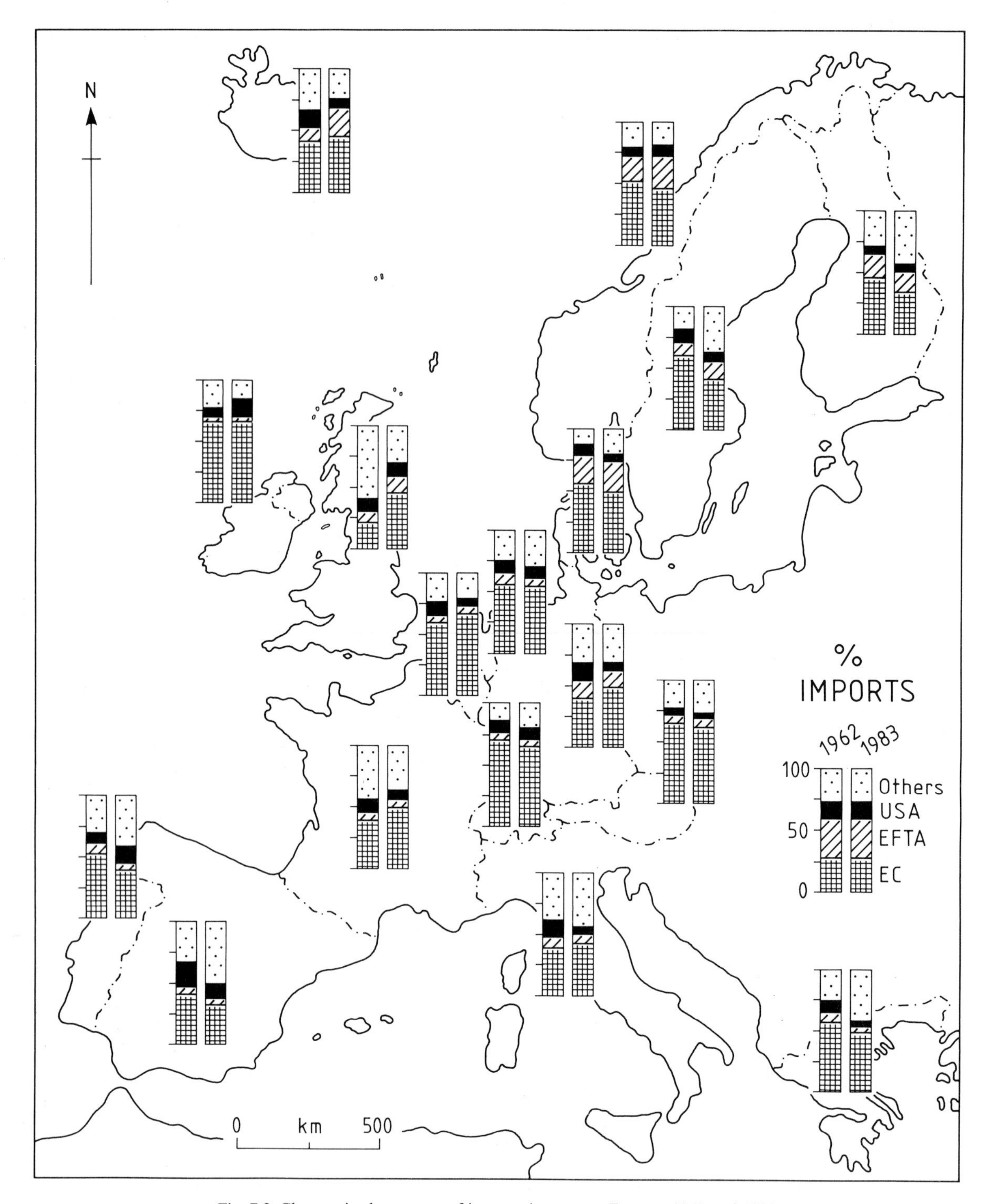

Fig. 7.2 Changes in the pattern of imports in western Europe, 1962 and 1983

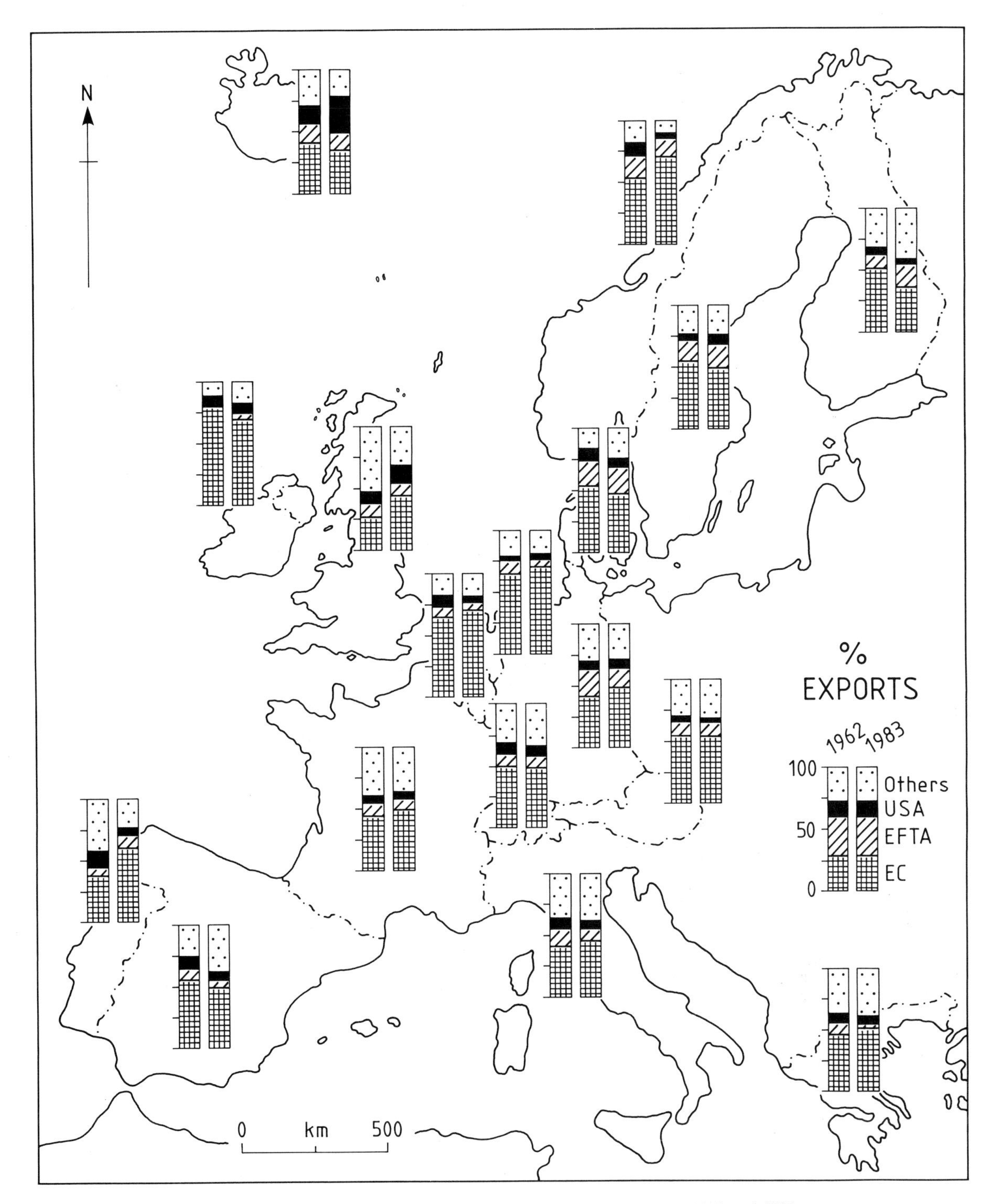

Fig. 7.3 Changes in the pattern of exports in western Europe, 1962 and 1983

on the import and export of goods, and all other measures having equivalent effect.

2. The establishment of a common customs tariff and of a common commercial tariff towards third countries.

3. The abolition, as between Member States, of obstacles to freedom of movement for persons, services, and capital.

4. The adoption of a common policy in the sphere of agriculture.

5. The adoption of a common policy in the sphere of transport.

6. The institution of a system ensuring that competition in the common market is not distorted.

7. The application of procedures by which the economic policies of Member States can be coordinated and disequilibria in their balance of payments remedied.

8. The approximation of the laws of Member States to the extent required for the proper functioning of the common market.

9. The creation of a European Social Fund in order to improve employment opportunities for workers and to contribute to the raising of their standard of living.

10. The establishment of a European Investment Bank to facilitate the economic expansion of the Community by opening up fresh resources.

11. The association of overseas territories and countries in order to increase trade and to promote jointly economic and social development.

The priority areas of establishing free movement of persons, capital, and services and common policies for agriculture, energy, and transport have made limited progress, as indicated in early sections of this book, and consequently the overall development of the European Community has been slow. Most progress has been made in trading arrangements, with the removal of internal restrictions, the creation of a common external tariff, and the association of overseas territories. Complete abolition of all restrictions on trade between member countries was aimed for by 1970, and in reality completed by 1968. The common external tariff governing imports from third countries was also completed in that year.

When the original Six was formed, France, Belgium, and the Netherlands still had colonies and dependencies and it was agreed that these should become associates for five years, to further their prosperity through economic, social, and cultural development. They were given preference in the Community markets and aid to the value of nearly 600 million units of account. By 1963, this agreement had been abolished and the Yaoundé Convention was signed by 17 African states and what is now the Malagasy Republic. The Yaoundé associates had a combined population of 17 million and the Convention ran from 1964 to 1969, with 620 million units of account of aid. Yaoundé II ran up to 1975 and aid was increased to 918 million units of account, most of which was used for infrastructure improvements, port facilities, and energy and agricultural improvement. In addition, the Arusha Convention had been signed in 1968 and associate status was extended to Kenya, Uganda, Tanzania, and Nigeria.

The relationship between the EC and the developing world was dramatically altered when the United Kingdom joined the Community. A large part of the United Kingdom's trade had been with the Commonwealth and links with Africa, the Caribbean, and the southern Pacific (ACP countries), were very strong. Hence, a precondition of UK entry was an agreement to cover these countries and, as the Yaoundé and Arusha Conventions were too limited in scope, it was agreed that a new agreement would be negotiated when they ceased in 1975. The result was the Lomé Convention, signed in 1975 between 57 African, Caribbean, and Pacific countries and to run in the first instance for five years. Following Warren (1977), the basic facts of the Convention were:

1. Under the agreement the Common Market gave free access to ACP industrial products.

2. It also gave free entry to all ACP farm goods which did not compete directly with the products of farmers of the then Nine. These agricultural products accounted for 84 per cent of the ACP countries' total agricultural exports.

3. The European Community had committed itself to aid for the ACP countries, totalling three billion units of account over the five-year period. Part of this was to go towards a new scheme which aimed to stabilize the export earnings from the main primary-commodity exports.

4. The ACP countries would have a much larger say in the operation of the European Deve-

lopment Fund. The Fund already handled over 10 per cent of the Community's aid.

5. Any ACP country was at liberty to opt out of the Lomé Convention if it became convinced that it was losing by it.

The Convention was of unequivocal importance to the ACP countries and the long-run importance of these concessions can hardly be over-estimated. The ACP was allowed to erect barriers against imports from the Nine to protect their infant industries, without losing their own free access to the European Community. The method of approach adopted was generally welcomed (White, 1976; Warren, 1977), and in 1978, for example, Community exports to the 57 ACP countries were worth £8.5 billions, whilst ACP exports to the Community totalled £7.2 billions.

The first Lomé Convention, which was worth $2.1 billion in aid, expired at the end of 1979 and was replaced by Lomé II in 1980. Negotiations between the EC and the then 58 ACP countries were never easy, but an agreement worth $4 billion was reached. Finally, the third Lomé Convention was signed in December 1984, between the 10 EC members and 65 ACP countries, to run from March 1985. The total aid package was $6.8 billion, which was well short of the 8 billion which ACP members felt was necessary to match Lomé II. However, there were two improvements over the second convention: first, an agreement was made to look into ways of combating the spread of deserts in African states; and secondly, there was a commitment to give ACP members access to surplus EC farm produce. In April 1985, it was announced that one of the remaining independent black African countries—Angola—was to join the pact.

The geographic rationale of the European Community is somewhat clearer than it is for EFTA. The member countries are contiguous, except for the stretches of water which separate Greece from the EC heartland, Britain from the mainland, and Ireland from Britain, and most are located within the core area of economic development in western Europe. Despite this, marked spatial variations are seen to exist between members when the indicators of employment, trade, and level of living used in the study of EFTA are again analysed (Table 7.2). Gross domestic product per person in 1982 varied from $13 698 in Luxembourg to $3711 in Greece, and there are wide disparities between members in such facilities as cars, telephones, and televisions. In terms of employment, Greece still employs around 30 per cent of its working population in agriculture, compared to figures of just 2.9 per cent in Belgium and 2.7 per cent in the United Kingdom. The service sector is the largest employer in all the member countries, reaching a peak of 66.3 per cent in the Netherlands. Trade statistics also reveal notable differences in the Ten. All countries continue to increase their total exports and imports, but the percentage share of exports going to other EC countries varies from 43.8 in the United Kingdom to 72.2 per cent in the Netherlands. Exports to EFTA also vary, from 2.7 per cent of total exports in Greece to 22.5 per cent in Denmark. The pattern for imports is similar, with Italy and Ireland importing the smallest and largest percentages of total imports respectively from the European Community, and Denmark importing the largest share of total imports from EFTA (24.8 per cent).

Given this diversity between the member countries, there is no doubting the importance of the Ten in world trade. As early as 1966, the Community was the largest exporter in the world, overtaking the United States, and, in 1983, exports to ACP countries totalled 15 907 million

7.1 The convention of Lomé 1975 (ECC)

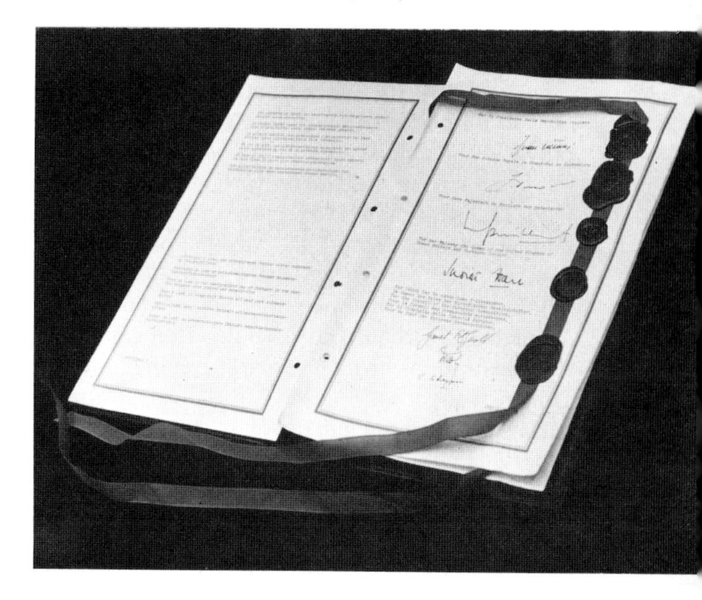

ECUs, just 2.5 per cent of total exports from the Ten. Including the three new members, exports from the European Community increased from \$39 720 million in 1962 to \$475 770 million in 1983, an increase of over 1000 per cent. Similarly, imports into the Community increased by 1033 per cent over the same time period. Over 80 per cent of exports are manufactured goods, whereas a reasonable proportion of imports consists of raw materials like oil, iron-ore, fruit and wood-pulp, one-third of which comes from the third countries.

As with EFTA, the focus of EC trade has changed between 1962 and 1983 (Table 7.3). Intra-EC trade has risen substantially and the percentage share of EC imports from member countries increased from 40.0 per cent in 1962 to 50.2 per cent in 1983, just as the percentage share of exports increased from 43.1 to 52.4 per cent. On the other hand, trade with EFTA has experienced a relative decline. In 1962, 9.7 per cent of EC imports came from EFTA countries; by 1977, this figure had decreased to 8.5 per cent, although by 1983 it had risen again, to 10.1 per cent. With regard to exports, the share declined from 14.5 to 10.6 per cent over the 20 year period.

The changing patterns of imports and exports in the individual countries of the European Community are summarized in Figs. 7.2 and 7.3. The percentage share of imports from the EC into the individual members of the Ten has increased in all cases but three; these are Denmark, where imports from the Community fell from 56.7 per cent of total imports in 1962 to 48.1 per cent in 1983; Greece, where the corresponding fall was from 56.5 to 48 per cent; and the Netherlands, where only a slight decline took place. Notable increases have occurred in the case of the United Kingdom, from 22.3 per cent in 1962 to 45.6 per cent in 1983, and in France, from 39.2 to 49.8 per cent. In contrast, the percentage share of imports from EFTA into the members of the Ten has either stagnated or declined in six out of ten cases, the three exceptions being Denmark, Ireland, and the United Kingdom. Figure 7.3 indicates similar variations between countries in the pattern of exports. Eight out of the ten EC members have increased their percentage share of total exports going to the Community; the two exceptions this time being Denmark and Ireland, who recorded losses of 5.2 and 10.3 per cent respectively between 1962 and 1983. The United Kingdom increased its percentage share of exports going to the Community from 26.1 to 43.8 per cent, the largest increase of any member country. Conversely, the share of toal exports going to EFTA countries has declined in eight out of ten cases; the two exceptions again being Ireland and Denmark, the latter having increased its share of total exports to EFTA from 20.9 to 22.5 per cent.

The European Community is the world's largest single trading bloc, but between 40 and 50 per cent of the trade of the individual member countries is with one another. As Parker (1981, p. 148) states 'while its world trade has been increasing very rapidly, its internal trade is increasing three times as fast, and while it is annually becoming more dependent on imports of certain commodities, it is rapidly attaining a high degree of self-sufficiency in many others'.

This chapter has demonstrated that both the political organization and patterns of trade have changed dramatically in post-war western Europe. However, the process of political and economic integration remains incomplete and the accession of Spain and Portugal into the EC in 1986, for example, will cause numerous problems for trade as the Community becomes swamped with large amounts of unwanted food. Nevertheless, it was announced in June 1985 that a determined assault on national controls on trade and travel in the EC would take place, in the hope that they would be fully removed by 1992. On a larger scale, it is also becoming clear that the Soviet Union is looking for a trade agreement between COMECON and the EC, to replace the bilateral agreements that individual east European countries may have with the EC. Therefore, the topic of trade in western Europe is important and very dynamic.

References

Blacksell, M. (1977). *Post-war Europe: a political geography*. Dawson, Folkestone.

Clout, H. D., Blacksell, M., King, R., and Pinder, D. (1985). *Western Europe: geographical perspectives*. Longman, London.

EFTA. (1975). *EFTA trade, 1973*. Geneva.

OEEC. (1955). 'Convention for European economic co-operation, Article 11'. *European Yearbook*, Vol. 1. Nijhoff.

Parker, G. (1981). *The logic of unity*. Longmans, London.
Robinson, A. H. (1973). *European institutions*. Stevens, Sarasota.
Warren, B. (1977). 'The EEC, the Lomé Convention and imperialism'. In: Nairn, T. (Ed.), *Atlantic Europe? The radical view*. Transnational Institute.
White, G. (1976). 'The Lomé Convention—a lawyer's view'. *European Law Review*, **1**, 197–212.

8

Tourism

Tourism represents one of the most important growth industries in western Europe and is closely related to post-war social and economic advances, which have raised living standards and provided people with more leisure time and paid holidays. The importance of international tourism as an economic force cannot be underestimated since, apart from being a potential source of foreign currency, it can help to eradicate trade deficits and surpluses, improve a country's infrastructure, personalize affluence in regions of poverty, and assimilate and destroy local cultures (Cosgrove and Jackson 1972). Although economic and political factors strongly influence trends in international tourism, it is socio-psychological factors which give the industry its own dynamism (OECD 1977).

According to the World Tourist Organization, there was a total of 285 million international tourists in 1980, an increase of 1040 per cent on the 1950 total of just 25 million. Western Europe accounts for over 50 per cent of international tourist arrivals, and receipts from tourism increased by 28 per cent between 1976 and 1977 alone. Indeed, the tourist industry in western Europe has been experiencing an upturn since the mid-1970s, continuing a trend that was severely interrupted by the 1973/4 oil crisis. However, growth rates have since been more modest and competition more fierce.

Before analysing tourist trends in western Europe, one should note that in any study of international tourism certain interrelated problems will be encountered (Williams and Zelinsky 1970). The first concerns the actual definition of an international tourist. There is no universally accepted definition of such a tourist, but the one normally used is that put forward by the United Nations Statistical Commission of 1963: an international tourist is 'any person visiting a country other than that in which he has his usual place of residence, for any reason other than following an occupation remunerated from within the country visited.' The term applies to those who travel for domestic reasons, pleasure, health, and meetings, and who are stopping for a period of 24 hours or more in a country or area other than that in which they usually reside. Therefore, the definition does not include immigrants, residents in a frontier zone, persons domiciled in one country or area and working in an adjoining country or area, transport crews and troops, and travellers passing through a country or area without stopping.

A second problem relates to the reliability and comparability of information sources. The lack of a universally accepted definition of a tourist makes it difficult to obtain comparable statistics on a world-wide or international basis, but a potentially more serious problem is that a standard method of actually counting the number of tourists does not exist. Data are normally based on frontier checks and on flight and disembarkation cards, which require passengers to state the purpose of their visit. In the absence of this method, data are based on hotel registrations. The two methods are not comparable and the latter excludes certain types of tourists, such as campers and visitors staying in private houses. In addition, duplication occurs if people move from one hotel to another.

A third problem is concerned with territorial definition. Official statistics take and have taken little account of tourist movements between associated territories such as Spain and the Canary Islands, France and French North Africa, and Ulster and the Irish Republic. Scandinavia presents a problem in that little effort is made to record tourist flows between Denmark, Sweden, Norway, and Finland. Tourist movements within one nation are also difficult to analyse because very few regional statistics are available. Therefore, in many instances there is little evidence of regional concentrations of tourists, although some idea can be obtained through a study of airport statistics.

8.1 West European trends and patterns, 1960–80

In western Europe, the scale of tourist activity is inversely related to population and economic potential (White 1976). The centre of population and economic potential (Chapters 2 and 6), has been shown to lie within a triangular area with its apices at Aachen, Maastricht, and Liège and consequently the number of international tourists would be expected to rise with increasing distance away from this core area. Evidence of a core-periphery relationship is afforded by numerous authors (see Christaller 1964; Peters 1969; Young 1973; Clout *et al.* 1985), who have demonstrated how tourism tends to be more important in the less-developed and often rural areas, removed from the urbanized population concentrations. A core–periphery form of migration takes place as people are attracted to the mountains, the woods, and the beaches. Young (1973) has shown that it is those countries with a low standard of living which often offer the most potential for tourism, and the climate and location of Mediterranean Europe, for example, have brought Iberia, Italy, and Greece an influx of wealth. Indeed, tourism is fundamental to their economies (Naylon 1967; Parsons 1973; Pacione 1977).

An analysis of trends in the destination of tourists (Table 8.1) shows that western Europe received 161 million international tourists in 1980. Spain emerged as the clear leader, topping 40 million visitors in 1981, a major increase on the 1951 figure of 1 million. Tourism earns approximately 50 per cent of Spain's foreign currency, and is largely responsible for the country's rapid economic growth since 1950 (Naylon 1967; Pacione 1977). Pearce and Grimmeau (1985)

Table 8.1 *International tourist movements to west European countries, 1960–80*

					Growth rates (%)		Estimated GDP per capita	
To:	1960	1970	1973	1980	1960–70	1970–80	1969	1981
Spain	6.11	22.66	34.56	39.65	271	75	872	4 749
France	5.61	13.70	10.43	30.10	144	120	2 783	10 470
Italy	9.10	12.72	13.16	22.19	40	74	1 548	6 074
Austria	4.59	8.86	11.36	13.88	93	57	1 687	9 066
United Kingdom	1.67	6.73	7.72	12.39	303	84	1 976	7 887
West Germany	5.48	7.71	7.47	11.29	41	46	2 512	10 735
Switzerland	4.95	6.84	6.82	10.65	38	56	2 965	15 076
Belgium	3.89	4.17	7.44	6.70	7	61	2 372	9 709
Greece	0.34	1.41	2.85	4.80	315	240	858	3 711
Netherlands	1.48	2.40	2.77	3.80	62	58	2 196	9 617
Portugal	0.35	3.34	1.65	3.76	854	13	529	2 114
Irish Republic	2.04	1.82	1.29	1.70	−11	−7	1 169	4 857
Totals	45.61	92.36	107.52	160.91	103	74	—	—

Notes:
1. Tourist numbers are in millions.
2. GDP per capita in US dollars.
3. In addition to these figures, there were an estimated 11.3 million (1970), and 13.20 million (1980) international tourists to the Scandinavian Passport Control Area.

Sources: UN Statistical Yearbooks, and *OECD: Tourism policy and International Tourism.*

attributed this dramatic growth to four main factors:

1. Proximity to the growing markets of western Europe.
2. The existence of a sunny climate and sandy beaches sought by these markets.
3. The relative cheapness of Spain as a destination.
4. The development of package tourism, and official encouragement and assistance from Franco's government.

The same authors demonstrated the spatially concentrated nature of both tourist accommodation and demand along the Mediterranean coast and in the Balearic and Canary Islands. Five provinces accounted for over 50 per cent of tourist accommodation capacity in 1981—Balearic Islands (18 per cent), Gerona (11.2), Barcelona (10), Malaga (7), and Alicante (6)—and the only interior province of significance was Madrid (5.2). In terms of demand, the Balearic Islands could be distinguished from the rest of Spain by two factors: first, the sheer volume of bednights* recorded (35 per cent of the nation's total); and secondly, the overwhelming dominance of European tourists there (91 per cent). Europeans also dominated the popular mainland coastal provinces, and overall nearly 90 per cent of Spain's visitors are west European (Fig. 8.1), with France supplying over one-third of the grand total.

Tourism has had an irreversible effect on Spain's economy, society, and landscape (Clout *et al.* 1985), and whole regions, like Benidorm and Formentera, have been transformed. The industry has benefited from successive government schemes, including loans to hoteliers and tax concessions. Between 1964 and 1966 alone, a total of $600 million (20 per cent of tourist revenue) was spent on spreading tourism to remote coasts, especially in the south, and to inland towns. Whilst the number of tourist arrivals increased throughout the 1970s, except for a slight interruption caused by the oil crisis of 1973/4, the importance of tourism to Spain's economy actually declined, from accounting for 34 per cent of the value of exports of goods and services in 1970 to 20.3 per cent in 1980 (Table 8.4). This is a reflection of two factors: first, the general diversification of the Spanish economy in the post-Franco era; and secondly, the growth in competition from rival, often Mediterranean, countries.

Spain is followed by France and Italy as the second and third ranking destinations for foreign tourists. In both countries, tourists are attracted by scenery, art, and gastronomy. Whilst the numbers have continued to increase in Italy, France experienced a period of stagnation in international tourism in the mid-1970s. It would appear that the higher hotel prices and the less ubiquitous sunshine caused the package operators to favour Spain and, on a smaller scale, Portugal and Greece. Therefore, package operators are a significant factor in helping to explain variations between countries.

Table 8.1 indicates that six countries experienced a relative decline in their share of international tourists visiting western Europe between 1970 and 1980—Austria, Switzerland, West Germany, Belgium, the Netherlands, and Portugal. Since the early 1960s, Austria has overtaken Switzerland and West Germany in terms of tourist arrivals, reflecting its low-cost appeal to package-tour companies. In fact, Austria attracts over 60 per cent of its tourists from West Germany. Other countries reflect considerable diversity in their tourist pulls. The United Kingdom has sustained a moderate growth rate and the estimated total of 12.39 million international tourists in 1980 is inflated by a large inflow of visitors from North America; indeed, one in five tourists came from the USA and Canada (Fig. 8.1). The Netherlands and Belgium benefit from large inflows of businessmen, and Portugal and Greece are comparative newcomers to package tourism. Portugal experienced a downturn in the number of foreign visitors during the mid-1970s, due to the oil crisis and internal political problems, but with government support the number of arrivals increased by 47 per cent between 1976 and 1977. Political factors have also contributed to the decline of foreign visitors to the Irish Republic, despite widespread promotion on the continent. In addition to these countries, Table 8.1 shows that there was an estimated total of 13.2 million international tourists visiting Scandinavia in 1980. This part of western Europe attracts a disproportionately high percentage of its visitors from North America, reaching figures of 20 per

*A bednight is one night spent by one person in tourist accommodation.

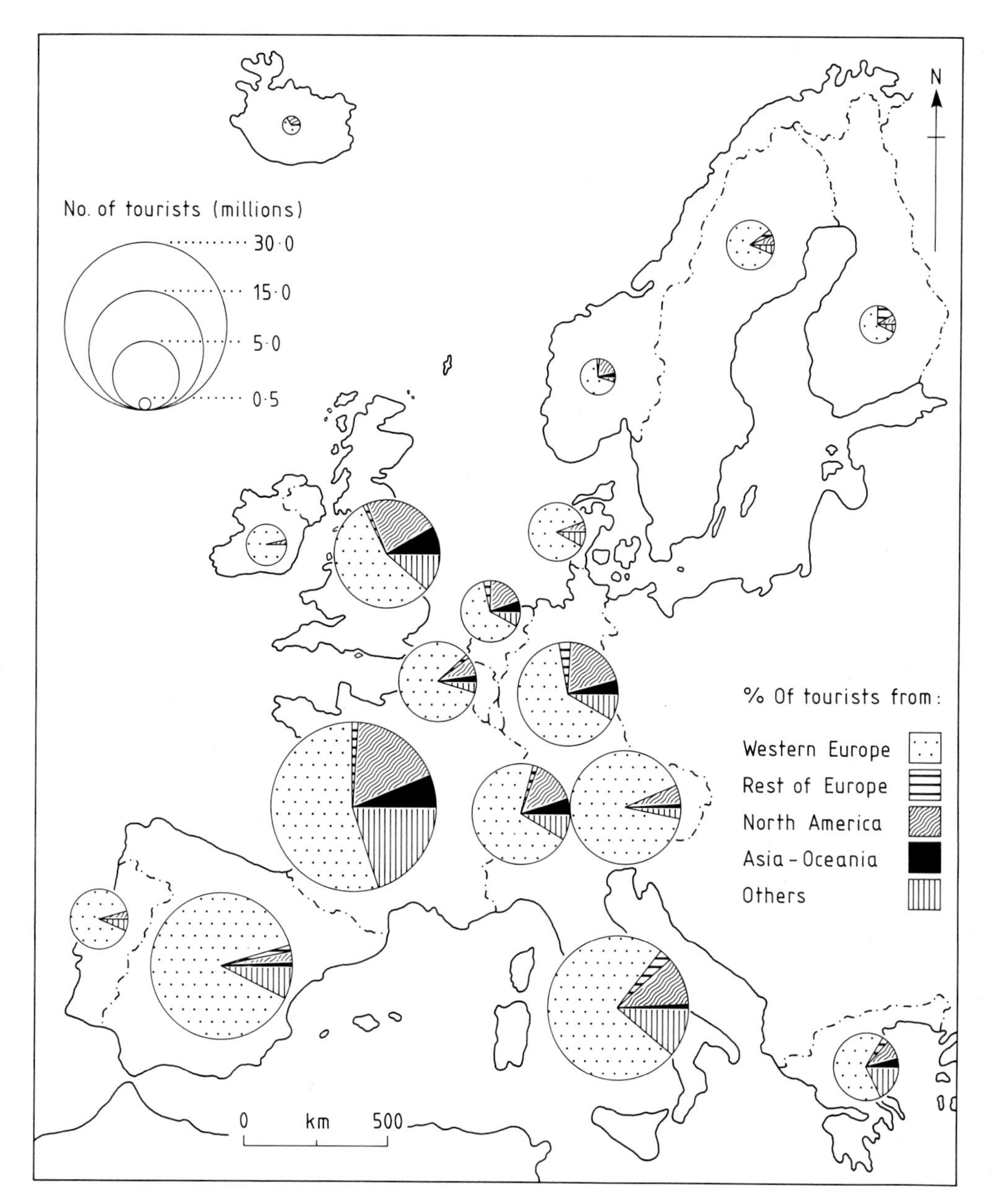

Fig. 8.1 Tourist arrivals in western Europe, 1980

cent for Norway and 33 per cent for Iceland.

The most significant correlation in Table 8.1 is derived from columns five and seven and suggests that in the 1960s tourist growth rates were largely a function of price levels (using gross domestic product as an indicator of tourist price levels). Of the five highest growth tourist destinations, three had gross domestic product figures in 1969 of less than $1000 per capita. This implies that the countries with lower standards of living attract the largest number of tourists. This relationship is less strong by 1980, although there is still evidence of low-cost economies experiencing the highest tourist growth rates.

Tourist departures are less comprehensively enumerated than arrivals, but the data in Table 8.2 indicate that in 1980, 71 per cent of all outgoing west European tourists originated in

Table 8.2 *International tourist movements from selected European countries to destinations in western Europe, 1960–80*

	Estimated number of tourists			Growth rates (%)		Pop$^{n.}$	% tourists to pop$^{n.}$	GDP per capita
From:	1960	1970	1980	1960–70	1970–80	1980	1980	1981
West Germany	10.65	29.02	36.48	173	26	61.2	60	10 735
United Kingdom	6.09	10.68	14.42	75	35	55.9	26	7 887
Netherlands	3.01	6.29	11.66	109	85	14.0	83	9 617
France	7.44	16.62	10.98	123	−34	53.4	21	10 470
Belgium	2.52	4.53	9.75	80	115	9.8	99	9 709
Switzerland	4.05	4.40	5.21	9	18	6.5	80	15 076
Spain	0.89	1.72	4.44	93	158	37.0	12	4 749
Italy	1.80	4.06	4.11	126	1	56.8	7	6 074
Portugal	0.54*	2.84*	3.50*	?	?	9.8	36	2 114
Austria	2.86	4.20	2.48	47	−41	7.5	33	9 066
Greece	0.25	0.49	0.52	96	6	9.3	6	3 711
Totals	40.10	86.85	103.55	117	19	321.2	32	—
Scandinavia	2.30	3.32	5.38	44	62	—	—	—

Notes:
1. Tourist numbers are in millions.
2. GDP per capita in US dollars.
3. Population in millions.

*Figures incomplete.

Sources: UN Statistical Yearbooks, and *OECD: Tourism Policy and International Tourism*

four countries: West Germany, France, the Netherlands, and the United Kingdom. West Germany, with a total of 36.5 million outgoing tourists, dominates the overall pattern and is a long way ahead of the second country, the United Kingdom. Allowing for the incomplete nature of the data presented in Table 8.2, there appears to be a positive correlation between gross domestic product and tourist departures. Countries with high gross domestic product figures have large outflows of tourists, whereas the lower income countries of Greece, Spain, Portugal, and Italy have the least number of tourist departures. Although every country witnessed an increase in the number of tourist departures between 1960 and 1970, two countries—France (−34%) and Austria (−41%)—experienced a decline in departures during the 1970s.

If a comparison is made of Tables 8.1 and 8.2, it is possible to ascertain some insight into tourist exchange patterns. In 1980, France, Spain, and Italy recorded a net gain of over 18 million tourists (Table 8.3), emphasizing the importance of tourism to their economies. Four other countries, including Scandinavia as one country, had net gains of over four million tourists. Portugal achieved a more even balance, with a net gain of just 0.26 million tourists, whereas the Netherlands and West Germany recorded losses of over 7 million tourists. West Germany, with 36.5 million tourist departures and just 11.3 million tourist arrivals in 1980, registered a high deficit of

Table 8.3 *Tourist exchange patterns, 1980*

	Tourist Arrivals (millions)	Tourist Departures (millions)	Balance (millions)
France	30.10	10.98	+19.12
Spain	22.50	4.40	+18.10
Italy	22.19	4.11	+18.08
Austria	13.88	2.48	+11.40
Scandinavia	13.20	5.38	+7.82
Switzerland	10.65	5.21	+5.44
Greece	4.80	0.52	+4.28
Portugal	3.76	3.50	+0.26
Belgium	6.70	9.75	−3.05
United Kingdom	11.29	14.42	−3.13
Netherlands	3.80	11.66	−7.86
West Germany	11.29	36.48	−25.19

Source: OECD: Tourism Policy and International Tourism

25.2 million tourists, representing a net loss to the economy of approximately $14 000 million.

The expansion of tourism in western Europe is due in part to improvements in the different modes of transport and in particular to the efficiency, comfort, safety, and cost of travel. Road transport accounts for the largest inflow of tourists in six countries: Austria (93%), West Germany (88%), France (86%), Italy (74%), Denmark (65%), and Spain (66%). However, it is air transport which has contributed most to the overall expansion in international tourism. This mode of transport is particularly important in countries receiving sea-borne tourists, accounting for two-thirds of tourists entering Greece, 61 per cent entering the United Kingdom, 42 per cent entering Portugal, and 40 per cent entering Sweden. In 1980, there were nearly 23 million passengers on non-scheduled charter flights and these were largely divided between four countries: Spain (32%), the United Kingdom (19%), West Germany (14%), and Scandinavia (9%). The three major flows are between the United Kingdom, West Germany, and Scandinavia on the one hand, and Spain on the other, although there has been an upward trend in the number of charter flights between the first three countries and Greece.

Benefits derived from international tourism are considerable and a cost/benefit analysis of tourist flows in western Europe shows that tourist expenditure in 1980 exceeded tourist receipts in seven countries. Not surprisingly, these countries, with the notable exception of Ireland, are located within the core area of economic activity and boast some of the highest standards of living in western Europe. In 1980, six countries had estimated tourist receipts in excess of $6000 million, with Italy recording the highest total of $8213 million (Table 8.4). Indeed, figures show a marked increase on those for 1970, when only five countries registered receipts of over $1000 million.

A more realistic way of assessing the importance of international tourism to the economies of western Europe is to examine the industry's contribution to the total value of exports of goods and services. For 1980, figures ranged from 20.3 per cent in Spain, and 20.0 per cent in Austria, to just 2.4 per cent in Sweden, 2 per cent in Belgium and Luxembourg, and 1.7 per cent in the Netherlands (Table 8.4). A distinct pattern is discernible and it is in the peripheral countries, such as Spain, Portugal, Greece, and Austria, where tourism contributes most to the value of exports of goods and services. These particular countries export

Table 8.4 *International tourist receipts and expenditure, 1960–80 (millions US dollars)*

	International tourist receipts			International tourist expenditure			Surplus of receipts over expenditure			International tourist receipts as % of value of exports of goods and services		
	1960	1970	1980	1960	1970	1980	1960	1970	1980	1960	1970	1980
France	500	1 189	8 197	263	1 057	6 001	+237	+132	+2 196	7.3	5.5	5.1
West Germany	398	1 024	6 565	631	2 493	20 598	−233	−1 469	−14 033	2.8	2.5	2.8
Austria	232	999	6 442	61	323	3 124	+171	+676	+3 318	15.4	23.3	20.0
Spain	297	1 681	6 968	50	138	1 229	+247	+1 543	+5 739	25.4	34.0	20.3
United Kingdom	473	1 039	6 922	521	924	6 410	−48	+115	+512	2.9	3.4	4.2
Italy	642	1 639	8 213	94	727	1 907	+548	+912	+6 306	12.0	8.7	8.5
Switzerland	365	905	3 149	148	427	2 357	+217	+478	+792	12.4	11.3	9.3
Netherlands	132	421	1 662	127	598	4 664	+5	−177	−3 002	2.5	2.8	1.7
Belgium/Luxembourg	110	348	1 810	138	492	3 272	−28	−144	−1 462	2.4	2.4	2.0
Greece	51	194	1 734	19	55	190	+32	+139	+1 544	12.8	15.9	18.9
Denmark	107	314	1 337	74	273	1 560	+33	+41	−223	5.4	7.0	5.5
Norway	51	158	751	55	152	1 310	−4	+6	−559	2.7	3.3	2.7
Sweden	67	144	962	148	427	2 236	−21	−338	−1 274	2.0	1.7	2.4
Finland	17	129	682	39	95	590	−22	+34	+92	1.5	4.5	3.9
Portugal	26	240	1 149	13	98	291	+13	+142	+858	5.1	15.6	16.8
Ireland	111	186	574	42	100	742	+69	+86	−168	17.1	11.9	5.7

Sources: UN Statistical Yearbooks, World Travel Statistics, and *OECD Tourism Committee Report 1977 and 1981*

8.1 The Ardennes Forest: a popular recreational spot (ECC)

small numbers of their own tourists and are thus left with a comparatively large surplus of receipts over expenditure. In contrast, the sizeable tourist receipts in West Germany, the Netherlands, Belgium, and Luxembourg are offset by tourist expenditure; in the case of West Germany the net debt was $14 033 million in 1980, a 850 per cent increase on the 1970 figure. The situation in the five countries with a net debt in tourism appears to be worsening, as indicated by figures for the 1960–80 period (Table 8.4).

Since the 1950s, the countries of western Europe have increasingly recognized that the tourist industry can be made a tool of governmental economic policy (White 1976). However, there is a wide range of policy objectives in western Europe and no common pattern of tourist policy exists. This is rather surprising when it is realized that, for the purpose of tourists movements, western Europe constitutes a unified market. Many controls on international movements, such as visas and passports, have been removed and major tourist flows take place. Yet, tourist policies remain national in character, despite their potential in overall regional development policy within western Europe. The only common objective is to offer all sectors of society the opportunity of a holiday.

Spain, the most important country for tourism, devotes equal attention to tourism for relaxation, with a concentration in coastal and mountain areas, and business tourism, which gravitates towards the urban areas and the major cities in particular. National incentives, to attract foreign visitors and to keep native tourists at home, have been used and were particularly important after the collapse of the package holiday, following the oil crisis of 1973/4. Attempts have been made in more recent years to control the growth of tourist facilities and spread the demand away from the congested Mediterranean coast and towards the less-developed parts. Similar policies have been used in Italy, Greece, Portugal, and the United Kingdom; and Belgium recognized the importance of international tourism to its economy by including it in the country's five year plan for 1976–80.

Certain governments have also attempted to redistribute tourists, by the use of such regional policies as loans, special funds, and specific projects to aid tourism in problem areas. Ireland has attempted to encourage leisure tourism in areas unsuited to agriculture and industry, whilst the United Kingdom has given priority to the development areas. France is using tourism to revitalize its regional economies and both Austria and Greece have recognized the need for a more balanced distribution of tourists, in order to relieve the pressure on those regions with a traditionally high tourist intake.

White (1976) claims that international planning is necessary for tourism in western Europe. This is because the benefits from this important sector of the economy do not always remain within the affected areas. Employment opportunities created by tourism, for example, are often seasonal in nature and hence attractive to migrant workers rather than to the resident population, and national tourism schemes diminish the importance of local ownership and control.

As a brief summary to this section, it is clear that one is dealing with an important, but highly fluctuating, sector of the economy. Four major points seem to emerge from a study of tourism trends between 1960 and 1980:

1. There has been a continued southerly drift of the centre of gravity of international tourism in western Europe, towards the Mediterranean coastlands.
2. The dominant tourist flows are from high income areas to low income areas. This observation appears to be valid on a regional as well as a national basis.
3. Tourism has risen in importance since the early 1960s and the average length of tourist movements continues to increase.
4. The mass-market tourist flows, aided by package tours, have significantly boosted the economies of certain west European countries, notably Spain, Greece, and Portugal.

8.2 Tourism in Austria and Switzerland

The predominantly Alpine environment of Austria and Switzerland places immediate constraints on the levels of human activity and the patterns of agriculture, industry, communications, and population. Statistics reveal that 70 per cent of Austria and 58 per cent of Switzerland are comprised of mountain chains, and that two-thirds of the surface area of Austria and one-fifth of the surface area of Switzerland are covered by snowy peaks, glaciers, forests, and lakes. Thus, the physical landscape offers opportunities for outdoor recreation and tourism is often the only profitable economic activity in the mountains. Indeed, tourism in the Alps has been a vital industry to the economies of Austria and Switzerland for over a century, and in Austria in particular it has played a large part in the country's post-war economic revival.

Historically, tourism was more important in Switzerland than in Austria, as development of the Austrian Alps was retarded by poor communications and set-backs to the German economy (Cosgrove and Jackson 1972). Between 1945 and the mid-1960s, Switzerland just managed to maintain its dominance, despite unprecedented development in Austria's tourist industry. Since the early 1970s, Austria has been queen of Alpine tourism and continues to attract many winter tourists who have become familiar with the long-established ski runs in Switzerland. A brief analysis of border-crossings of incoming tourists supports this pattern of change and whilst Austria can boast a continual pattern of increase between the early 1960s and 1983, Switzerland appears to have been influenced by the slump in world tourism and actually experienced a decline in border crossings between 1972 and 1976.

Foreign exchange earnings from international tourism are very important to both Austria and Switzerland. In Austria, they compensate for up to 90 per cent of the country's trade deficits, and in Switzerland tourism ranks third, in terms of receipts, in the country's trade balance. Table 8.4 indicates that in 1980 there was a surplus of receipts over expenditure in both countries, although earnings from foreign tourism were four times as high in Austria. Tourist receipts, as a percentage of the value of exports of goods and services, continued to rise in Austria, from 15.4 per cent in 1960 to 23.3 per cent in 1976, before falling to 20 per cent in 1980, and this is in contrast to the downward trend in Switzerland, from 12.4 to 9.3 per cent over the twenty-year period.

In 1983, Austria and Switzerland received 19.9 and 11.1 million tourists respectively. In each case, over 60 per cent of the arrivals were from foreign lands, although Table 8.5 indicates that internal tourism is more important in Switzerland

Table 8.5 *Tourism in Austria and Switzerland, 1983 (millions)*

	Tourist arrivals	% of arrivals		No. of bednights	No. of beds available	% of bednights		W. German (as % of foreign)
		Foreign	National			Foreign	National	
Austria	19.9	72.9	27.1	115.8	1.14	75.5	24.5	69.2
Switzerland	11.1	62.2	37.8	33.6	0.28	59.0	41.0	32.8

Sources: OECD (1984): Tourism policy and international tourism, Statistisches Jahrbuch Der Schweiz, 1984, and *Statistisches Handbuch Für Die Republik Österreich, 1984*

(comprising 37.8 per cent of total arrivals and 41 per cent of total bednights) than it is in Austria. Tourists remain in Austria for an average of 6.8 bednights, just over twice the average length of stay in Switzerland.

If one analyses the origins of foreign tourists in Austria and Switzerland (Fig. 8.1), western Europe supplies 88 and 72 per cent of the respective totals and thus totally dominates the pattern. Only nine per cent of Austria's foreign tourists come from outside the European continent; the figure for Switzerland is 27 per cent, with a majority of those coming from North America. A more detailed breakdown of the origin of west European tourists to Austria and Switzerland in 1983 is presented in Figs. 8.2 and 8.3. In both countries, West Germany supplies the largest numbers; 66 per cent in Austria and 47 per cent in Switzerland. The second and third largest suppliers to Austria are the Netherlands, with 9.4 per cent, and the United Kingdom, with just 5.9 per cent. Switzerland exhibits a more widespread pattern and, apart from West Germany, attracts between 5 and 11 per cent of its west European

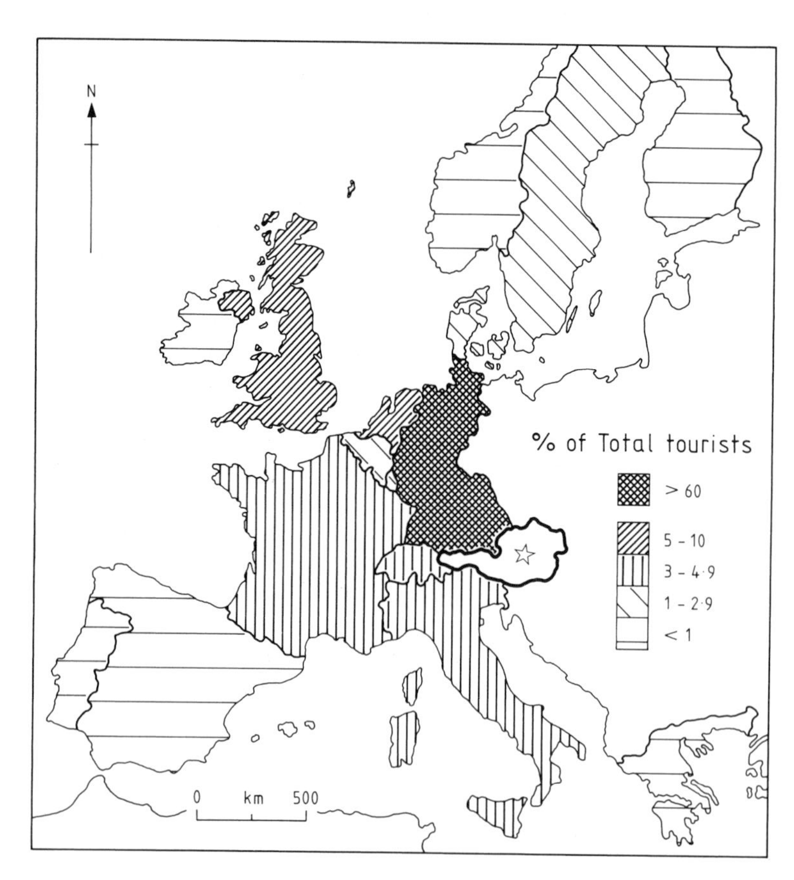

Fig. 8.2 Origin of west European tourists visiting Austria, 1983

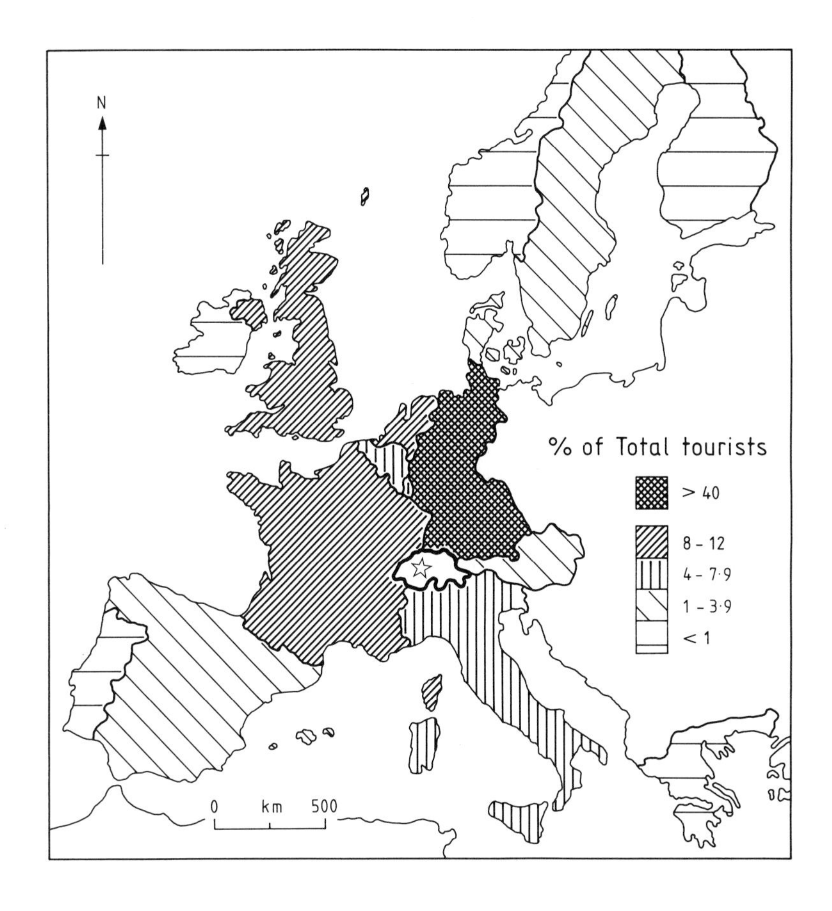

Fig. 8.3 Origin of west European tourists visiting Switzerland, 1983

tourists from each of the following: the Netherlands (8.8), the United Kingdom (10.8), France (9.8), and Italy (7.4). The number of international tourists visiting Austria appears to decrease with increasing distance away from the country; this simple relationship does not hold for Switzerland.

An examination of bednight statistics for the *Länder* of Austria and regions of Switzerland reveals some interesting points about the regional concentration of tourism. In Austria, the western *Länder* receive the largest number of tourists (Fig. 8.4), and Tirol, with just over one-third of the total number of bednights spent in Austria in 1983, is easily the most popular *Land*. Interestingly, Tirol is attractive to the foreign tourists and not the national tourists, who spend most bednights in Steiermark, a *Land* which ranks only seventh amongst foreign tourists. In 1983, the western *Länder* of Tirol, Salzburg, and Karnten accounted for 67 per cent of all tourist bednights, and 75 per cent of foreign tourist bednights spent in Austria.

This pattern of concentration is not reproduced in Switzerland and Fig. 8.5 depicts a more even spread of tourist bednights. The most popular region is Grisons in the extreme south-eastern corner of the country and this is followed first, by the Lake Geneva region, and secondly by central Switzerland. The latter two regions are less important amongst the national tourists, who seem to favour the southern and south-eastern regions of Grisons, Ticino, and Valais.

The origins of foreign tourists visiting the regions of Austria and Switzerland are listed in Table 8.6. Nearly 60 per cent of Austria's foreign visitors come from West Germany, which also supplies one-third of Switzerland's total. Eight of the nine *Länder* are dependent on West Germany for the majority of their tourists. The one exception is Vienna, an urban *Land* which attracts its visitors from all parts of the world and accounts for 35 per cent of the total number of bednights spent by North American tourists in Austria. The Swiss regions attract varying proportions of their

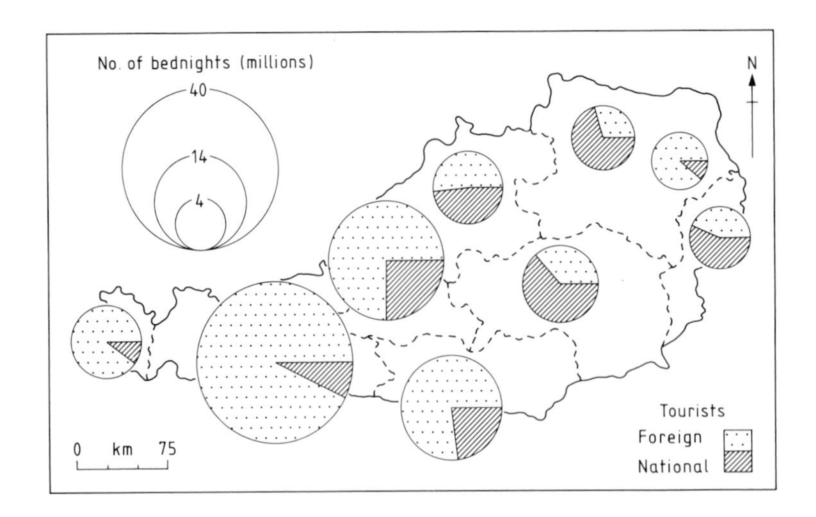

Fig. 8.4 Distribution of tourist bednights in Austria, 1983

foreign visitors from West Germany, ranging from 56.4 per cent of foreign bednights in Grisons to just 9.2 per cent in Lake Geneva. American visitors account for over one-fifth of Central Switzerland's tourist bednights and they also favour the Zürich, Bernese Mittelland, and Lake Geneva regions. In addition, Table 8.6 emphasizes the importance of French tourists to Valais and the Jura, and British tourists to the Bernese Oberland and central Switzerland.

At a more localized level, certain tourist centres are especially important. In Austria, tourists are found primarily in the west and extend as far east as the Carinthian lakes and the Gastein valley; the only tourist areas in the east are Mariazell and Semmering-Weschel (Schadlbauer, 1975). Popular winter tourist resorts include Innsbruck, St. Anton, Söll, and Sölden, in Tirol, and Saalback, St. Gilgen, and Obertauern, in Salzburg. Vienna, Salzburg city, Innsbruck, and the Carinthian lakes are popular locations for summer tourists. In Switzerland, more than 50 per cent of all tourists are welcomed into 15 major tourist centres, which provide just one-third of the country's hotel accommodation. These 15 'top spots' are widely dispersed, although located at accessible points on the road and rail network, and include centres for winter and summer sports, such as St. Moritz, Arosa, and Davos. Also popular are lakeside centres like Interlaken, Locarno, and Geneva and the major cities of Basle, Zürich, Berne, and Geneva, which are important for art, culture, and conference meetings.

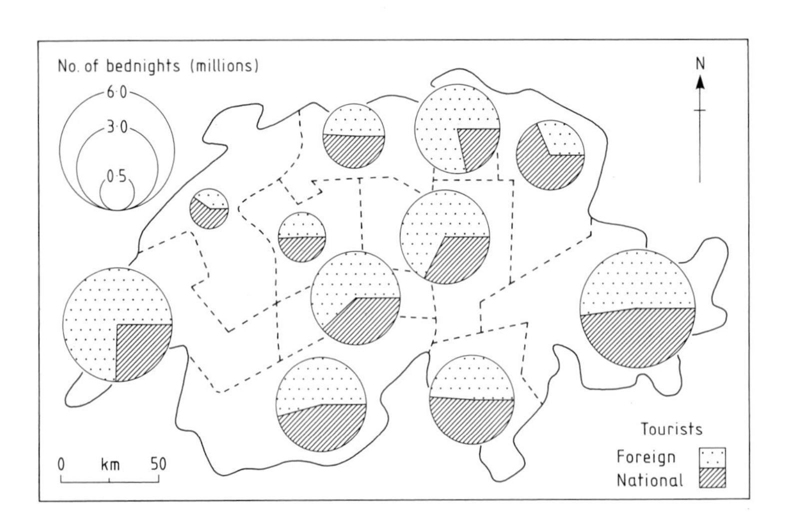

Fig. 8.5 Distribution of tourist bednights in Switzerland, 1983

Table 8.6 *Origins of foreign tourists in the regions of Austria and Switzerland, 1983 (percentage of total bednights spent by foreign tourists)*

To: Austrian *Länder*	Origins of foreign tourists							
	Belgium	West Germany	France	Great Britain	Italy	Netherlands	USA	Others
Burgenland	0.6	85.3	0.8	0.6	1.7	2.2	0.9	7.9
Kärnten	2.0	71.8	1.4	2.0	2.3	11.1	1.2	8.2
Nieder-Osterreich	1.3	59.2	3.0	2.2	5.7	9.0	4.7	14.9
Ober-Osterreich	1.7	68.3	2.7	5.1	2.6	4.9	2.4	12.3
Salzburg	2.0	58.6	3.1	4.3	3.2	7.5	5.8	15.5
Steiermark	3.2	59.3	3.2	2.8	5.5	5.6	2.7	17.7
Tirol	3.2	59.2	3.7	7.7	1.3	10.8	4.0	10.1
Vorarlberg	2.1	72.1	5.4	2.9	0.6	6.5	1.5	8.9
Vienna	1.3	24.7	4.1	4.9	10.2	2.5	13.8	38.5
Austria	2.4	58.5	3.4	5.2	3.0	8.3	4.7	14.5
To: Swiss regions	Belgium	West Germany	France	Great Britain	Italy	Netherlands	USA	Others
Grisons	6.2	56.4	6.7	5.7	3.2	5.7	7.0	9.1
Bernese Oberland	2.1	35.6	5.9	23.9	1.4	6.4	10.9	13.8
Central Switzerland	2.7	33.9	3.6	14.0	1.8	4.3	22.2	17.5
Ticino	3.5	51.2	2.8	6.1	9.4	7.4	6.3	13.3
Valais	9.6	40.4	14.8	9.4	2.9	5.9	8.1	8.9
Lake Geneva	5.2	9.2	13.1	10.7	5.2	2.8	12.9	40.9
Bernese Mittelland	1.6	25.6	6.4	4.1	6.7	3.5	13.8	38.2
Jura	4.7	27.8	15.6	4.5	5.3	6.5	7.4	28.3
Eastern Switzerland	1.8	49.5	5.0	4.7	5.0	5.2	7.7	21.1
Zurich	1.1	15.8	3.0	5.8	5.2	2.7	16.4	47.9
North-western Switzerland	3.4	31.1	5.8	8.2	6.2	6.2	9.2	29.9
Switzerland	4.2	32.5	7.6	10.2	4.2	4.8	12.0	24.5

Sources: Statistiches Jahrbuch Der Schweiz, 1984, Statistiches Handbuch Fur Die Republik Osterreich, 1984

Tourism has also had a dramatic impact on many of the more remote alpine valleys of Austria and Switzerland and 'within the past 25 years or so, most alpine communities have changed from virtual dependence on near-subsistence agriculture to almost total dependence on tourism' (Kariel and Kariel 1982, p. 2). The same authors developed three models to conceptualize the effects of tourism on mountain regions. One of these was set within the framework of diffusion theory and outlines the various social and cultural changes which accompany the transition from a rural–agrarian society to an urban–technological one (Fig. 8.6). When applied to four communities taken from a west to east transect through the Austrian portion of the Alps, the model adequately depicted the manner in which changes associated with the growth of tourism proceeded. Indeed, evidence was presented to show how the way of life in these communities was coming to resemble that of the urban–technological society. Kariel and Kariel (1982) also found that residents in the sampled communities emphasized the positive rather than negative effects of tourism, although they expressed a strong tie to the land and a feeling for their home community.

Both Austria and Switzerland are aware of the special role of tourism in their national economies, and regional development programmes pay particular attention to tourist potential, especially in rural areas where industrial development is largely impossible. The Austrian government has attempted to stabilize the country's economic situation by investing in tourist facilities, in the hope that receipts from international tourism will contribute to a better balance of payments equili-

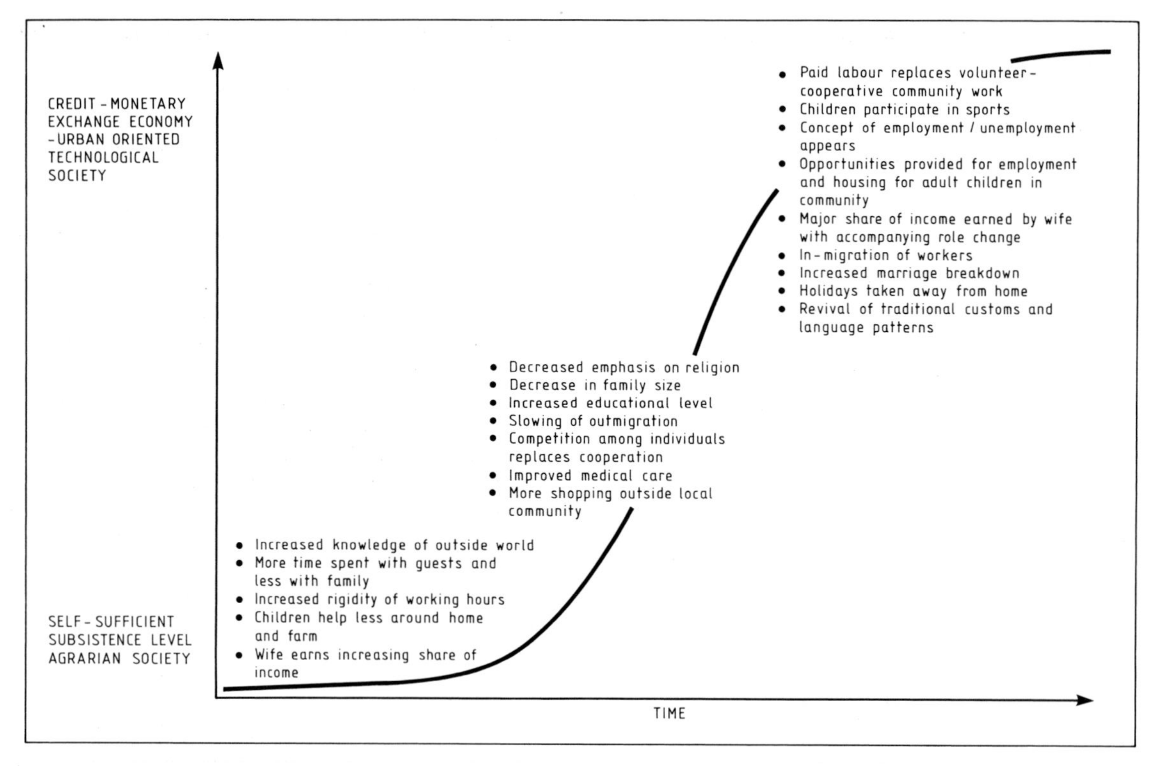

Fig. 8.6 Diffusion of socio-cultural changes resulting from tourism (Kariel and Kariel 1982)

brium. Bodies responsible for tourism exist at the Federal, *Länder*, and Commune levels (OECD 1978), and their objectives include:

1. The improvement of services.
2. The provision of additional types of tourist entertainment which are independent of weather conditions.
3. The improvement of the marketing of tourist services.
4. The improvement of vocational training.
5. The protection of rest centres from environmental aggression.

In an attempt to spread tourism to all parts of Austria, farmers have been encouraged to offer accommodation facilities, or create new ones, and low interest loans and cheap publicity have been offered as incentives. Although the role of tourism in Austria's economy is now declining, as more Austrians take holidays abroad, a Tourism Promotion Programme was launched in 1980 for a ten-year period. This had three main objectives: first, to devote greater effort to attract tourists; secondly, to attempt to increase demand in remoter areas; and thirdly, to provide more support for basic organizations and services.

The Swiss government has become increasingly concerned about the country's tourist industry and in 1973, the Advisory Commission for Tourism was established by the Federal Council to determine a pattern for Swiss tourism. In the following year a Federal Act made a block appropriation of Swiss francs 500 million available to the mountain regions for promoting development and improving economic structures. Tourism was to benefit indirectly through infrastructure improvements. Government aid has been available since 1976, through the Société de Crédit Hôtelier, and the Swiss National Tourist Office awarded a special grant of one million francs in 1976 and 1977, to finance publicity campaigns in the major generating markets. A total of Swiss francs 99 million was invested in tourism in 1977 and it was during this year that the Swiss plan for tourism was completed. The plan was submitted to the Federal authorities in 1978 and subsequently used as the official guide for future tourist development. In 1983, the Fédéral Service de tourisme was transferred to the Federal Department for Public Economy, emphasizing the continued importance of tourism to the country's development plans.

Thus, tourism is a vital industry in the Alpine economies of Austria and Switzerland and one of continued importance. As the number of international tourists visiting Austria continues to increase, and as Switzerland shows signs of a tourist revival, tourism figures strongly in the development programmes of the respective countries and one of the objectives of regional policy is to achieve a more equitable distribution of tourist destinations.

References

Christaller, W. (1964). 'Some considerations of tourism locations in Europe'. *Regional Science Association Papers*, **12,** 95–105.

Clout, H. D., Blacksell, M., King, R., and Pinder, D. (1985). *Western Europe: geographical perspectives*. Longman, London.

Cosgrove, I., and Jackson, R. (1972). *The geography of recreation and leisure*. Hutchinson, London.

Kariel, H. G., and Kariel, P. E. (1982). 'Socio-cultural impacts of tourism: an example from the Austrian Alps'. *Geografiska Annaler*, **64B,** 1–16.

Naylon, J. (1967). 'Tourism: Spain's most important industry'. *Geography*, **52,** 23–40.

OECD. (1977). *Tourism policy and international tourism*. Paris.

OECD. (1978). *Tourism policy and international tourism*. Paris.

OECD. (1984). *Tourism policy and international tourism*. Paris.

Pacione, M. (1977). 'Tourism: its effect on a traditional landscape in Ibiza and Formentera'. *Geography*, **62,** 43–7.

Parsons, J. J. (1973). 'Southwards to the sun: the impact of mass tourism on the coast of Spain'. *Association of Pacific Coast Geographers, Yearbook*, **35,** 129–46.

Pearce, D. G., and Grimmeau, J. P. (1985). 'The spatial structure of tourist accommodation and hotel demand in Spain'. *Geoforum*, **16,** 37–50.

Peters, M. (1969). *International tourism*. Hutchinson, London.

Shadlbauer, F. (1975). 'Austria'. In Clout, H. D. (Ed.) *Regional development in western Europe*. Wiley, London.

White, P. E. (1976). 'Tourism and economic development in the rural environment'. In Lee, R. and Ogden, P. E. (Eds.) *Economy and society in the EEC*. Saxon House, Farnborough.

Williams, A. V. and Zelinsky, W. (1970). 'On some patterns in international tourist flows'. *Economic Geography*, **46,** 549–67.

Young, G. (1973). *Tourism: blessing or blight*. Penguin, Harmondsworth.

9

Social Well-being

During the 1960s and early 1970s, most west European countries experienced substantial rates of economic growth. Despite this improved situation, it is doubtful whether the benefits reached all segments of society and concern in more recent years has been focused upon the 'social well-being' of people in these countries. Indeed, the economic crisis of the mid-1970s caused unrest in Europe, especially in peripheral regions, and led to a shift in government priorities, away from the promotion of economic growth to a concern for an individual's 'quality of life'.

Increasing concern for the 'social well-being' of people has prompted policy-makers and academics to call for improved systems of measurement of the more socially-orientated dimensions of the socio-economic system to complement existing statistics, and to produce more comprehensive data banks; good examples of the latter are found in Scandinavian countries (Knox 1984). With a gradual rise in the publication of social statistics, geographers have begun to examine spatial variations in social conditions. The objective of this chapter is to examine variations in social well-being at various levels within Europe. Before that is possible, it is necessary to develop a suitable method of study.

9.1 Development of a methodological framework

Various terms have been used in geographical studies of well-being, such as 'quality of life', 'level-of-living', 'social welfare' and 'satisfactory life', and there has been much debate over which term is best (Smith 1973 and 1977; Knox 1974a and 1975; Coates, Johnston and Knox 1977, Pacione 1982). The term 'social well-being' will be adopted in this book because it is widely accepted in geography as a generic term for the family of overlapping concepts listed above. However, not too much importance is placed on the actual label and, as pointed out in an OECD report (1976), the proof of the pudding is in the eating and the specification and operationalization of the concept is more important.

One of the chief proponents of a more socially-oriented discipline was D. M. Smith, who argued for a geography of social well-being (1973) and outlined a welfare approach which could provide a useful framework for human geography (1977). Well-being can be defined as the satisfaction of the needs and wants of the population. The growing awareness of the social and political health of a nation was manifested in the Social Indicators Movement and a search for indicators other than those based on GNP. A social indicator is an aggregate or composite measure of well-being, or some element of it (Knox 1975). The development of social indicators has been reviewed by many people (eg. Kamrany and Christakis 1970; Knox 1974b), and the aim of such indicators is to monitor levels of well-being at various points in time, in order to improve levels of information required for policy decision-making. A social indicator on health, for example, could be obtained by aggregating various statistics on health into one meaningful summary statistic. Indicators on various aspects of life, amenable to quantification, can thus be obtained and combined into an overall measure of social well-being.

Geographers are necessarily interested in territorial social indicators which relate to any geographical section of a nation, such as a region or urban area. However, certain problems are apparent:

1. The geographer is often restricted to government-defined boundaries, for which data are available.
2. Data on social indicators are often only available at the national level and this necessarily hides spatial inequalities in social well-being.
3. The number of factors affecting and creating disparities in well-being is enormous and there has been little standardization in terms of those social indicators which are considered most appropriate.
4. It is difficult to assess whether 'objective' and 'subjective' indicators will produce contrasting patterns of social well-being (Pacione 1982). The former are hard measures describing the environment within which people live and work, whereas the latter concentrate upon the way in which people perceive and evaluate conditions around them.

In order to develop a set of social indicators to analyse spatial variations in social well-being in Europe, it is necessary to follow four basic steps:

1. The selection of geographical units of analysis.
2. The specification of general areas of social concern for which indicators are to be developed.
3. The collection of a relevant set of basic statistics from which a summary statistic can be created.
4. The aggregation of basic statistics to obtain a limited number of meaningful social indicators.

A brief review of two attempts to provide a methodological framework for the development of a set of social indicators will be given, in the hope that they will make the analysis of social well-being easier to follow. The selected studies are those by the United Nations and the OECD, both of considerable importance to countries in western Europe.

Studies of international variations in social well-being were pioneered by the United Nations, which designed a general 'level-of-living index' for its members as early as 1954 (United Nations 1961 and 1966). 'Level-of-living' was defined as the actual living conditions of the people and the concept was to be analysed in terms of a series of *components*, each intended to represent a class of human needs (such as health, nutrition, and housing), the satisfaction of which could be measured indirectly by a series of *indicators* (Knox 1975). The committee of experts recommended seven components to measure level-of-living and these are listed, along with their respective indicators, in Table 9.1. Items closely related to, but not direct measures of level-of-living were added to the list, to provide necessary background information for the interpretation of level-of-living; these included population and labour force characteristics, income expenditure, and communications and transport.

The indicators used were necessarily elementary in nature because of the wide range of cultures and conditions in the member countries. A single level-of-living index for each country was the main attraction of this study, although the index could be disaggregated to give scores for each component. A more detailed analysis of the United Nations approach is beyond the scope of this chapter and those interested are referred to the United Nations reports (1961 and 1966) and to Drewnowski and Scott (1968).

In 1970, at a Council of Ministers meeting, the OECD explored the possibility of developing a set of social indicators to 'better focus and enlighten public discussion and decision-making' (OECD 1976, p. 157). The starting point was the listing of 24 social concerns, which were comparable and common to most member countries. A social concern was defined as 'an identifiable and definable aspiration or concern of fundamental and direct importance to human well-being' (OECD 1973, p. 8). The 24 fundamental concerns were grouped into eight primary goal areas (Table 9.2), which together would provide an overall perspective on social well-being. The identification of goal areas implied value judgements about what is important to human well-being but as Smith states (1973, p. 72), 'value judgements are involved in any selection of criteria of social well-being'.

The list of social concerns acted as a preliminary to the specification of a set of social indicators, which was designed to reveal the level-of-living for each social concern and so monitor changes in these levels over time (OECD 1976). Actual social indicators developed were to be viewed as a step towards the measurement of

Table 9.1 *General design of the United Nations level-of-living index*

	Component		Indicator
1	Nutrition	a	Calorie intake per head
		b	Protein intake per head
		c	Proportion of calorie intake derived from cereals, roots, tubers and sugars
2	Shelter	a	Quality of habitation
		b	Density of occupancy
		c	Independence of occupancy
3	Health	a	Access to medical care
		b	Mortality—due to parasitic and infectious diseases
		c	Proportional mortality ratio
4	Education	a	School enrolment ratio
		b	School output ratio
		c	Pupil–teacher ratio
5	Leisure and Recreation	a	Average leisure time
		b	Daily newspaper circulation
		c	Incidence of radio and television sets
6	Security	a	Incidence of violent deaths
		b	Proportion of population covered by unemployment and sickness benefits
		c	Proportion of population covered by retirement schemes
7	Surplus Income	a	Surplus to satisfaction of basic physical and cultural needs

Source: Knox (1975, p. 26)

social well-being. The OECD also recognized the need to include information on how people actually feel about the quality of their lives; however, the collection of data on such 'subjective' social indicators is still in an early stage of development.

Partly as a result of the efforts of the OECD, several west European countries have issued reports based on available data sources relating to social well-being. The United Kingdom began the trend in 1970 with regular publication of *Social Trends*, a package of statistical information relating to social conditions. This was followed by reports in such countries as France (1973), West Germany (1973), the Netherlands (1974), Italy (1975), Sweden (1975), and Spain (1975). As problems associated with the comparability of the data are sorted out, international comparisons of such information will help to shed light on how alternative schemes and institutional frameworks can effectively contribute to improving levels of social well-being.

9.2 Variations in social well-being in western Europe

Western Europe is characterized by spatial disparities in social well-being. Attention in this section is focused on the intensity of these disparities, at three different levels of analysis: national, regional, and urban. The means of measuring and portraying the patterns are relatively crude and approximate, but in the early stages of any such investigation the identification of disparities is more important than methodological perfection (Coates *et al.* 1977).

(i) *National variations in social well-being.* In Chapter 1 it was shown how an analysis of certain individual indicators of social well-being revealed distinct inequalities between the countries of western Europe. What is required, however, is some overall measure of social well-being and an index, based on the methodology of Knox (1974a), was developed by Ilbery (1984). This study represented a preliminary attempt to assess the extent of broad spatial variations in social well-being in Europe generally and it should be stressed that it was exploratory in nature and not intended as a definite statement of well-being. The analysis was restricted by the lack of suitable and comparable data at the national level, and certain aspects of social well-being were necessarily excluded.

Table 9.2 *List of social concerns common to most member countries*

A *Health*
1 The probability of a healthy life through all stages of the life cycle.
2 The impact of health impairments on individuals.

B *Individual development through learning*
1 The acquisition by children of the basic knowledge, skills, and values necessary for their individual development and their successful functioning as citizens in their society.
2 The availability of opportunities for continuing self-development and the propensity of individuals to use them.
3 The maintenance and development by individuals of the knowledge, skills, and flexibility required to fulfil their economic potential and to enable them to integrate themselves in the economic process if they wish to do so.
4 The individual's satisfaction with the process of individual development through learning, while he is in the process.
5 The maintenance and development of the cultural heritage relative to its positive contribution to the well-being of the members of various social groups.

C *Employment and quality of working life*
1 The availability of gainful employment for those who desire it.
2 The quality of working life.
3 Individual satisfaction with the experience of working life.

D *Time and Leisure*
1 The availability of effective choices for the use of time.

E *Command over goods and services*
1 The personal command over goods and services.
2 The number of individuals experiencing material deprivation.
3 The extent of equity in the distribution of command over goods and services.
4 The quality, range of choice, and accessibility of private and public goods and services.
5 The protection of individuals and families against economic hazards.

F *Physical environment*
1 Housing conditions.
2 Population exposure to harmful and/or unpleasant pollutants.
3 The benefit derived by the population from the use and management of the environment.

G *Personal safety and the administration of justice*
1 Violence, victimization, and harassment suffered by individuals.
2 Fairness and humanity of the administration of justice.
3 The extent of confidence in the administration of justice.

H *Social opportunity and participation*
1 The degree of social inequality.
2 The extent of opportunity for participation in community life, institutions, and decision-making.

Source: OECD (1973, pp. 14–17)

A total of 27 measurable variables were selected for this analysis (Table 9.3). The unweighted variables were representative of seven major constituents of social well-being: health, education, housing, material well-being, economic growth, demographic structure, and leisure and recreation. Therefore, the data matrix to be analysed consisted of 27 variables for each of the 26 European countries, excluding Iceland but including Turkey. The data were synthesized in two successive stages.

In the first stage, correlation analysis was used to examine the degree of relationship between the original 27 variables. Many significant correlations were produced and as with Knox's study (1974a), closer examination of the correlation matrix revealed a number of important intercorrelations amongst the variables. This suggested that some general spatial patterns associated with social well-being existed. These patterns were resolved using the technique of principal components analysis, the second stage in the synthesis of the data. Principal components analysis has been discussed by various people (see Child 1970), and it suffices to say that the technique produces a new and smaller set of variables (known as components) which are derived from the original variables and which account for a large proportion of

Table 9.3 *Social well-being in Europe: list of variables*

1	Number of persons per square kilometre, 1979.
2	Percentage of population 0–15 years, 1979.
3	Percentage of population 15–64 years, 1979.
4	Percentage of population 65 and over, 1979.
5	Number of births per 1000 inhabitants, 1979.
6	Infant mortality per 1000 live births, 1979.
7	Total fertility rate, 1979.
8	Percentage of households with five or more persons, 1979.
9	Percentage of total population classed as urban, 1980.
10	Number of doctors per 10 000 inhabitants, 1976
11	Number of people per hospital bed, 1970–7.
12	Number of phamarcists per 10 000 inhabitants, 1975–7.
13	Number of people per physician, 1977.
14	Gross national product per person, 1977, ($US).
15	Percentage of gross national product spent on education, 1978.
16	Number of persons per room, 1976.
17	Percentage of public expenditure spent on education, 1978.
18	Percentage of population economically active, 1980.
19	Percentage of working population employed in agriculture, 1980.
20	Percentage of working population employed in services, 1979.
21	Consumption of steel per inhabitant, 1979 (kg).
22	Consumption of energy per inhabitant, 1979 (kg coal equiv.).
23	Number of cars per 1000 inhabitants, 1979.
24	Number of telephone owners per 1000 inhabitants, 1979.
25	Number of television receivers per 1000 inhabitants, 1979.
26	Consumption of newsprint per inhabitant, 1977 (kg).
27	Cinemas per 100 000 inhabitants, 1979.

Data Sources:
Eurostat Basic Statistics of the Community, 1980/1.
European Marketing Data and Statistics, 1980/1.
Comecon Data 1981. Vienna Institute of Comparative Economic Studies.
United Nations Statistical Yearbook, various years.
World Development Report, 1982. The World Bank.

Source: Ilbery (1984, p. 292)

the variation in the original data. The components are identified by their *loadings*, or the correlation coefficient between the new variable and the original one, and are interpreted as measuring the major dimensions of social well-being extracted from the data.

The analysis produced four significant and unrelated components, the relative importance of each being determined by the 'percentage of total variance' column in Table 9.4. The total variance is equal to the number of variables used in the analysis (27) and as component one accounts for the equivalent of 14.7 of the 27 variables, it extracts: $\frac{14.7}{27} \times 100 = 54.43$ per cent of the total variance. This makes it by far the most important component.

Component one is identifiable with certain socio-geographic processes and was labelled a general socio-economic well-being component, with high loadings on six of the seven constituents of social well-being (Table 9.5), the one exception being education. The main characteristic of component one was that it contained contrasting

Table 9.4 *Relative importance of the first four components*

Component	Component variance	% of total variance	Cumulative % of total variance
1	14.69	54.43	54.43
2	4.03	14.92	69.35
3	1.76	6.50	75.85
4	1.34	4.95	80.80

Source: Ilbery (1984, p. 294)

Table 9.5 *The important loadings for component one*

Variables	Loadings
Percentage of population aged 0–15 years	0.93
Percentage of population aged 15–64 years	−0.65
Percentage of population aged 65 years and over	−0.90
Number of births/1000 inhabitants	0.94
Infant mortality/1000 live births	0.93
Total fertility rate	0.92
Number of persons/room	0.91
Percentage of households with 5 or more persons	0.91
Percentage of total population classed as urban	−0.71
Number of doctors/10 000 inhabitants	−0.62
Number of people/hospital bed	0.79
Gross national product/person	−0.86
Number of people/physician	0.71
Percentage of population economically active	−0.51
Percentage of working population in agriculture	0.96
Percentage of working population in services	−0.80
Consumption of steel/inhabitant	−0.57
Consumption of energy/inhabitant	−0.67
Number of cars/1000 inhabitants	−0.82
Number of telephone owners/1000 inhabitants	−0.80
Number of television receivers/1000 inhabitants	−0.91
Consumption of newsprint/inhabitant	−0.74

Source: Ilbery (1984, p. 294)

groups of variables. This is known as a bi-polar component and can be described in terms of a continuum of well-being; variables with high positive loadings represent one pole of the continuum (in this case lower levels of social well-being), and variables with high negative loadings represent the other pole (higher levels of social well-being).

Therefore, component one suggested that major inequalities exist in social well-being in Europe and in order to assess the spatial importance of this new variable (component) its distribution was mapped using the component *scores* (Fig. 9.1). A component score for any one country, on the first component, is obtained by multiplying the standard score for each original variable in that country by the appropriate loading for the first component; fortunately the scores are produced as part of the standard principal components package.

In Fig. 9.1, high negative scores indicate a 'better' state of social well-being, whereas high positive scores indicate a 'worse' state. The most prominent feature of the map is the concentration of high negative values in the 'core' countries of north-western Europe, reaching a peak in Sweden. As with many other aspects of the human geography of western Europe covered in this book, a core–periphery relationship is discernible, with the European periphery exhibiting the lowest levels of well-being and comprising the countries of Ireland, Portugal, Spain, Greece, and Turkey, together with all east European countries except East Germany and Czechoslovakia. In terms of component one, the countries experiencing the extremes of social well-being are Sweden (highest) and Turkey (lowest).

A similar exercise could have been conducted for the remaining significant components, but these were far less meaningful than component one and accounted for a very small percentage of the total variance (Table 9.4). Despite the comprehensive nature of component one, all significant components are concerned with aspects of social well-being in Europe. Therefore, what is required is some overall measure, or index, of social well-being and this was achieved by following the methodology developed by Knox (1974a). In the first instance, use was made of the component structure to select a small number of 'diagnostic' variables from the set of original variables. The choice of diagnostic variables is arbitrary, but Knox stated (p. 13) that each should:

1. be highly associated with at least one of the components (a loading of ±0.60 or more), and have a large proportion (75 per cent or more) of its variance involved in the component structure;
2. be as normative as possible, where a change along a quantified scale can be recognized as either 'good' or 'bad'; and
3. in combination, reflect the character and composition of the significant components.

On the basis of these criteria, six diagnostic variables were selected:

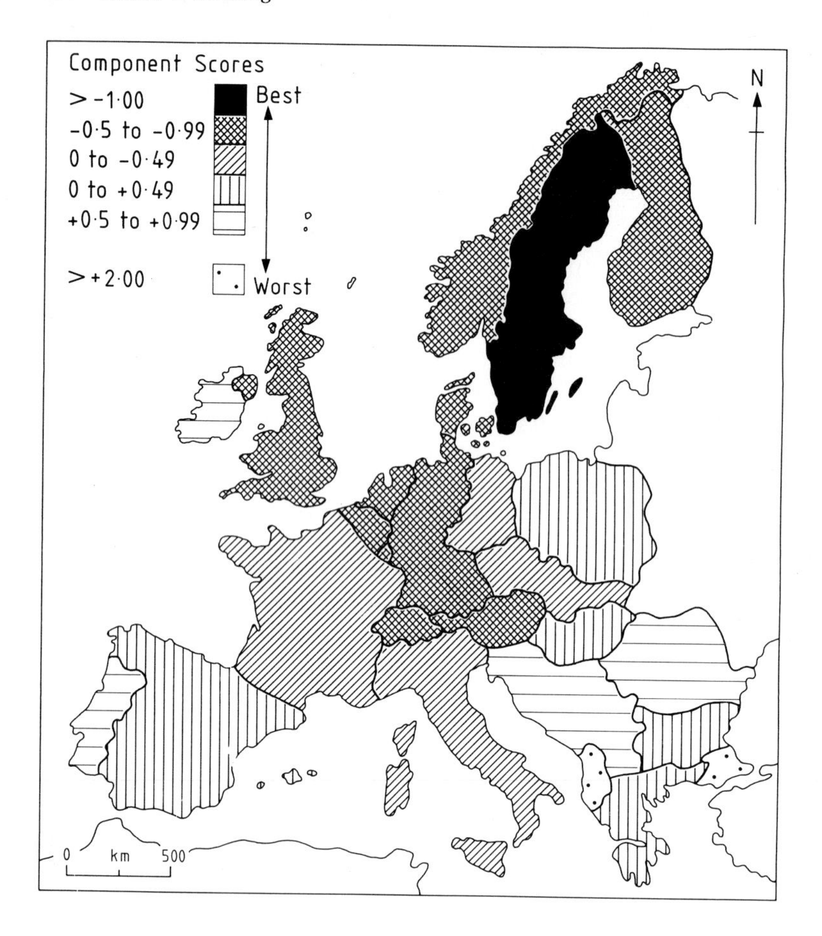

Fig. 9.1 Component one: general socio-economic well-being in Europe. (*Source:* Ilbery 1984, p. 295)

1. percentage of households with five or more persons.
2. percentage of population 0–15 years.
3. number of births per 1000 inhabitants.
4. infant mortality per 1000 live births.
5. percentage of working population employed in agriculture.
6. number of television receivers per 1000 inhabitants.

Once the variables had been selected, an index of social well-being was obtained by substituting the appropriate values into Knox's formula:

$$I_j = 100. \sum \left(\frac{R_j}{N.C.} \right)$$

where I_j = social well-being index of country j,
R_j = sum of individual rank-scores, for country j,
N = number of diagnostic variables (6),
C = number of cases (26).

For each of the six diagnostic variables, the countries were placed in rank-score order, with lower ranks regarded as 'bad' in terms of social well-being. The results of this ordering and the final social well-being index for each country are presented in Table 9.6. High index values are indicative of low levels of social well-being and results ranged from a high of 98.72 in Turkey to a low (best) of 10.90 in Sweden. The spatial distribution of index values is shown in Fig. 9.2 and a distinct pattern of inequality emerges. Once again a core area is in evidence, although smaller in extent than the one derived from the analysis of component one. Sweden and West Germany have the highest levels of social well-being in Europe and together with Belgium, Switzerland, Denmark and the United Kingdom form a heartland area. Luxembourg, Austria, Norway, Finland and the Netherlands are relegated to the semi-periphery, where they are joined by France, Italy,

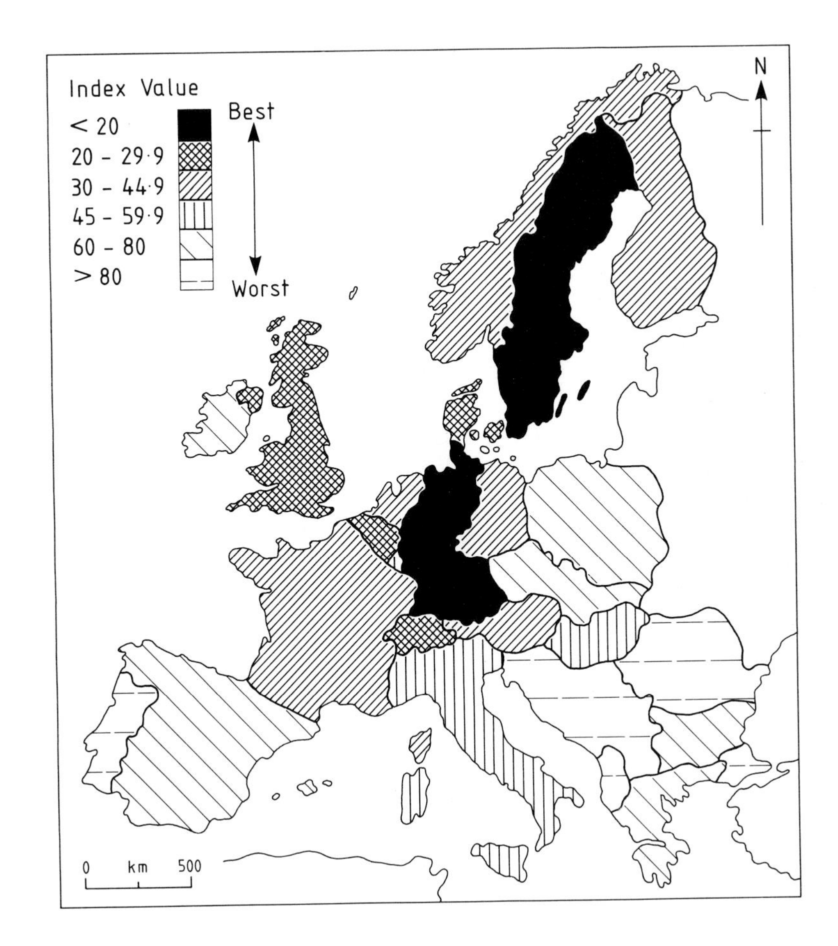

Fig. 9.2 An index of social well-being in Europe. (*Source:* Ilbery 1984, p. 296)

East Germany and Hungary. Index values continue to rise quite rapidly with increasing distance from the core area and seven countries (Ireland, Greece, Portugal, Turkey, Yugoslavia, Albania, and Romania) have values of over 70 and are undoubtedly part of the European periphery.

Unfortunately, the index is dependent on the rather subjective choice of diagnostic variables and could conceal important differences in the national characteristics of social well-being. For example, Norway and Luxembourg have similar index values, but a comparison of the rank-scores on the six diagnostic variables (Table 9.6), shows significant differences. Consequently, the two countries do not necessarily share the same social well-being traits and may belong to different groups. To overcome this difficulty, Ilbery (1984) used cluster analysis to classify the 26 countries of Europe on the basis of the four dimensions of social well-being derived from the principal components analysis. The classification produced eight groups or 'types' of countries, with two groups each containing just one country, and is indicative of the marked spatial variations in social well-being which exist in Europe. Noticeable differentiating characteristics were recognised and in preference to examining each group in detail, the results have been succinctly summarized in Table 9.7. The distribution of the eight groups is portrayed in Fig. 9.3, and a strong core-periphery contrast still exists, although the groupings are different from those produced for the index of social well-being.

Although relatively straightforward, the methodology adopted in this section suffers from certain problems, which should be borne in mind when analysing the patterns of social well-being. In the first instance, the representativeness and relative importance of the selected variables should be considered; the set of indicators chosen

Table 9.6 *Diagnostic variables and the social well-being index*

Country	Diagnostic variables												Social well-being index
	2	Rank	5	Rank	6	Rank	9	Rank	19	Rank	25	Rank	
West Germany	19.2	2	9.5	1	13.5	13	9.2	2	6.0	5	337	5	17.95
France	22.7	13	14.1	14	10.1	7	15.5	10	8.8	10	292	10	41.03
Italy	22.8	15	11.8	7	15.3	16	21.5	19	14.2	14	231	17	56.41
Netherlands	23.2	17	12.5	8	8.7	5	20.4	18	4.6	3	292	8.5	38.14
Belgium	20.7	6	12.6	9	11.1	8	16.1	12	3.0	2	293	8.5	29.17
Luxembourg	18.5	1	11.2	2	13.0	11.5	17.6	14	6.3	6	245	16	32.37
United Kingdom	21.8	10	13.1	11	12.8	10	14.5	8	2.6	1	394	1	26.28
Ireland	30.6	24	21.5	24	12.4	9	35.3	24	19.2	17	233	18	74.36
Denmark	21.6	9	11.6	5.5	8.8	6	11.6	5	7.0	7	358	3	22.76
Greece	23.3	18	15.7	17	18.7	17	22.1	20	30.3	22	147	23	75.00
Spain	26.5	21	16.1	18	15.1	15	24.6	21	18.8	16	253	14	67.31
Portugal	27.9	23	16.3	19	38.9	23	27.3	22	28.6	21	122	24	84.62
Turkey	39.9	26	34.9	26	80.0	26	63.3	26	60.7	25	58	25	98.72
Norway	22.8	15	12.8	10	8.6	3.5	10.8	3.5	8.5	9	288	11	33.33
Sweden	19.7	3.5	11.6	5.5	7.3	1	7.1	1	5.6	4	374	2	10.90
Switzerland	19.7	3.5	11.3	3	8.6	3.5	15.7	11	7.4	8	312	7	23.08
Austria	21.1	8	11.5	4	14.8	14	14.6	9	10.7	11.5	282	12	37.50
Finland	20.8	7	13.3	12	7.6	2	18.0	15	11.8	13	316	6	35.26
Yugoslavia	26.6	22	17.8	20.5	40.4	24	30.5	23	48.5	24	189	21	86.22
Bulgaria	22.2	11	15.3	16	22.0	19.5	19.6	17	24.8	19	196	20	65.71
Czechoslovakia	22.8	15	17.8	20.5	19.0	18	16.9	13	14.6	15	256	13	60.58
East Germany	22.4	12	14.0	13	13.0	11.5	10.8	3.5	10.7	11.5	342	4	35.58
Hungary	20.2	5	15.0	15	24.0	21	14.0	6	22.1	18	249	15	51.28
Poland	24.1	19	19.5	23	22.0	19.5	14.4	7	27.1	20	216	19	68.91
Romania	25.1	20	18.6	22	31.0	22	19.5	16	31.0	23	163	22	80.13
Albania	36.3	25	30.0	25	70.0	25	52.4	25	61.0	26	19	26	97.44

Diagnostic variables: 2 Percentage of population 0–15 years
5 Number of births/1000 inhabitants
6 Infant mortality/1000 live births
9 Percentage of households with 5 or more persons
19 Percentage of working population employed in agriculture
25 Number of television receivers/1000 inhabitants

Source: Ilbery (1984, p. 297)

needs to be broad enough to include all the major life concerns of the population. Secondly, many of the selected variables will be highly correlated and to weight them equally could be misleading (Gehrmann 1978); the use of principal components analysis helps to solve the weighting problem by replacing the original variables with a small number of orthogonal components. Thirdly, there is the assumption that the aggregation of a series of measures of aspects of social well-being will produce a meaningful summary statistic. Finally, there is the common problem of the ecological fallacy, whereby resultant correlations do not necessarily reflect the concerns of all individuals within the areal unit selected for study.

In relation to this last point, an interesting comparison with the results in this section, based on objective social indicators, is provided by the Eurobarometer surveys of the EC countries. These are public opinion polls conducted on behalf of the Commission of the European Communities. To obtain a subjective assessment of the people's feelings of 'well-being' or 'satisfaction', 9689 homes were surveyed in the ten EC countries (Eurobarometer 1982). Feelings of satisfaction were shown to relate to income and political ideologies and three groups of countries were deemed to exist. Denmark and the Netherlands were well clear of other member countries and were most satisfied. Ireland, Luxembourg, Belgium, West Germany, and the United Kingdom

Table 9.7 *A typology of social well-being characteristics in Europe*

Group	Characteristics
A West Germany, Belgium United Kingdom, and Netherlands	(i) Most densely populated countries in Europe, highly urbanized but with low number of persons per room. (ii) Lowest employment in agriculture in Europe, but above-average employment in services. (iii) Above-average GNP, except the UK, with some of the highest proportions of GNP spent on education. (iv) Good scores on material well-being, but poor cinema provision. (v) Favourable demographic characteristics, with below-average number of young people, births, infant mortality, and fertility. (vi) Mixed standard of health-care, but rather poor provision of hospital beds, physicians and pharmacists.
B France, Luxembourg, Denmark, Switzerland, and Austria	(i) Good demographic structure, with below-average number of young and old people. (ii) Mixed rates of urbanization, little overcrowding, and low agricultural employment. (iii) Amongst highest GNP in Europe. (iv) Mixed levels of health-care, education, and energy and steel consumption. (v) Above-average number of cars and telephones, but mixed on televisions and newsprint consumption.
C Italy, Greece, and Spain	(i) Average demographic conditions, although high fertility and below-average infant mortality. (ii) Lower levels of urbanization than A or B, although above-average (except Greece) tendency for overcrowding. (iii) Poor on economically active population, but higher agricultural employment (14–30 per cent) than in A or B. (iv) Below-average GNP, leading to less spent on education provision, and steel and energy consumption. Above-average cinema facilities. (v) Mixed health-care, good on doctors, but below-average on hospital beds and physicians.
D Ireland, Portugal, and Albania	(i) Poor demographic structure, with high birth rates, fertility and infant mortality (except Ireland), leading to some of the highest concentrations of young people. (ii) Below-average levels of urbanization, but some of the worst overcrowding in Europe. (iii) Medically good on pharmacists, but poor on other aspects of health-care. (iv) Below-average GNP, leading to poor steel and energy consumption and to low rates of car, television and telephone ownership. Yet, above-average expenditure on education.
E Norway and Finland	(i) Favourable demographic conditions, with lowest infant mortality (after Sweden). Good on births and number of young people, but mixed on number of old people. (ii) Lowest population density in Europe, leading to low urbanization rates; high economically active rates and below-average employment in agriculture. (iii) Very good health-care: excellent on pharmacists and provision of beds, average on doctors and poor on physicians. (iv) High GNP, with above-average proportions spent on education. Good material well-being, but poor cinema provision.
F Sweden	(i) Number one ranked country on eight of the variables, covering many constituents of well-being. Second or third ranked on a further six variables. (ii) Good on material well-being and leisure, with above-average cinema provision. (iii) Weakest area is in health-care: very poor on pharmacists and only average on doctors and physicians. (iv) Lowest population density, after Norway andFinland.

Table 9.7 (continued)

Group	Characteristics
G Yugoslavia, Bulgaria, Hungary, Czechoslovakia, East Germany, Poland, and Romania	(i) Fairly weak demographic structure, especially in terms of fertility and infant mortality, but above-average working population. (ii) Mixed levels of urbanization, with above-average number of persons per room but little overcrowding (households with 5 or more persons). (iii) High economically-active rates, with above-average agricultural employment (except East Germany and Czechoslovakia) and below-average service employment. (iv) Very mixed on material well-being, with some of highest and lowest steel and energy consumption figures. Poor provision of cars, telephones and newsprint. (v) Mixed health and education, with below-average GNP.
H Turkey	(i) Lowest ranked country on half of original variables. (ii) Ranked 21st or worst on most of remaining variables, thus very serious social well-being problems. (iii) Best performance comes on education, with above-average percentage of public expenditure devoted to it. Ranked 15th (below-average) on percentage GNP spent on education.

Source: Ilbery (1984, pp. 298–9)

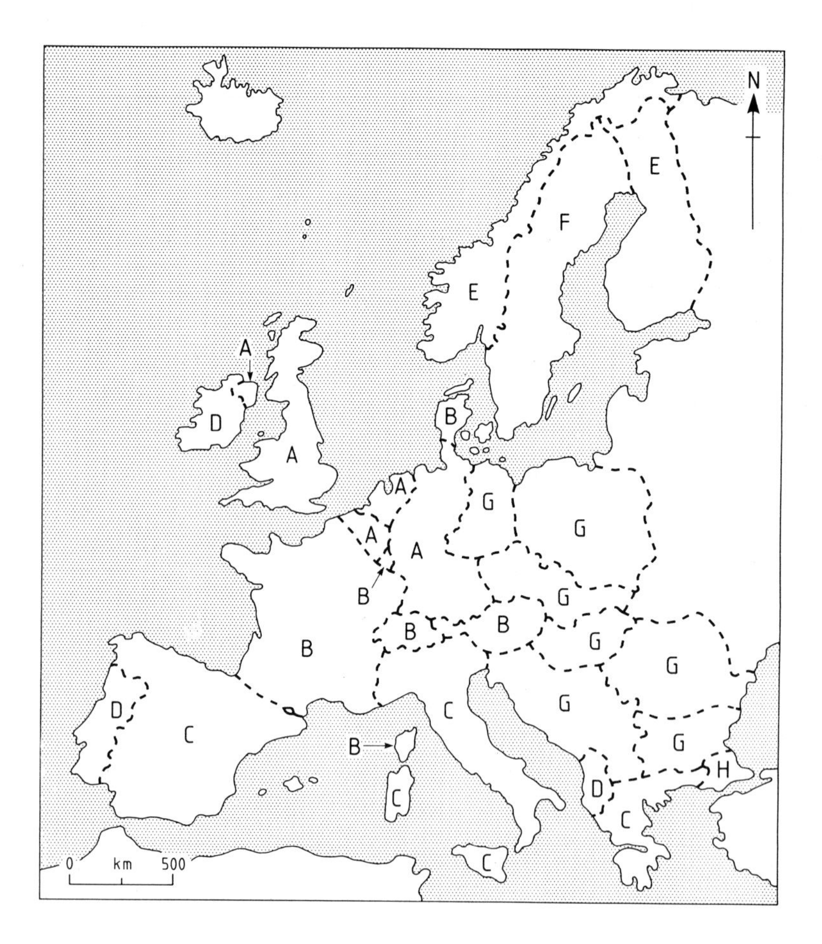

Fig. 9.3 Social well-being in Europe: A classification of countries according to 27 socio-economic characteristics. (*Source:* Ilbery 1984, p. 298)

formed group two and also had above-average Community scores for levels of satisfaction. However, France, Italy, and Greece formed a group of less-satisfied countries. Therefore, there exists a reasonable correlation between objective and subjective assessments of well-being, with two notable exceptions: first, Ireland, which is much more satisfied than objective measures of well-being would suggest; and secondly, France, which is less satisfied.

(ii) *Regional variations in social well-being.* Just as variations in social well-being have been shown to exist at the national level, European studies have highlighted their existence at the regional level. However, once again it is necessary to emphasize the lack of comparability in national statistical series, which has inhibited cross-national studies on the regional scale (Knox 1984). One of the earliest studies of regional disparities in such conditions contributing to social well-being as affluence, health, education, and opportunity of employment was that by Coates and Rawstron (1971) in Britain. In terms of income levels, for example, an island of prosperity, stretching from Essex to Worcestershire and from Sussex to Leicestershire, was found to exist; income declined in a regular manner away from this island, with the peripheral regions of northern Scotland having incomes far below the national average. A later study by Knox (1974a) analysed the variations in level-of-living between all the counties and county boroughs of England and Wales in 1961, on the basis of 53 unweighted variables. Unlike Coates and Rawstron, Knox derived a single 'level-of-living index' for the 145 counties and county boroughs and the result was a marked regional pattern of inequality.

Studies have also been conducted in other west European countries, including France (Knox and Scarth 1977), Northern Ireland (Goodyear and Eastwood 1978) and Switzerland (Walter-Busch 1983). The former has shown considerable concern over the social well-being of its inhabitants and this led to the appointment, by President Giscard d'Estaing, of the first Minister for the Quality of Life in 1974. Extensive surveys have demonstrated the severity of regional variations in the 'quality of life' in France (Le Point 1974; Knox and Scarth 1977), and one of the main objectives of the newly appointed Minister was to reduce these great regional disparities. The first major survey (Le Point 1974), analysed existing disparities between the 95 *départements* of France on the basis of 48 variables, each relating to one of six basic components of the quality of life: health, environment, social stability, material well-being, economic growth, and culture and recreation. The 48 variables were weighted according to their relative importance and an overall index was obtained for each *département* by combining the scores for these variables.

The pattern of regional inequality is shown in Fig. 9.4 and, following Coates *et al*, (1977), three main zones, running from west to east across the country, can be differentiated:

1. *The poorest zone*, stretching from Brittany to Lorraine and including the lower Loire valley, Normandy, the industrial area in the north, Lorraine, and the suburbs to the east of Paris. This zone contains industrial and rural areas of social backwardness and malaise. The industrial areas are characterized by economic difficulties and inadequate and crowded housing conditions, whilst the rural areas are slow to adjust to modern technology and change.
2. *The richest zone*, in sight of the Alps and the Pyrenees and stretching from the *Landes* of Aquitaine, through the Mediterranean coastlands to Lyon and the Alps of Savoy. Paris and its outer eastern suburb of Yvelines form an outlier and the only area in the north which enjoys excellent cultural, medical, and economic facilities. The favoured areas in the south benefit from the outstanding recreational facilities of the Alps and new industries based on modern hydro-electric power resources.
3. *The intermediate zone*, between the northern and southern zones and embracing the Massif Central, Burgundy, and the eastern part of the Paris Basin.

In a similar survey, Knox and Scarth (1977) found an almost identical zonal pattern of inequality, based this time on 41 unweighted socio-economic variables. However, they felt that an aggregate index could possibly 'mask important differences in the regional characteristics in the quality of life' (p. 11). In an attempt to highlight these differences, the technique of cluster analysis was used to classify the *départements* on the basis of the various 'indicators' of the

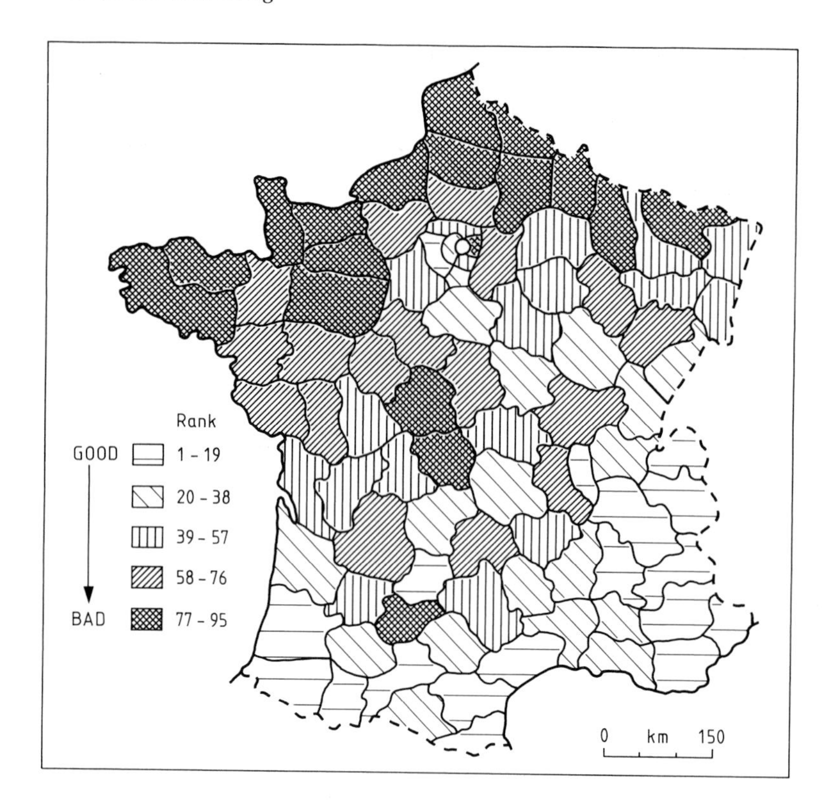

Fig. 9.4 Social well-being in France, 1970. (*Source:* Coates, Johnston and Knox 1977, p. 69)

quality of life. Nine groups of *départements* emerged, each with a strong regional component and noticeable differentiating characteristics. The characteristics of the nine groups of *départements* (A-I) are briefly summarized in Table 9.8 and their spatial expression portrayed in Fig. 9.5.

(iii) *Urban variations in social well-being.* There have been relatively few attempts to analyse urban variations in social well-being within the countries of western Europe (Gehrmann 1978); this is particularly true at the intra-urban level. However, marked spatial variations in social well-being exist at both the intra- and inter-urban levels, a fact which has been verified by certain studies in the United Kingdom.

At the inter-urban level, a pioneer study, conducted by Moser and Scott (1961) and based on 57 different socio-economic variables, illustrated how patterns of inequality existed in many aspects of social well-being in the 157 towns and cities of England and Wales with a population of over 50 000. A more recent study, by Holtermann (1975), highlighted dramatic spatial variations in deprivation between the 87 518 enumeration districts (EDs) which comprised the urban administrative districts of Britain in 1971. Holtermann initially analysed the worst 5 per cent of EDs in Great Britain on the basis of 18 separate indicators of deprivation, which could be grouped into components such as housing, employment, and socio-economic structure. Results indicated that 'Clydeside consistently has a very much more than proportionate share of the worst 5 per cent of EDs on nearly all kinds of deprivation, and so to a lesser extent does the rest of Scotland' (p. 37). In addition, Inner London has more than its fair share of most types of housing deprivation.

In an attempt to measure 'multiple deprivation', Holtermann identified those EDs which fell in the worst 15 per cent of all EDs on *each* of three criteria:

1. the level of male unemployment;
2. the percentage of households which lacked exclusive use of basic amenities such as hot water, a fixed bath, and an inside toilet; and
3. the percentage of households classed as overcrowded.

Table 9.8 *Socio-economic characteristics of the nine groups of départements in France*

Type		Characteristics
A	Western France and the fringes of the Massif Central. (25 *départements*)	(i) a generally rural environment, with low density of population and rural depopulation (ii) houses are small, overcrowded, and often lacking in modern facilities (iii) salaries are low and below the national average (iv) poor educational provision and achievement (v) alcoholism is a noticeable problem
B	Northeastern France (15 *départements*)	(i) high rates of infant mortality and low life expectancy (ii) low levels of educational achievement (iii) high incidence of tuberculosis, but fewest hospital beds in France (iv) very high unemployment, due mainly to the decline in coal-mining, textiles, and iron and steel in the industrial towns (v) slightly above-average rates of car-ownership and provision of cinemas
C	The Paris Basin (12 *départements*)	(i) high incidence of social pathologies—suicide, divorce, delinquency, theft (ii) below-average rates of unemployment (iii) above-average housing conditions, car-ownership, taxable incomes, and numbers of telephones and televisions (iv) good communications, large and highly mechanised farms, and relatively prosperous industrial towns
D	The metropolitan suburbs of Paris (6 *départements*)	(i) very high levels of educational achievement, incomes, and in-migration (ii) good housing amenities and high rates of telephone ownership (iii) poor scores on provision of hospital beds, road accidents, divorce, crime, and overcrowding in the house and in the classroom (iv) a most varied area
E	Mediterranean France (7 *départements*)	(i) best group in terms of quality of life (ii) high life expectancy, high levels of health-care provision, and in-migration (iii) high levels of telephone ownership, and well-appointed homes (iv) low rates of suicide and mortality from alcoholism (v) high levels of unemployment, especially in rural areas, and high divorce rates
F	Southern France (I) (16 *départements*)	(i) low levels of car ownership and job availability (ii) good scores on education, housing conditions, and mortality from alcoholism (iii) all *départements* experienced population increase between 1968 and 1975 (iv) close proximity to recognizable urban centres or growth poles
G	Southern France (II) (12 *départements*)	(i) below-average scores on educational attainment, salaries, and television ownership (ii) above-average scores on social pathologies such as alcoholism, divorce, suicide, and delinquency (iii) eight of the 12 *départements* are losing population (iv) small market towns rather than large urban centres
H	Corsica	(i) extremely low salaries, and low levels of car and television ownership (ii) low levels of educational attainment, and high rates of infant mortality (iii) high rates of crime and delinquency, but low rates of suicide and alcoholism (iv) small farm units and unfavourable terrain
I	Paris	(i) high provision of health-care facilities, research centres, hotels and restaurants, cinemas, libraries, roads, and newspapers (ii) high educational attainment, and life expectancy (iii) high rates of divorce, crime, and car accidents (iv) well above-average salaries (v) congestion and overcrowding

Source: Adapted from Knox and Scarth (1977)

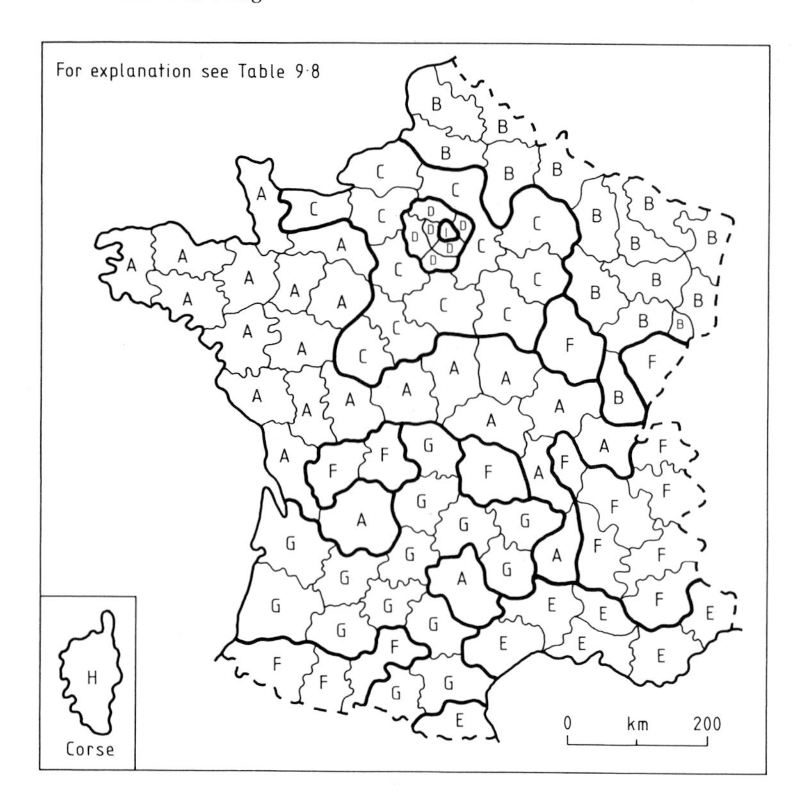

Fig. 9.5 Quality of life in France, 1973: A classification of *départements* according to 41 socio-economic characteristics. (*Source:* Knox and Scarth 1977, p. 12)

The geographical distribution of EDs with more than one kind of deprivation is once again dominated by the major conurbations, especially Clydeside and Inner London. Glasgow, with 578 deprived EDs, suffered more than the London boroughs or the third worst city, Birmingham, which had 170 deprived EDs.

Coates *et al.* (1977) used the results of Holtermann's study to calculate quotients, which enabled a comparison to be made of the proportional share of multiple deprivation experienced in the regions and conurbations of Britain. The results are listed in Table 9.9 and a location quotient of one is obtained if the percentage of deprived EDs in any one area is the same as the national percentage, whereas a value of over one indicates a greater than national share. Location quotients ranged from 0.13 in the relatively non-deprived Outer London boroughs, through 1.76 for the West Midlands conurbations and 2.1 for the Inner London boroughs, to a high of 6.19 for the Clydeside conurbation. Indeed, Scotland had far more than its fair share of multiple deprivation, with a location quotient of 3.21.

Table 9.9 *Localization of 'multiple deprivation' within the regions and conurbations of Britain*

Regions	Location Quotient
North	1.11
Yorks and Humberside	0.74
North-West	0.81
East Midlands	0.84
West Midlands	1.10
East Anglia	0.20
South-East	0.66
South-West	0.25
Wales	0.28
Conurbations	
Tyneside	1.60
West Yorkshire	1.37
Merseyside	1.07
South-East Lancashire	1.15
West Midlands	1.76
Outer London Boroughs	0.13
Inner London Boroughs	2.10
Clydeside	6.19

Source: Coates, Johnston and Knox (1977, p. 76)

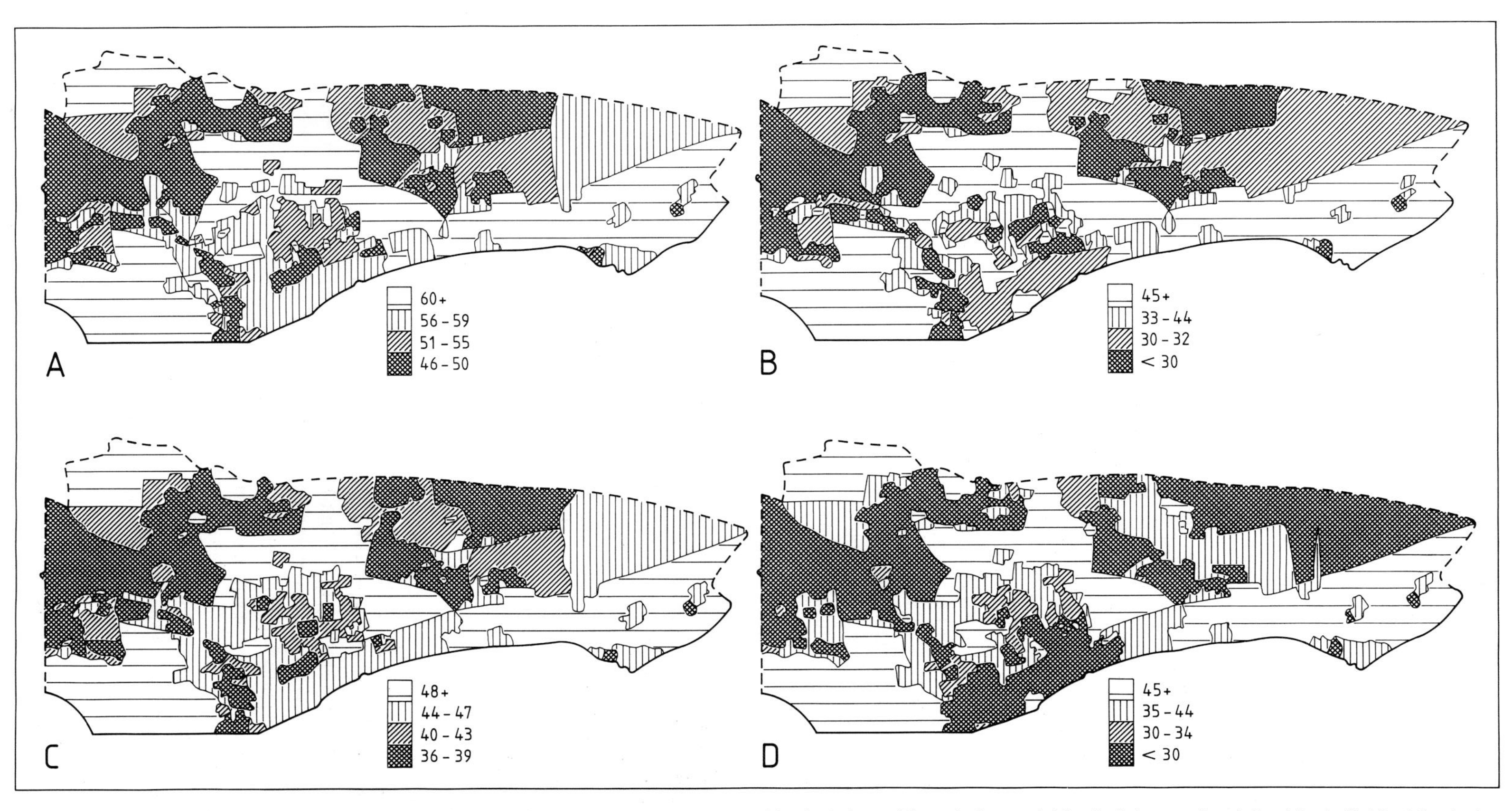

Fig. 9.6 Social well-being in Dundee, 1974: (A) an index of level of living based on 'objective' data; (B) an index on 'objective' data and weighted by individuals' priority preferences; (C) an index based on levels of satisfaction; (D) an index based on levels of satisfaction weighted by individuals' priority preferences. (*Source:* Coates, Johnston and Knox 1977, p. 77)

At the intra-urban level, a rather novel approach to the measurement of territorial variations in social well-being was that presented in MacLaren's (1975) study of Dundee. MacLaren, using questionnaire data rather than published data, identified quite marked spatial variations in well-being (Fig. 9.6). With the use of data on both objective indicators of well-being and peoples' feelings and preferences, MacLaren was able to devise four indices of social well-being in Dundee:

1. An index of level-of-living based on 'objective' data.
2. An index based on 'objective' data and weighted by individuals' priority preferences.
3. An index based on levels of satisfaction.
4. An index based on levels of satisfaction and weighted by individuals' priority preferences.

The territorial variations in social well-being displayed by these indices are shown in Fig. 9.6 (A-D) and, although noticeable differences between the four maps were discernible, there was a general consensus of agreement regarding areas of low and high levels of living. Dundee is not the most typical of urban areas in Western Europe, but MacLaren's work is important for two reasons: first, it combines both objective and subjective indicators of social well-being in one study; and secondly, it could form the basis for detailed studies of the larger European cities.

This preliminary survey has highlighted national inequalities in social well-being in Europe. The findings remain tentative and are necessarily influenced by the availability of comparable data and the selection of variables. National studies of social well-being naturally hide even greater regional and local variations, some of which have been outlined in the latter sections of this chapter, and these cannot be satisfactorily examined until data are collected at, and social indicators developed for, these more localized levels of analysis. Despite this Knox (1984) has shown that there is little evidence of 'convergence' of inequalities in social well-being in Europe; at best the 'status quo' is being maintained and at worst 'divergence' is to be expected as the core–periphery system permits the continued expansion of the prosperous regions at the expense of the less developed ones.

References

Child, D. (1970). *The essentials of factor analysis*. Holt, Rhinehart and Winston, London.

Coates, B. E. and Rawstron, E. M. (1971). *Regional variations in Britain: selected essays in economic and social geography*. Batsford, London.

Coates, B. E., Johnston, R. J., and Knox, P. L. (1977). *Geography and inequality*. Oxford University Press, Oxford.

Drewnowski, J. and Scott, W. (1968). 'The level-of-living index'. *Ekistics*, **25,** 226–75.

Eurobarometer. (1982). *Satisfaction with life and the feeling of happiness*. Commission of European Communities, Brussels.

Gehrmann, F. (1978). ' "Valid" empirical measurement of quality of life?' *Social Indicators Research*, **5,** 73–109.

Goodyear, R. M. and Eastwood, D. A. (1978). 'Spatial variations in levels of living in Northern Ireland'. *Irish Geography*, **11,** 54–67.

Holtermann, S. (1975). 'Areas of urban deprivation in Great Britain: an analysis of 1971 census data'. *Social Trends*, **6,** 33–47.

Ilbery, B. W. (1984). 'Core-periphery contrasts in European social well-being'. *Geography*, **69,** 289–302.

Kamrany, N. M. and Christakis, A. N. (1970). 'Social indicators in perspective'. *Socio-Economic Planning Sciences*, **14,** 207–16.

Knox, P. L. (1974a). 'Spatial variations in level of living in England and Wales'. *Transactions of the Institute of British Geographers*, **62,** 1–24.

Knox, P. L. (1974b). 'Social indicators and the concept of level of living'. *Sociological Review*, **22,** 249–57.

Knox, P. L. (1975). *Social well-being: a spatial perspective*. Clarendon Press, Oxford.

Knox, P. L. (1984). *The geography of western Europe: a socio-economic survey*. Croom Helm, London.

Knox, P. L. and Scarth, A. (1977). 'The quality of life in France'. *Geography*, **62,** 9–16.

Le Point. (1974). 'Où vit-on heureux en France'. *Le point*, **68,** 35–50.

MacLaren, A. C. (1975). *Spatial aspects of relative deprivation*. Mimeographed paper presented to the Urban Geography Study Group of the Institute of British Geographers, Reading.

Moser, C. A. and Scott, W. (1961). *British towns*. Oliver and Boyd, London.

OECD. (1973). *List of social concerns*. The OECD Social Indicator Development Programme, 1. Paris.

OECD. (1976). *Measuring social well-being*. The OECD Social Indicator Development Programme, 3. Paris.

Pacione, M. (1982). 'The use of objective and subjective measures of quality of life in human geography'. *Progress in Human Geography*, **6,** 495–514.

Smith, D. M. (1973). *The geography of social well-being in the United States*. McGraw-Hill, New York.

Smith, D. M. (1977). *Human geography: a welfare approach*, Edward Arnold, London.

United Nations. (1961). *International definition and measurement of levels of living: an interim guide*. UN, New York.

United Nations. (1966). *The level of living index*. Report No. 4, UN Research Institute for Social Development, Geneva.

Walter-Busch, E. (1983). 'Subjective and objective indicators of regional quality of life in Switzerland'. *Social Indicators Research*, **12,** 337–92.

10

Regional Policy

Regional policy is largely concerned with minimizing the spatial imbalances in economic development and prosperity between regions. The situation is not straightforward and is made more difficult by the fact that disequilibria between the regions of some countries are worse than others and areas needing special help are not strictly comparable; for example, the poorest areas in West Germany are as wealthy as the richest areas in Ireland.

Regional problems can be classified into two groups:

1. those arising from the fact that the increase of work and that of people wanting it are not in the same place; and
2. those arising because, although the work and the workers are in the same place, it is for some reason the wrong place.

In the first group, policy-makers have to decide whether it is better to move the work or the workers and in the second group, one has to decide what is the right spatial pattern for population and economic activity. The most difficult task in both groups is to devise ways of dealing with these problems. Approaches to regional policy vary greatly from country to country and are necessarily influenced by political as well as by economic and other considerations (Sant 1974; Hayward and Watson 1975; Clout 1976 and 1981; Hall 1976; Gillingwater and Hart 1978; Yuill *et al.* 1980).

The aim of this chapter is to examine the various policies which have been employed to reduce regional disparities and problems. In the first instance, the different solutions to regional problems in western Europe will be discussed. The scale of analysis is then reduced and attention is focused, first, on regional policy in the European Community and, secondly, on one problem region in particular, the Mezzogiorno of southern Italy.

10.1 Differing approaches to regional policy in western Europe

A number of regional problems exist in western Europe and evidence of regional disparities has already been given in earlier chapters. Marked spatial variations have been shown to exist in such indicators as incomes (Fig. 1.3), population densities and migration losses (Figs. 2.1 and 2.8) and levels of social well-being (Fig. 9.2). One further indicator of variations in economic opportunities between regions is unemployment. Allowing for the ambiguities and imperfections of international data on unemployment rates, Fig. 10.1 clearly shows a pronounced pattern of national and regional inequality. Unemployment in the EC rose from 2.5 per cent in 1973 to 11.8 per cent in 1985, when over 13 million people were out of work. Figures range from a low of 1.9 per cent in Luxembourg to 18.1 per cent in Ireland, and the regional dimension has been exacerbated, with marked differences between the ten strongest (less than 5 per cent) and ten weakest regions (greater than 20 per cent). Indeed, it is not only the fringes of western Europe which are experiencing increasing rates of unemployment, as the problem is also acute in the more central and industrial regions and in regions with strong production levels. The gap between regions, in terms of unemployment rates, is likely to increase throughout the 1980s because of differences in population pressure.

The core–periphery model has been widely

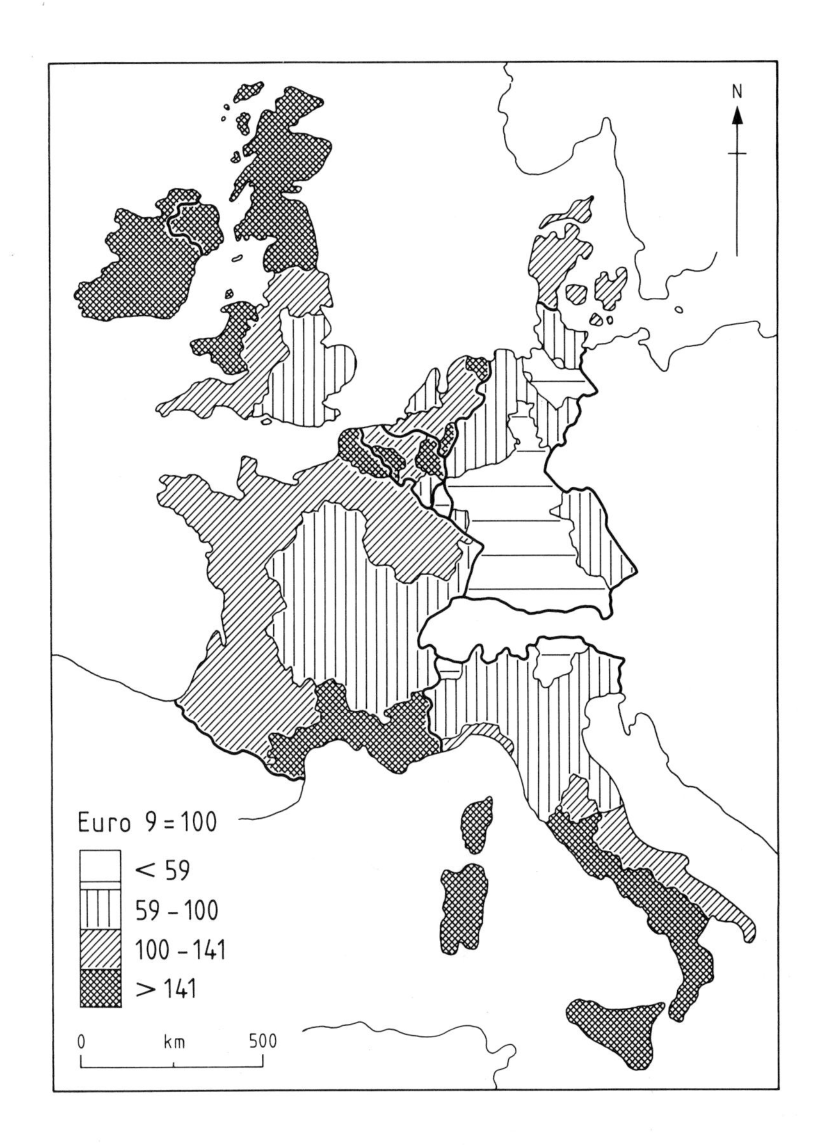

Fig. 10.1 Unemployment in the European Community, 1981

adopted as being appropriate for discussing the regional problem in western Europe. However, Pinder (1983) has warned of the oversimplified view it can provide, as there is much diversity within both the core and the periphery, and 'peripheral areas' are not confined to the fringes of western Europe but are also found in the more central areas. This led Holland (1976a) to devise a more complex typology of 'regions' in western Europe, which comprised:

1. *Over-developed regions*, where a large proportion of the national population and economic life are found and where social deprivation and environmental decay are problems. Paris, London, the Randstad, and north Italian cities are good examples.

2. *Neutral regions*, on the immediate fringes of the over-developed areas and characterized by population growth, high employment and income, and less congestion. South-east England and the outer parts of the Paris Basin come into this category.

3. *Depressed regions*, which comprise former innovative areas now experiencing relative or absolute decline. Classic examples in this group would include old coal-mining districts (e.g. Ruhr, Saar, Limburg, and Wallonia) and textile areas (Lancashire, Flanders).

4. *Under-developed regions*, where industrial

capitalism is not established and where the economy is still agricultural/rural-based. The peripheral parts of Europe represent such regions.

5. *Intermediate regions*, which display a mixture of conditions (employment, income, and migration) found in the four other regions. Naples is a very good example of a large, decaying urban area in an under-developed region (Spooner 1984).

6. *Frontier regions* (Blacksell 1977; Clout *et al.* 1985), where there is inadequate cross-border infrastructure and differences in income and law. The most notable are those in Austria and West Germany which abut onto east European neighbours.

Therefore, different types of problem region exist in western Europe, from those based on agriculture and coal-mining to those experiencing industrial decay and urban congestion. Consequently, regional policy has varied and still varies spatially. In some countries, attempts have been made to stimulate conditions in the more prosperous core areas, whilst in others the poorer and often peripheral areas have been given attention. It is virtually impossible to generalize about regional policy in western Europe and the situation has been aptly summarized in a statement from Brown and Burrows (1977 p. 193), 'the fundamental ingredients of regional policy have been mixed in very different proportions and total quantities in different countries, according to the nature of their problems and their political and social structures and traditions'. However, Hall (1976) believes that most west European countries are faced with similar problems, such as depopulation, unemployment, congestion, and housing shortages, and the solutions are only different because the political outlooks of the major west European countries have remained sharply differentiated in the post-war period. As Clout (1981) suggested, different problems are perceived to exist in different countries and it is this perception which has led to policies being operated in different ways, over different time periods.

Therefore, national approaches to regional development in western Europe differ and Clout (1981) classified the reasons for this into three main groups:

1. The nature and intensity of problems differ in the various regions of western Europe, as for example in the empty lands of northern Europe and the congested urban area of Paris. These problems are perceived in different ways by government institutions and planning bodies and as Thirlwall (1974) noted, the type of regional policy must differ according to the type of regional problem being experienced.

2. National systems of administration, government, and planning vary enormously at the institutional level, from the more favourable centralized traditions in countries like the United Kingdom, the Netherlands, Spain, and France, to the socio-democratic traditions in Scandinavia and the federal traditions in West Germany, Austria, and Switzerland. Regional policy is usually a modification of national systems and policies, and is consequently a reflection of unique national conditions.

3. National systems vary in their stage of evolution. The long-established regional planning systems in the United Kingdom, the Netherlands, and France contrast with the more recent attempts at regional policy in Austria, Switzerland, and Iberia.

Using Brown and Burrows' (1977) classification of regional problems, it is possible to analyse some of the diverging regional policies which exist in western Europe. Areas receiving assistance for regional development up to the early 1970s from the various schemes in operation in different parts of western Europe were mapped by Clout (1976) and are shown in Fig. 10.2. The map indicates that similar schemes exist in different countries, but the origin and emergence of these schemes are diverse and complicated.

The policy of moving workers to available jobs has not been very popular in western Europe, mainly because it involves a movement to the big cities which is, in general, politically and socially undesirable. However, examples of such a policy can be found in inter-war Britain and post-war Netherlands and Sweden. Workers were encouraged to move in the United Kingdom in 1928, when a problem of severe localized unemployment existed, and the Netherlands had a policy of encouraging and assisting migration to the more prosperous areas. During the 1950s and 1960s, movement in search of work in Sweden was given positive assistance. This necessarily meant a north–south drift of people to the city regions, a

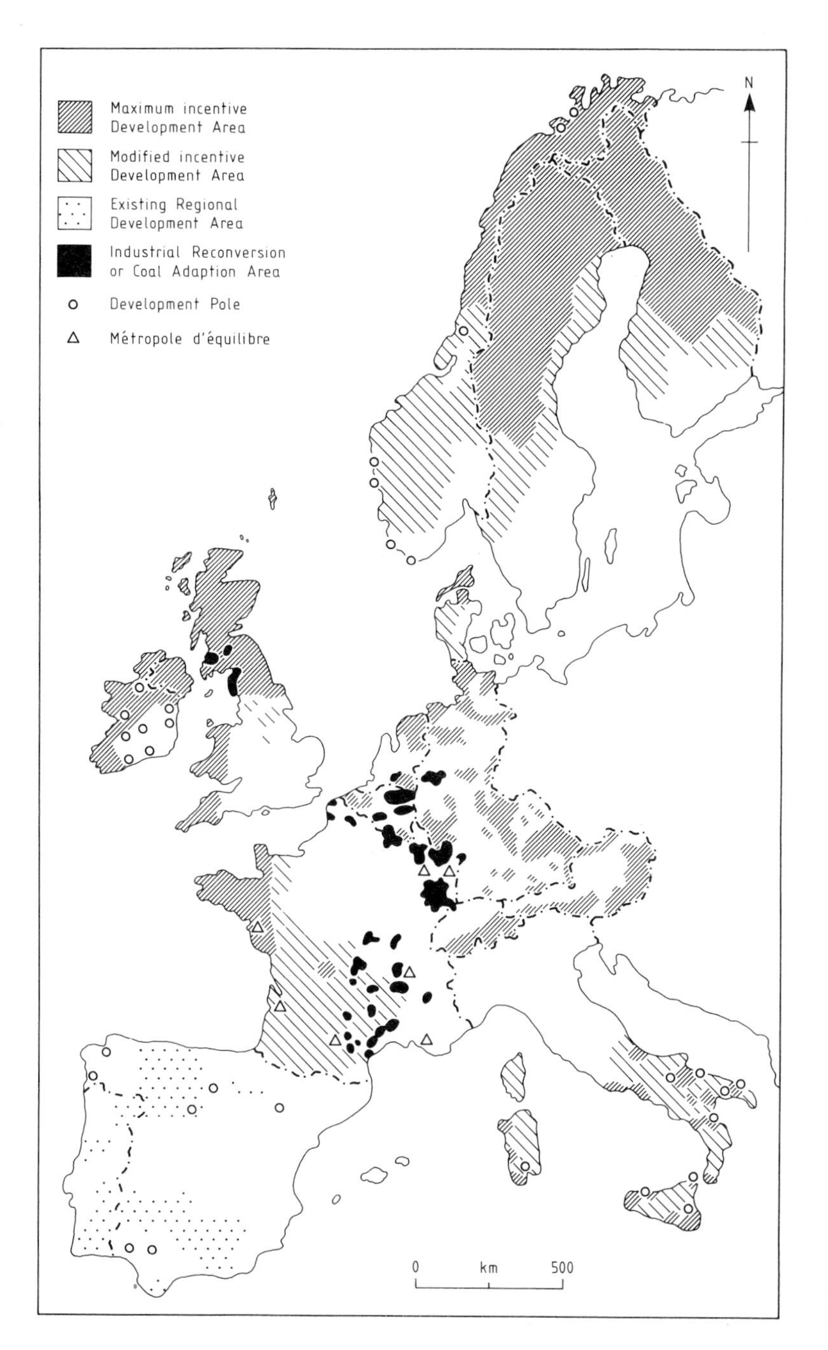

Fig. 10.2 Problem areas in western Europe. (*Source:* Clout 1976)

trend which added to the problems of depopulation in the northern areas (Stenstadvold 1975).

In contrast, policies aimed at moving work to the workers are more numerous. All types of strategies and instruments have been used to encourage industry to move to assisted regions, ranging from capital grants and subsidies on capital expenditure and wage bills to tax concessions, cash inducements, training grants, negative controls, and growth points and centres. Countries have varied enormously in the degree to which they have implemented these policies, but one feature common to practically all west European countries is the frequency with which policies have consistently changed.

A fashionable strategy in western Europe in the 1960s was the selection of growth points or growth centres, aimed at encouraging the development of backward areas and relieving the congestion of major metropolitan areas, namely the capital cities. The nature and objectives of this strategy differed from country to country and one of the most noticeable differences was in terms of the size of proposed growth points. Extremes of this size range can be found in Scandinavia and France.

Scandinavia represents quite an extreme version of the core–periphery contrast and the governments of Sweden, Finland, and Norway have been concerned with the locational disadvantages of the northern regions. However, growth point policy has varied between Norway on the one hand, and Sweden and Finland on the other. The main concern of policy objectives in Norway has been to preserve the scattered settlements on the islands, along the coast, and in the valleys and mountains (Stenstadvold 1975). In 1965, a number of test 'growth-centres' were proposed, only to be later replaced, with a change of government, by six 'development regions'. Therefore, emphasis was placed on growth areas rather than growth points and the wealth generated from oil and gas in the late 1970s has helped to maintain the policy of protecting a settlement pattern inherited from the previous period of industrial development (Hansen 1983). This contrasts with policy in Sweden and Finland, where the growth-centre concept was accepted and national systems of central places were identified. In Sweden, the first system was presented in 1970 and concerned itself with metropolitan areas, at the expense of peripheral areas. The system was modifed in 1972, when three levels of centre below the large city regions were proposed: the primary centres, which included the metropolitan areas contained in the 1970 system; the regional centres; and the municipal centres. Consequently, each enlarged commune contained a centre, but the aim was not to promote equal opportunities for all types of settlements and areas, and policy was firmly orientated instead towards growth-points.

Brown and Burrows (1977) have outlined some of the problems associated with growth-point strategy in northern Europe. The authorities in Sweden estimated that, in order to be viable, a growth centre should have a minimum population of 30 000, but in an area of very low population densities this would comprise the population of an extremely large area. In addition, most of the towns of viable size were located on the coast and inland areas were losing population rapidly. Although similar aids were originally given to coastal and inland growth-centres, it was later necessary to give preferential assistance to the inland centres.

The rather small growth points in northern Scandinavia contrast markedly with the large *métropoles d'équilibre* (balancing metropolitan areas) in France. Eight *métropoles* were designated in 1964, as part of the fourth National Plan, and the objective of these counter-weights was to slow down the rate of migration to the Paris region. The centres selected were already large provincial cities and represented logical places for the generation of economic growth in their respective regions (Fig. 10.3). As well as aiding the regeneration of old industrial areas, the *métropoles* were designed to help the prosperous industrial areas and support the poor hill-farming areas. Assistance given to these centres was mainly in the form of infrastructure investment, followed by investment grants for office and similar development. The idea of extending provincial growth centres beyond the *métropoles* led to a further seven centres being selected in 1967 to receive financial help for attracting tertiary employment: Besancon, Clermont-Ferrand, Dijon, Limoges, Montpellier, Poitiers, and Rennes.

Growth-point policy was partially successful in France, in so far as the *métropoles* grew rapidly in the 1960s and at a greater rate than Paris. However, certain snags occurred with this policy (Hall 1976; Brown and Burrows 1977; Clout 1981; Hudson and Lewis 1982). The *métropoles* were effectively creating further 'Paris and its desert' situations as they were growing at the expense of their own hinterlands, causing the rural areas to remain outside their sphere of influence. The problems of lower-order centres were being ignored and the costs of further development in the *métropoles* had exceeded those for similar development in smaller centres. In addition, investment in the *métropoles* had to compete with investment in Paris.

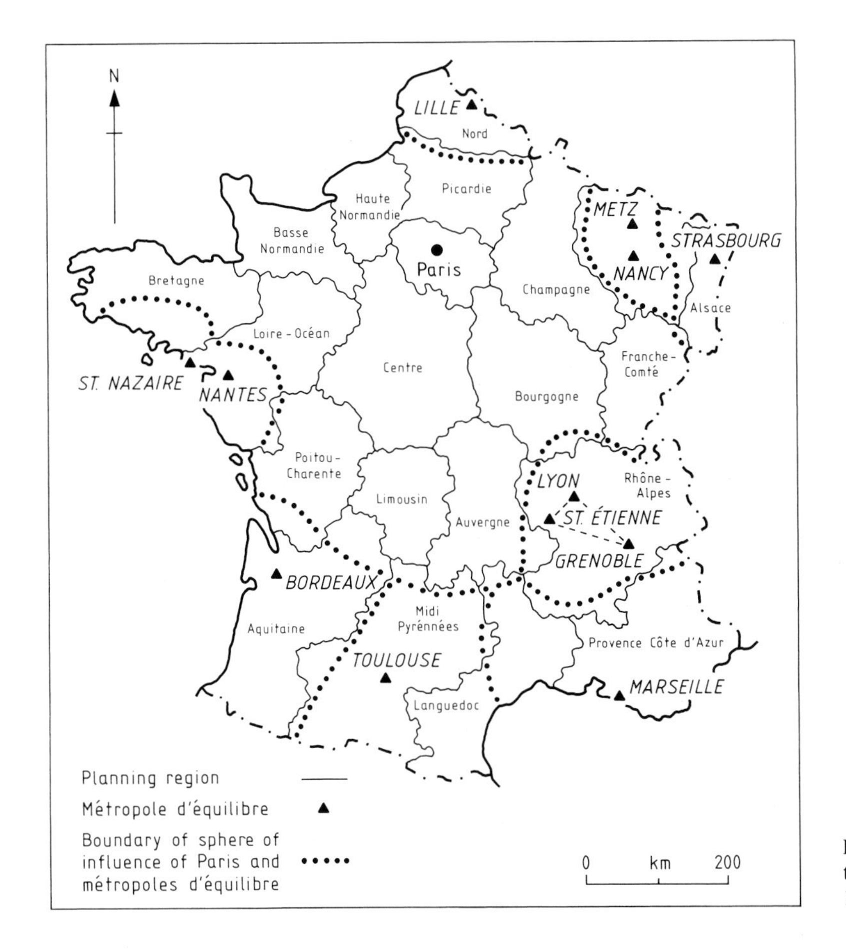

Fig. 10.3 French planning regions and the *métropoles d'équilibre* (*Source:* Hall 1976)

In the early 1970s, the virtues of small and medium-sized towns were realized and more attention was paid to improving housing, transport, job provision, and cultural facilities in towns of less than 100 000 people. Nevertheless, Hall (1976) believes that growth-point developments in France are unlikely to modify the strong underlying trend of the concentration of people in the major urban centres. Indeed, the ideas of growth points and centres are difficult to operate in practice and as Brown and Burrows state (1977 p. 184), 'the more a country is urbanized already, and provided with reasonably serviceable infrastructure, the less the strategy, in its original form, is called for'. Detailed case studies of the Fos-sur-Mer growth pole in the Marseille metropolitan area of France (Kinsey 1978 and 1979; Bleitrach and Chenu 1982; Hudson and Lewis 1982) have demonstrated the major difference between intentions and outcomes. Whilst productivity had increased, the area itself was not benefiting. This was because the industry attracted was capital- and not labour-intensive and geared to meet external rather than local demand. This point will be returned to later.

The instruments for carrying through the spatial strategy of regional development are numerous and include infrastucture improvements, cash inducements, and negative controls. The first two, often found in association, have been used to initiate regional development in the older industrial and poorly developed agricultural regions. In Italy, infrastructure improvements have been used in combination with cheap loans, grants, and tax concessions in an attempt to attract industry to the south, and in the town of Emmen in Holland, some of the infrastructure expenditure has been used to provide building grants. To ensure the full use of the infrastructure, it has often been necessary to provide incentives such as grants, concessions and loans at low rates of interest.

West Germany has used both infrastructure improvements and financial incentives as part of its policy for regional development. As in other west European countries, the objective of regional planning in West Germany is regional equality, but the major responsibilities for carrying out policy measures are given to the 10 *Länder* and not the central government. Planning has been essentially concerned with the remote rural districts and three types of area requiring assistance were recognized in the late 1960s: development areas, development centres, and the frontier zone. Development areas were themselves frontier zones and most were located along the zonal border with East Germany. It was in these areas that infrastructure improvements and financial incentives were especially important. Improved communications were necessary because the areas were remote from the main lines of transport and cash inducements, in the form of loans and grants, were needed to help private investment, notably in industry. Assistance was highest in the frontier areas, with investment grants as high as 25 per cent (Burtenshaw 1974); in addition, these areas received special tax benefits and federal contracts. Infrastructure improvements were also important in relation to the 312 German development centres. These centres were, in general, small market towns which had a surplus of agricultural labour. With better communications and services, there was a period in West Germany (1959–66), after full employment had been obtained, when employers eagerly moved to these small towns.

In an attempt to give closer co-ordination of development programmes, regional action programmes were initiated in 1969–70, for which a joint planning effort over five years was to be made. Twenty-one regions, covering 59 per cent of the area of West Germany and 33 per cent of the total population, were defined and development was to be focused on commercial and industrial centres with a hinterland of at least 20 000 people. Up to 20 per cent of investment costs for new workplaces were covered, and up to 10 per cent of the cost of securing workplaces. Burtenshaw (1974) commented on the encouraging results of the regional action programmes and estimated that over 750 000 new jobs would be created by the end of the programme period. However, the economic recession has increased regional disparities in West Germany and the tenth Planning Framework (1981–5) has redrawn the assisted areas and aid has become more spatially concentrated (to 49 per cent of the territory and 30 per cent of the population), just as there is greater selectivity in the designation of enterprises qualifying for assistance (Ulrichjung 1982).

Negative controls have been employed to restrict the ability of firms to go anywhere but the assisted areas. Essentially, firms have been stopped from locating in already congested areas and conurbations (Jarrett 1975). This was first expressed in the system of Industrial Development Certificates, introduced in the United Kingdom in 1947. Control of office building followed in 1964 and collectively the two measures played an important role in regional policy in the United Kingdom, especially during the 1960s. In the 1970s, there was a relaxation of such negative policies, except in the more prosperous regions, and this contrasts with the rest of western Europe which, with the exception of France, avoided negative-type controls until this decade.

France followed a similar pattern to the United Kingdom and control of industrial building in Paris in 1955 was supplemented by control of office building in 1969 (Moseley 1980; Flockton 1982). In Italy and the Netherlands, negative controls became important in the 1970s. The Italians introduced a scheme in 1971 whereby industrialists in congested areas were required to give notice of major expansion plans and to submit to a 25 per cent surcharge on construction costs if the scheme violated the national plan. In 1975 the Netherlands also restricted development in the conurbations by introducing a licensing system on industrial and commercial building in the Randstad area, with a tax in addition in Rotterdam.

Scandinavia has witnessed a series of negative controls since 1970 and schemes to decentralize offices of the central government were formulated in Sweden in 1970, in Finland in 1974, and in Norway in 1975/6. Sweden introduced a negative measure in 1971, when firms proposing to expand in the main urban areas of the south—Göteborg, Malmö, and Stockholm—were obliged to consult with the government. This was essentially a queuing device and an opportunity to present firms with information on alternative centres. However, it was an ineffective system of building licences and it was not until 1974 that a Royal Commission recommended a comprehensive system of control, based on similar lines to those introduced in the United Kingdom in 1947. Norway and Finland followed suit and in the mid-1970s considered a system based on industrial development certificates.

From this brief review of the instruments of regional policy, it would appear that no one set of measures is adequate when dealing with regional disparities and that a combination of measures is preferable. Brown and Burrows (1977 p. 191) summarize the situation quite well when they state, 'a combination of incentives with disincentives or negative controls is necessary for implementing a regional policy'.

Despite the vast differences in regional policy in the countries of western Europe, Allen (1974) listed a number of instructive similarities. The major ones are as follows:

1. All west European countries increased their expenditure on regional policy, especially during the 1960s and 1970s (up to 1979). However, there have been relatively few attempts to quantify the costs and benefits of regional policy. In the United Kingdom, Brown (1972), Moore and Rhodes (1973 and 1976), and Keeble (1976) have indicated an increase in employment and a change in the distribution of new industrial building in favour of the assisted areas, and in the Netherlands, Hendriks (1974) estimated that up to 50 per cent of the increase in industrial and service employment in the Groningen area was due to assistance. Stenstadvold (1975) commented on the success of regional policy in Scandinavia, and Brown and Burrows (1977) concluded that such policy had a substantial effect in western Europe, quoting examples of the movement of manufacturing industries out of Paris and the stemming of migration into Stockholm, the Dutch Randstad, and south-east England.

2. The policies have changed frequently and after the war the United Kingdom, for example, shifted from large development areas to small development districts and back to development areas, and from growth area policy to grants and depreciation allowances. With the exception of French controls on development in the Paris region, most countries avoided the type of negative controls which the United Kingdom operated for industrial and office building until the 1970s. Instead they relied heavily on inducements such as grants, loans, and provision of infrastructure.

3. With the exceptions of France and the United Kingdom, and to a lesser extent Ireland (Bannon 1973), there has been little policy for service mobility (Alexander 1979). Where it does exist, it is weak and tends to take the form of decentralization, as for example in the movement of government offices to development areas. A possible reason for this relative lack of policy is the generally held view that services are unproductive. Service mobility has been considered in France and Switzerland and is based on the notion that a firm is a set of functions, with each function having different locational requirements, and that there is no reason why these functions should not be spatially spread according to their location needs. The Swiss have been working in terms of five functions—processing, administration, distribution, training, and head office—and Allen believes that this is a process which needs stimulating.

4. In all countries, regional policy has been affected by political factors and until relatively recent years there was an almost complete lack of measures for declining areas, despite strong social and economic arguments for such measures. The need for declining policies is especially acute if growth-area policies are being pursued in the 'core' regions.

5. The policies have often lacked realism and are deficient in considering the time scale of regional development. Regional development is a slow process and no major European problem area has been removed from the designated list (Fig. 10.4). Policies have aimed at solving regional problems relatively quickly; the Cassa, for example, was created in southern Italy in 1950 to resolve the 'southern' problem, and its envisaged life span was 10 years. However, the Cassa existed until 1980, with increased funds, and the north-south disparities have not been removed (see Section 10.3).

6. Regional policy failed to consider the secondary impact of firms on the area of designation. The demand for materials by a new plant is often met by firms outside the region and not by local markets. Consequently, there has been a low

multiplier effect, a common problem in southern Italy and Scotland. Lever (1974), in a study of a panel of firms in Scotland, revealed that 80 per cent of their materials and semi-finished inputs came from outside Scotland.

One observation to emerge from this last point, and some of the others, is the limited success of regional policy in western Europe. This was highlighted during, and since, the economic recession of the mid-1970s, when it was realized that the regional planning orthodoxies of the 'growing' 1960s were increasingly inappropriate to the control of capitalist forces during the 1970s and 1980s (Holland 1976b; Hudson and Lewis 1982; Pinder 1983). The major phase of post-war growth favoured large-scale and multi-national enterprises, which made wide-spread use of development funds to obtain good sites in assisted areas. As Massey (1979) explained, regional problems reflected the evolution of business organization and capital's manipulation of space. A new spatial division of labour emerged, with low-skill functions in the assisted (peripheral) areas and high-level R & D functions reserved for the advanced (core) regions. Consequently, local needs in the poorer areas were ignored as the newly attracted branch plants failed to develop links with other enterprises in the area or use local raw materials. This worked against regional policy objectives and actually reinforced the structural weaknesses in assisted areas, especially as branch plants were most suspect to closure during recession periods.

Hudson and Lewis (1982) emphasized the need for a new, radically different method and theoretical basis for regional planning in western Europe. This revolved around the idea of increased regional self-management, signs of which can be detected in such countries as Belgium, France, Italy, Portugal, Spain, and the United Kingdom. Spain is a particularly good example because regional policy has been transformed over the last ten years, first by the economic recession, and secondly by the return to democracy after Franco's death in 1975. The new constitution of 1978 permitted substantial devolution of sovereignty to newly-created regional governments, although national unity and inter-regional solidarity remained as important objectives (Hebbert 1982). A 'bottom up' approach to regional development was adopted, encouraging local initiatives, rather than a 'top down' approach which characterized the pre-democracy Franco era.

As a result of the limited success of regional policy measures, many governments have 'cut back' their areas of assisted status; for example, after 1982 the assisted areas in the United Kingdom accounted for only 25 per cent of the workforce, compared with 40 per cent in the mid-1970s (Clout, *et al.* 1985). Less money has been spent on regional aid by individual governments since 1979 and emphasis has shifted from area-specific to more general attempts to create jobs. Some of these observed changes will be further developed in section 10.2.

10.2 Regional policy in the European Community: the European Regional Development Fund (ERDF)

The Treaty of Rome stated that the Community should design policies to reduce regional inequalities and that the development of backward areas, such as southern Italy, north Holland, and west and south-west France, was a necessity. However, the European Community had little effect on regional policy in member countries and, until recent years, attempts to develop a common regional policy were slow and weak. Regional disparities which existed when the original EEC was established have not been reduced and if anything have tended to grow (Holland 1976a; Armstrong 1978). With such wide disparities in prosperity within the Community, such as income differentials of up to 600 per cent between the lower Rhine industrial district and Eire (Blacksell 1977), the desirability of a common policy was widely recognized, but political rather than economic factors restricted the rate of progress.

Before 1971, little was accomplished and funds for regional development purposes were small, as there was nobody specifically concerned with regional development. Early contributions to regional policy were made by four main organizations:

1. The European Investment Bank (EIB), set up under articles 129 and 130 of the Treaty of Rome specifically to help finance development projects in the less developed regions. Between 1958 and 1972, 86 per cent of the loans were allocated to regional development schemes, with

a large proportion going to the Mezzogiorno area of southern Italy; in 1983, two-thirds of the funds were still devoted to this purpose. The EIB, which is the world's second largest bank and the first development bank, has undoubtedly contributed to regional development. However, it is a bank and consequently rates of interest are charged on loans and investment grants, and other incentives are not provided. Pinder (1978) and the EIB (1983) have shown how 25 billion ECUs have been provided in the first 25 years, for a whole range of projects (85 per cent within the EC), and how the level of finance has trebled since 1973 and stood at 5.47 billion ECUs in 1983 alone. Since the 1960s, the pattern of investment has broadened and, although still biased towards southern Italy, a greater proportion of the funds has been channelled to the United Kingdom, Ireland, Greece, and France. In recent years, the EIB has encouraged the growth of loans to 'smaller' ventures, through a system of global loans. Over 50 per cent of loans now go to such ventures, as it is felt that new ideas often come from small firms, who are able to maintain competition and prevent monopolistic situations.

2. The European Social Fund (ESF), which supports schemes for the training and retraining of unemployed workers, but which has had little effect on regional problems because of a lack of revenue. However, with 13 million people unemployed in the EC, its role is of increased significance, especially as higher rates of assistance are available in the most seriously disadvantaged regions (Clout *et al.* 1985).

3. The Agricultural Guidance and Guarantee Fund (FEOGA), where revenue from the guidance section is spent on structural changes in the backward agricultural areas. Unfortunately, the guarantee policies offset the regional benefits of the guidance section.

4. The European Coal and Steel Community (ECSC), which led to the creation of new jobs and the retraining and re-employment of workers from the coal and steel industries of the original Six. The ECSC probably had more effect on regional problems in the 1960s than the three other bodies put together, but the spatial impact is obviously restricted to mining and steel-making areas.

It was only when negotiations to enlarge the Six were taking place in 1971 that regional policy became a 'live' issue. Britain insisted upon a regional development fund being established, as a precondition of her entry into the Community. This was agreed in principle during the Paris Summit Conference in 1972. Various proposals were forwarded at the conference, to counteract the differences in living standards and economic growth between the regions, and a pledge was made by the member countries to co-ordinate their regional policies, even though responsibility for formulating such policies would continue to rest with the national governments (Thirlwall 1974). The first steps taken to implement and operate a Community Regional Policy came in the form of three main proposals:

1. Part of the guidance section funds from FEOGA should be used for the creation of industrial jobs in rural areas.
2. A regional development corporation should be set up, having both advisory and technical assistance tasks.
3. A regional development fund should be created, to give grants and interest subsidies for industrial, service, and infrastructure projects in problem areas.

The Paris Summit led to the publication of the Thomson Report in May 1973, which announced the establishment of a Regional Development Fund (ERDF). The report identified different types of problem area, but as Clout (1981) observed, these were delimited on the basis of diverse sources of information, which were not strictly comparable, and thus should be treated with caution. Thomson, the Commissioner responsible for regional affairs, proposed 2250 million units of account for regional projects over the 1974–6 period (500 million units in 1974; 750 in 1975; and 1000 in 1976); and the criteria upon which regions were to receive aid were as follows (Commission of European Communities 1973):

1. A lower gross domestic product than the Community average, plus at least one from the following:
2. A heavy dependence on agricultural employment (greater than the Community average), or employment in declining industries (at least 20 per cent of local employment).

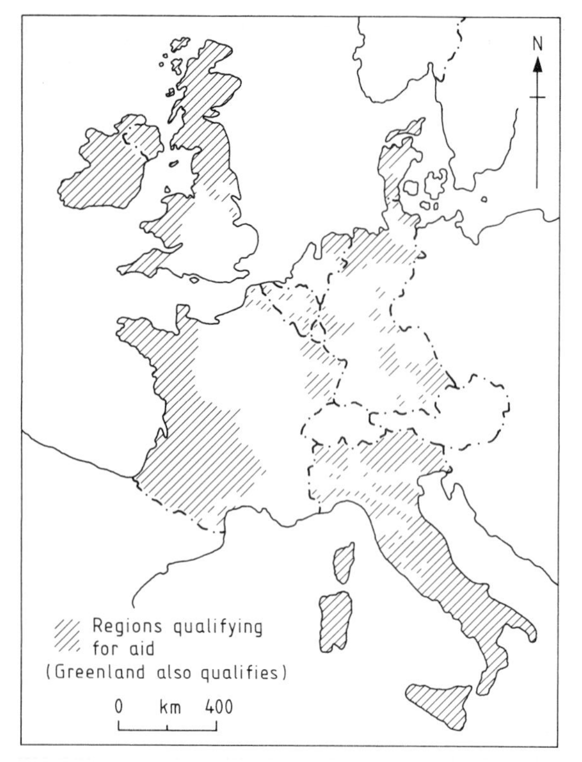

Fig. 10.4 Areas eligible for grants according to the Thomson Report

3. A persistently high rate of unemployment, at least 3.5 per cent.
4. A high rate of net emigration (10 per 1000 each year averaged over a long period).

Even if these criteria were satisfied, only certain types of development qualified. Projects had to be aimed at creating or maintaining jobs, require more than 50 000 units of account and be in the industrial or service sectors. The ERDF could contribute up to 15 per cent of the total cost of an industrial project or 30 per cent of an infrastructure project, but the contribution could not exceed 50 per cent of national aid expenditure on a given industrial investment.

Areas which satisfied the criteria laid down and thus qualified for help from the ERDF were extensive (Fig. 10.4). Core–periphery contrasts are again striking and the areas covered included the whole of Ireland, most of upland Great Britain, western France, parts of northern Germany and Denmark, most of Italy, and the former industrial regions of France, Belgium, and West Germany. Aid was to be made available to a whole range of problem regions, from the geographically and politically isolated to the depressed agricultural and industrial.

The ERDF was not intended as a substitute for a fully-blown regional policy and development programmes were only eligible for funds if they complemented existing national policies. Therefore, Community aid was not given on its own and the overall objective of a Common Regional Policy was to complement rather than replace national systems of aid. It was first necessary to harmonize national policies and then formulate common Community objectives. The view of the Commissioner was clearly demonstrated in the following statement from the 1973 report, 'the role of Community regional policy will progressively increase as the Community increases and improves its instruments of intervention, together with the co-ordination of national policies which will be undertaken in the light of the varying extent of regional problems' (Commission of European Communities 1973, p. 12).

The proposals in the Thomson Report received varying degrees of criticism from certain member countries and, because of political disagreement, there was difficulty in distributing the funds allocated to it. In terms of finance, West Germany would be the major contributor to the fund, having relatively few problem areas, and southern Italy the major beneficiary. In the ensuing debate, there was much discussion over the extent of assisted areas, the criteria used for definition and the ratio between likely contributions and receipts (Clout 1981). It was not until January 1975 that the ERDF became operative, when it was agreed to distribute 1300 million units of account, over a three year period, among the poorer regions (300 million in 1975 and 500 million each in 1976 and 1977). This figure was considerably lower than the 2250 million proposed in the Thomson Report and led to a decline in the number of eligible areas. In 1975 the ERDF allocated quotas for the 1975–7 period in the following proportions: Italy, 40 per cent; United Kingdom, 28 per cent; Ireland, 6 per cent (these were net beneficiaries); France, 15 per cent; West Germany, 6.4 per cent; the Netherlands, 1.7 per cent; Belgium, 1.5 per cent; Denmark, 1.3 per cent; and Luxembourg, 0.1 per cent (these six were net contributors). In the 1975 regulations the detailed 1973 proposals for establishing the ERDF were abandoned in favour of a simple method of defining eligible areas for Fund assistance as those parts of the Nine also eligible for the individual regional policies of member-states. This point will be discussed later.

During the late 1970s, a series of reports were published by the Commission of the European Communities. These had certain fundamental effects on Community regional policy, which were very well summarized and assessed by Armstrong (1978). Proposals for a revised Community Regional Policy for 1978, and subsequent years, were drafted in July 1977 (Commission of European Communities 1977a and b). The ERDF was given powers to finance interest rebates on a wide variety of Community loans, as well as continuing the system of grants. In addition, the Fund was able to contribute up to 50 per cent of the associated expenditure by public authorities on infrastructure projects in certain severely depressed areas, whereas the limit remained at 30 per cent elsewhere. The 1977 proposals also envisaged a strengthening of pressures on individual member states, to prevent them substituting ERDF assistance for their own regional assistance; member countries would be required to demonstrate clearly the ways in which Fund assistance enabled them to operate a more powerful regional policy.

However, Armstrong (1978) believed that Community Regional Policy remained in an 'embryonic' state throughout the 1970s and contained numerous weaknesses, including the following major ones:

1. The package of regional policy instruments was extremely limited and compared unfavourably with the wide range of measures used in France and the United Kingdom, for example. Armstrong stated that there was a need for 'a system of disincentives to be applied consistently throughout the relatively prosperous regions of the EC' (p. 518).

2. The ERDF suffered from a lack of funds allocated to it. The Commission envisaged a Fund of 750 million units of account for 1978, but the amount actually allocated was 581 million. This rose to 620 million units of account in 1979 and 650 million in 1980. As Armstrong noted, this represented less than 0.05 per cent of the combined gross domestic product of the member states. The Fund was virtually non-existent when compared to the EC budget, which was dominated by the CAP and the guarantee section of the FEOGA in particular.

3. The ERDF was inflexible. It did not concentrate its resources particularly well, primarily because shares of the Fund were determined in advance. The system of pre-determined quotas was modified in 1977, by introducing an extra non-quota allocation. This was distributed at the discretion of the Community, adding more flexibility, but unfortunately non-quota assistance constituted just 5 per cent of the resources of the ERDF. Whilst national quotas reflected the location of problem regions in the EC, the process of spatial concentration needed to be expanded.

4. The abandonment of the 1973 proposals for delimiting problem areas in favour of the 1975 system, where eligible areas were defined as those parts of the Community which were eligible for the individual regional policies of member states, was a retrograde step. As the individual countries used different methods to delimit problem areas, the eligible areas of the Fund were not compatible. This added to the inflexibility of the ERDF, which had lost its power to direct funds to desired locations.

5. Regional development policy had entirely ignored areas of urban deprivation. Urban areas represented a particular type of regional problem which required separate measures, but these were not forthcoming. Urban policy needed to be integrated with regional policy.

6. The co-ordination of regional policies was slow. This was not an easy task as it required detailed information on the objectives and methods of regional policies in the member states. Information sources have improved and member states have submitted an annual statement and statistical summary since 1975, and Regional Programmes since 1977. From 1979, the Commission proposed to draft a comprehensive report, every two years, on regional problems and policies within the European Community.

The early years of the 1980s have been characterized by the Commission reappraising the ERDF and arguing for changes in the distribution of aid. Reform was certainly necessary, especially as the Community's expenditure on regional matters was equal to only about one-tenth of the collective sum given by member states to their assisted areas (Klein 1981). In 1981 Greece joined the EC and the ERDF quotas were adjusted, giving the following proportions: Italy (34.9 per cent), the United Kingdom (23.4), Greece (15), and Ireland (5.9) were net beneficiaries; whilst France (13.2), West Germany (4.5), the Netherlands (1.2), Belgium (1.1), Denmark (0.9), and Luxembourg (0.1) were net contributors.

The efforts of the ERDF during its first nine years are summarized in Table 10.1. Although still accounting for just 0.08 per cent of the EC's GDP in 1983, the ERDF's budget exceeded 2000 million **ECU** for the first time, which was equivalent to 8.7 per cent of the Community's total budget. Between 1975 and 1983, 21 500 investment projects were supported, over three-quarters of which were related to infrastructure. The number of projects increased each year and stood at 3700 in 1983. During this year, 95 per cent of the funds were concentrated in five countries—Italy, the United Kingdom, France, Greece, and Ireland (Fig. 10.5).

Indeed, 20 per cent of the regions received 80 per cent of the assistance, and regions with

Table 10.1 *ERDF allocations, 1975–83*

Year	Amount (Mio ECU)	Annual % increase	Share in Community budget
1975	257.6	—	4.8
1976	394.3	53.1	5.6
1977	378.5	−4.0	4.9
1978	581.0	53.5	4.6
1979	945.0	62.7	6.1
1980	1165.0	23.3	6.7
1981	1540.0	32.2	7.3
1982	1759.5	14.3	7.6
1983	2010.0	14.2	8.7

Source: ERDF Annual report, 1984

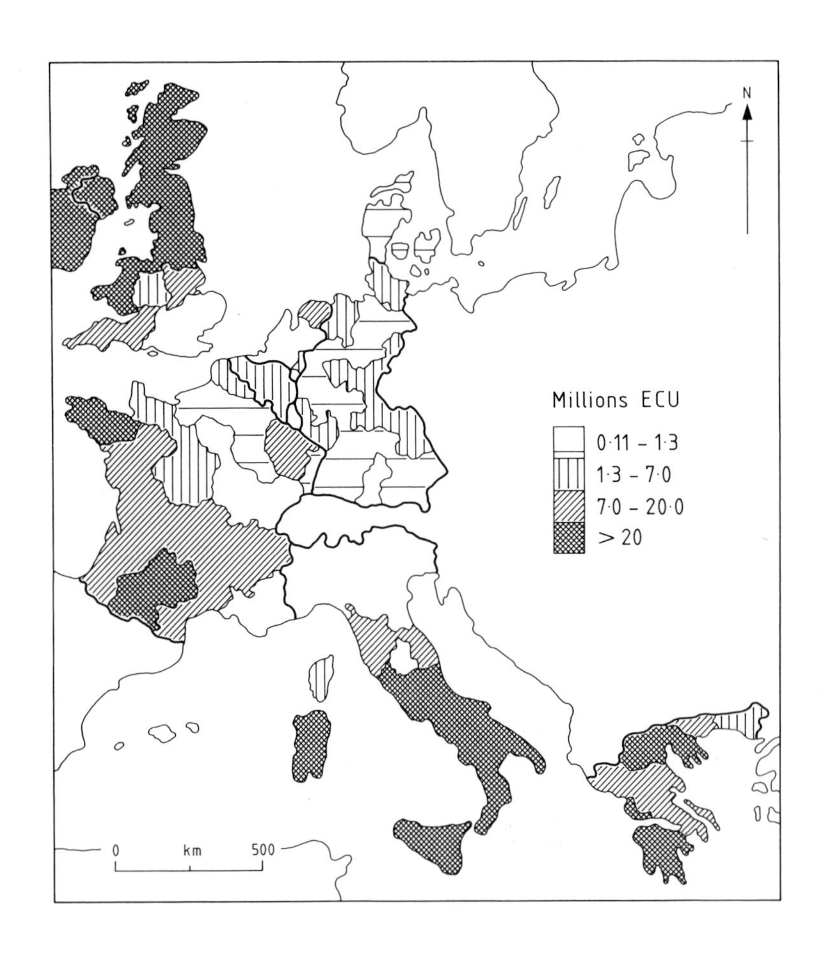

Fig. 10.5 Regional distribution of ERDF assistance in 1983. (*Source:* ERDF Annual Report 1984)

'priority status'—the Mezzogiorno, Ireland, Greece, Northern Ireland, Greenland, and the French Overseas Departments—accounted for 64.1 per cent of funds in 1983, compared with 57 per cent in 1975. The non-quota section of the ERDF had also achieved some success and five schemes were approved by 1982, involving frontier regions (Ireland, Northern Ireland), industries affected by world trends in, for example, steel (the United Kingdom, Italy, and Belgium), and areas affected by the policy of enlarging the EC (the Mezzogiorno and southern France).

Despite these efforts, the Commission remained unhappy. This revolved around the small size of the ERDF's budget, which was dwarfed by other funds and overwhelmed by world recession (Clout *et al.* 1985), and the inadequacy of the quota system, which defined the annual aid to which each country was entitled, irrespective of the projects proposed. For example, regions in the Netherlands and West Germany, with higher per capita GDP than the richest regions in Italy and the United Kingdom, were qualifying for aid. Regional policy measures in the EC were simply progressing too slowly, a problem that would be accentuated with Spain and Portugal's membership in 1986. Consequently, the Commission proposed certain changes in 1981 and 1983.

The first recommendation was to increase the spatial concentration of policy under the quota section, by giving aid to regions where per capita GDP was very low and long-term unemployment particularly high (Fig. 10.6). This would entail reallocation of the quota system in the following manner: Ireland (7.3 per cent), the Mezzogiorno (43.7), all of Greece except Athens and Thessalonika (16), Northern Ireland and parts of Scotland, Wales, north and north-west England (29.3), French Overseas Departements (2.4), and Greenland (1.3) (Greenland has voted to withdraw from the EC). The implication of this

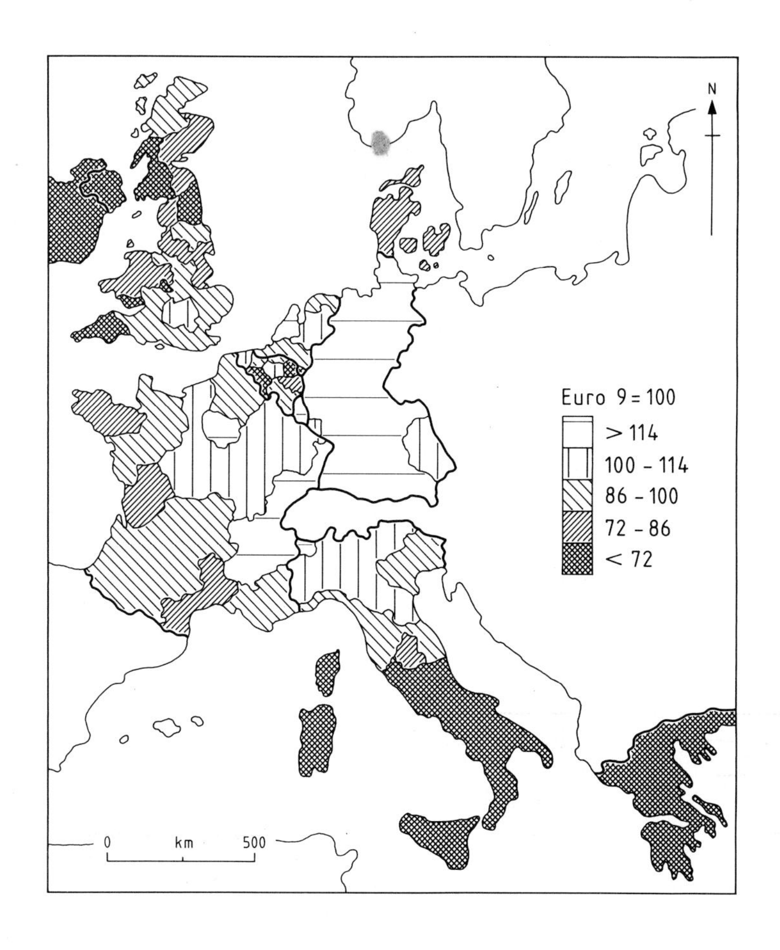

Fig. 10.6 Acuteness of regional problems in the European Community 1981

change is that West Germany, Denmark, mainland France, and the Benelux countries will no longer qualify for assistance from the quota scheme.

A second proposal was to increase the non-quota section from 5 to 20 per cent of the ERDF fund, for redeployment in regions which have been harshly affected by recent industrial decline (Clout *et al.* 1985). In addition, both quota and non-quota payments will not be directed to individual projects, but to broad-based development programmes over a period of years. More recently (November 1983) an amended proposal was forwarded by the Commission to the Council, which suggested the abolition of the distinction between the support measures (quota) section and the specific Community measures (non-quota) section of the ERDF. These should be replaced by quantitative guidelines for each member state, which range between a guaranteed lower limit and an upper limit, and Community programmes. Indeed, by the late 1980s, the Commission envisaged that at least 40 per cent of the ERDF's resources would be allocated to finance Community programmes and national programmes of Community interest. The changeover to programme financing is expected to allow greater selectivity and concentration of objectives and more emphasis to be placed upon productive investment, notably in small and medium-sized businesses (ERDF, *Annual Report* 1984).

Whilst these changes might represent a 'step in the right direction', it is clear that Community regional policy is not progressing at an adequate pace or able to control the growing disparities which exist in the Ten. A more radical policy seems necessary, with additional funds, if the problems of the enlarged Community are to be tackled in a meaningful manner.

10.3 Regional policy at the local level: the Mezzogiorno

Southern Italy is probably the most pronounced problem region in the European Community and one which has received most attention. The Mezzogiorno contains seven of the 19 administrative regions of Italy, (Fig. 10.7), and although accounting for 40 per cent of the total area, only one-eighth of the region is flat and suitable for productive agriculture. A distinct core-periphery relationship is recognizable in Italy and 'the backward, agricultural, peasant south contrasts markedly with the forward-looking, industrial north' (King 1981, p. 119). All the characteristics of 'economic infirmity' (Pacione 1976), are exhibited in southern Italy and these have been discussed at length by many authors (for example, Dickinson 1955; Rodgers 1970; Allen and MacLennan 1970; King 1970 and 1981; Mountjoy 1973; Pacione 1976 and 1982).

Indicators of underdevelopment are numerous and include migration, birth-rates, unemployment, and gross domestic product (Fig. 10.8). The Mezzogiorno contains 36 per cent of the Italian population, yet it accounts for only one-quarter of the gross domestic product. In 1981, gross domestic product per person was well below the national and EC averages (Fig. 10.8c), and parts of Calabria, Sardinia, and Sicily were experiencing great poverty. Per capita incomes in the south were only 40 per cent of those in the prosperous north in 1950, and by the early 1980s this figure had risen to only a little over 50 per cent.

The disparity between north and south is further exemplified by an analysis of birth-rates and rates of out-migration (Fig. 10.8a and b). The birth-rate remained high in the south throughout the 1950s and 1960s, and in 1976 it stood at 16.7 per 1000, compared to 11.6 in the north. Campania had the highest birth-rate in Italy in 1982 (16.9 per 1000), and this was 2.6 times as high as the region with the lowest rate, Liguria in the north (6.6 per 1000). Rural over-population was the result, causing a 'drift to the north' and emigration. Associated with this is the problem of unemployment (Fig. 10.8d). In 1950, agriculture employed nearly 60 per cent of the working population, but as people were being released from the land there was a lack of alternative employment. Consequently, labourers often worked for just 100–150 days per year. This enhanced the level of poverty, which was mixed with overcrowded housing conditions, poor sanitation, a lack of drinking water and a very high rate of illiteracy.

Minshull (1978 p. 231) summarized the problems as follows: 'a low level of economic activity; income levels well below other parts of the EC; poor housing conditions; and an overwhelming dependence on agriculture with resulting rural overpopulation.' Therefore, the south has

Fig. 10.7 Economic planning regions in Italy

remained remote, backward, and, until recently, neglected. In contrast, the north has benefited from a whole range of factors, including a favourable position in relation to the advanced regions of Europe, large markets, technology, capital availability, supplies of skilled labour, good communications, and sources of power.

A complex of factors interacting over time and space has produced this 'ecology of backwardness' (King 1981). Mountjoy (1973) divided these factors into three categories: geographical, historical, and social and economic. Geographical factors are plentiful and include the following:

1. Remoteness from the main centres of economic activity, due to the sheer physical distance and length of the country, combined with poor transport facilities.

2. The influence of latitude. The Mezzogiorno experiences a truly Mediterranean climate with very high temperatures in the summer. This would be suitable for many crops if water supplies were adequate, but the rainfall is seasonal and comes at the wrong time for agriculture. The summer drought is worsened by high evaporation rates and consequently agriculture, the main form of livelihood, suffers from harsh climatic conditions.

3. The very difficulty topography. The two peninsulas and two islands of the south inhibit spatial integration, and 85 per cent of the Mezzogiorno consists of mountains and hill country. The Appennines are deeply dissected, making communications difficult, and good land is confined to plains around Naples, Salerno, and Foggia.

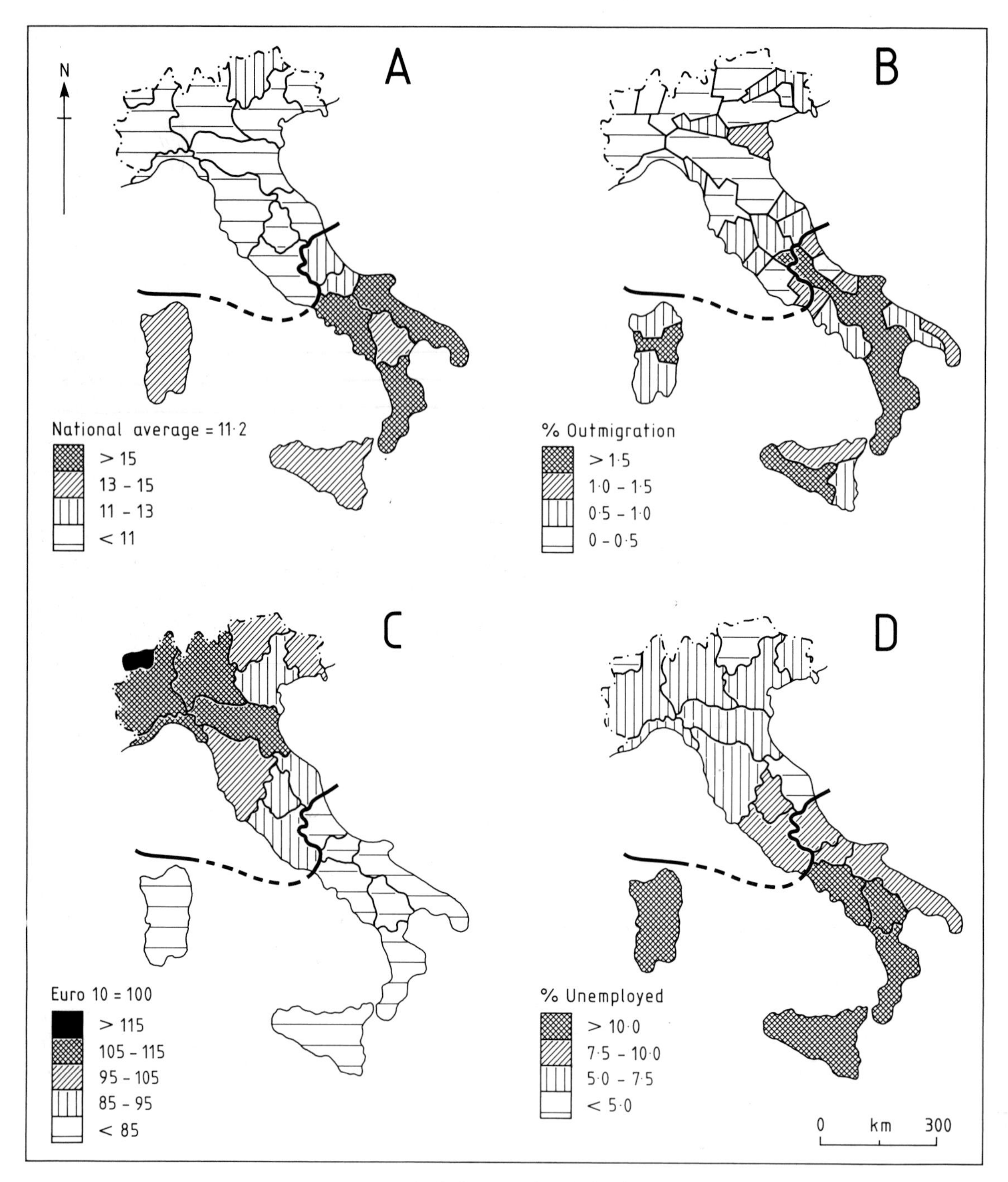

Fig. 10.8 Italy: indicators of underdevelopment

4. The unfavourable geology. Much of the area is characterized by steep slopes and hard dolomitic limestone, which is permeable and affords little soil cover. Other parts of the Mezzogiorno consist of claylands and these suffer from the extremes of being heavy and sticky in the winter, and hard and dry in the summer.

5. The location of southern Italy in the centre of the Mediterranean, providing an ideal meeting place for invaders and colonists, with the former

10.1 The deeply dissected schist country on the western flank of the granitic Sila Grande, Calabria, placing tremendous physical constraints on agricultural production (A.B.M.)

especially benefiting from the articulated coastline.

As a result of this last factor, the area became the centre of much activity in early times and was affected by a series of invasions. Early settlers destroyed timber supplies and traded more with north Africa and Greece than with northern Italy. The Normans formed the United Kingdom of south Italy and Sicily and this was separated from the states and dukedoms of the north; the north–south division was thus established in the twelfth century. The Bourbons established the Kingdom of Two Sicilies, but their administration was corrupt and little investment was made in communications and public utilities. Feudal serfdom, with large estates called 'latifundia', was the norm and represented a system which persisted into the twentieth century. Italy was unified in 1861, but by this time the damage had been done; agriculture in the south, based on the latifundia, was feudal and stagnant. A policy of protection after 1861 did no more than encourage the maintenance of inefficient agriculture in the south, allowing industry in the north to grow.

The situation in the Mezzogiorno was heightened by various social and economic factors:

1. The south lacked the necessary power resources of coal and hydro-electric power.
2. Industry was small-scale, inefficient and artisan in character. The unification of Italy meant that the south was open to industrial competition from the north and the gap between north and south actually increased.
3. The structure of agriculture encouraged inefficiency. Besides the large latifundia, there was much fragmentation and farms were divided and leased in small plots to the peasantry. By 1950, many holdings were less than two hectares. In addition, land tenure was very insecure and this had damaging effects.
4. There were growing population pressures, especially in the rural areas. The consequence was

an over-extension of arable activities and cultivation was pushed up the hill slopes, causing widespread soil erosion.

5. State action had little impact because of various loopholes. Attempts were made to provide credit facilities and ease taxation on the peasant farmers, but these benefits rarely reached the peasant.

6. The south was characterized by unemployment, poverty, overcrowding, and poor communications.

The serious nature of the situation in the Mezzogiorno caused the government to embark upon a long-term plan for regional development. King (1981) has divided Italian policy for the south into four time-periods:

1. 1945–50, a period of post-war reconstruction, with American Marshall Aid.
2. 1950–7, a pre-industrial phase characterized by emphasis on agricultural and infrastructural reform.
3. 1957–63, a phase of industrial development and growth-poles.
4. 1965 onwards, a phase of more integrated planning.

Over 300 laws were passed to deal with the problems of the south, but most failed because of bad administration and a lack of co-ordinated planning. However, in 1950 the 'Cassa per il Mezzogiorno' or 'Fund for the South' was established, a 10-year development plan. The Cassa was a linking organization between the government ministries in Rome and the local authorities in the south, and was designed to deal with work beyond the scope and resources of individual ministries. In the first instance, the objective of the Cassa was to promote land reform and infrastructure improvements. To do this, £600 million were allocated to the Fund for the 1950–60 period. In reality, the time-scale and amount of finance were far too restrictive and the life-span of the Cassa was perpetually lengthened, first to 1962, then to 1965, and finally to 1980. This extension was accompanied by the allocation of more funds, and between 1950 and 1970, £5000 million were spent on development. The Cassa was also allowed to seek loans from abroad and to attract private investment; in the first decade, 30 per cent of its funds came from the World Bank. The European Investment Bank was another important source of revenue, supplying £200 million in the 1960s.

Initially, the Cassa put emphasis on the rehabilitation of agriculture and infrastructure, with 77 per cent of the investment consigned to the former and 23 per cent to the latter. In 1950, three land reform laws were passed in Italy (King 1973), and these were largely concerned with the expropriation of large and inefficiently operated latifundia and the creation of smallholdings for the peasants. Over 700 000 hectares of land had been redistributed in Italy by 1965 (Clout 1976), with more than 70 per cent of this redistribution affecting the south. Nearly 115 000 families received land and a closer pattern of settlement, with new villages, was established. Unfortunately, land reform measures only affected 10 per cent of the Mezzogiorno.

Infrastructural improvements were associated with the land-reform movement and took various forms. Water supplies, drainage, irrigation, afforestation, mechanization, reclamation, and communications were all improved. The landscape was enriched with 125 000 new farmhouses and 80 000 barns (Clout 1976), 50 service centres, 7500 wells and 17 000 kilometres of roads (Minshull 1978). In addition, 300 farm co-operatives were started, agricultural training and education were considered necessary, packing, processing, and refrigeration units were established as were reservoirs, schools, hospitals, telephone systems, hydro-electric power stations, and marketing facilities for agricultural products.

These measures achieved some success and a change in land-use, from wheat and olives to citrus fruits, vegetables, and livestock, was noticeable in the better areas. Agriculture in the hilly areas was little affected, as farmers found it difficult to adjust to a new way of life. However, King (1973) thought that land reform was a one-dimensional and inflexible development programme, which had serious limitations as a regional development policy. With an economic boom in the north, King felt that the context of land-reform had changed faster than could the policy itself. Consequently, land-reform became obsolete and was a victim of its own timing.

The first extension of the Cassa, to 1962, allocated funds in a similar manner, with 69 per cent devoted to agriculture, 29 per cent to transport,

10.2 Land reform farms and farmhouses in the Metapontino (A.B.M.)

and 2 per cent to tourism (Pacione 1976). However, a broader-based strategy was required and although the need for industry was recognized in the Vanoni Plan of 1954, it was not included as part of development policy until 1957. In that year, the Industrial Areas Act allowed the Cassa to support the development of industrial zones in the south and by 1962, 60 per cent of the Cassa's investment was in industry, compared to none in 1950. Nevertheless, industrial policies had to overcome the problems of isolation, limited markets and raw materials, and a lack of capital investment and skilled labour. Measures designed to minimize these problems have been classified into three types by Pacione (1976):

1. *Compensation*, a range of financial and fiscal incentives to compensate industry for the difficulties, including tax exemptions, direct grants, easy term loans, transport concessions, and free advice and technical assistance.
2. *Stimulation*, direct measures to gain industrial development, through the movement of state controlled firms to the south and the use of the government contract-award system
3. *Concentration*, the idea of creating growth poles to encourage development.

In addition, there were the added incentives of abundant and relatively cheap labour, new fuel sources, mainly in the form of local natural gas, and better port facilities, to aid the importation of bulky goods.

Early results were rather disappointing, partly because of the lack of an industrial tradition in the south, and it was difficult to compete with areas to the north. Funds were again too restrictive and the incentives were of little value as there was nobody to promote industrial development in the Mezzogiorno. Large-scale private enterprises were not attracted and the greatest contribution to southern industrialization came from government-controlled companies, such as IRI (holding company for industrial reconstruction) and ENI (state hydrocarbons agency). By 1964–5, public companies were required by law to place 40 per cent of their investments in the Mezzogiorno, yet less than 20 per cent of private industrial investment had been directed to the south.

The Mezzogiorno experienced certain growth in heavy industry, mainly in petrochemicals, iron and steel, heavy engineering, and oil refining. Iron and steel works were established at Taranto and Bagnoli, and petrochemicals at Brindisi and Gela in southern Sicily. The steelworks at Taranto and petrochemicals in Gela provided over 6000 jobs and created 12 000 additional jobs in associated factories (Clout 1976). Unfortunately, the industry attracted was mainly capital-intensive, in an area of high unemployment. Light consumer-goods industries were not developed and a need for more sophisticated development policies was thus apparent.

The 1957 Industrial Areas Act introduced the idea of growth-centres and two levels of centre were introduced: first, *areas* of industrial development, which consisted of cities or urban regions possessing a population of over 200 000 and could demonstrate that an industrial vocation was already in existence; and secondly, *nuclei* of industrialization, with a population of less than 75 000 and containing small and medium-sized firms. By 1968, 42 areas and nuclei had been designated, covering 29 per cent of the area and 45 per cent of the population of the Mezzogiorno, but these tended to be scattered throughout the area, even though a policy of concentrated effort had been adopted.

This philosophy of concentrated effort in the south led to the designation of 82 agricultural development and connected zones, and 29 areas of tourist development, in addition to the 42 industrial growth-centres. Much overlap existed between the various zones and six major areas of overlap became designated growth-poles. These were Cheiti–Pescara; in lower Lazio and Campania; around Palermo; the Bari–Brindisi–Taranto triangle; Catania–Siracusa; and the area of Cagliari–Sulcis Iglesiente–Oristano in Sardina (Pacione 1976).

Substantial amounts of investment were channelled into these points and each pole was to develop a variety of factories and a range of services. The Bari–Brindisi–Taranto growth-pole in Apulia received much attention and eight industrial plants and 24 linked factories were planned, in an attempt to develop industry in this 'heel' of Italy. The growth-pole was fully integrated into the national motorway network, via the Autostrada del Sole, and besides the iron and steel works at Taranto, which with an annual capacity of 10 million tonnes is one of the largest in western Europe, there is mechanical engineering at Bari and Taranto and petrochemicals at Brindisi, producing plastic and ethylene. The European Commission produced a detailed development plan for this growth pole and invested over 150 000 million lire in industrial plant between 1973 and 1977 (Minshull 1978). Despite the concentrated effort in this area, numerous problems remained, such as a general shortage of adequate skilled labour, unemployment in Bari and a lack of jobs and subsidiary industries being generated by the iron and steel industry. Growth-pole policy in the Mezzogiorno has been criticized by King (1981) on a number of grounds:

1. The distinction between 'intensive areas' and 'diffused nuclei' became blurred.
2. No overall assessment was made of the relative suitability of different locations for certain types of industries.
3. No real co-ordination existed between the various areas and nuclei, with the exception of the Bari–Brindisi–Taranto triangle.
4. No interrelation existed between town planning and industrial development.
5. No new instruments were adopted to promote the location of northern private industry in the south.

The last phase of regional development in the Mezzogiorno, from 1965 onwards, has tried to adopt a more integrative approach. In 1972, the Cassa was authorized to spend £4800 million by 1977, which is more than it spent for the whole of the 1950–72 period. Among the policies envisaged was the formation of a special finance corporation to provide loans for development in the south. State-owned corporations were directed to locate 80 per cent of their future capital investment in the south and the Cassa was to concentrate on financing special projects to cover new industrial zones, expansion of large urban areas, and the safeguarding of natural resources (Clout 1976). Special grants and cheap loans were made available, in an attempt to attract small factories into the areas which had been experiencing rural outmigration. Between 1976 and 1980, one-fifth of Cassa's funds was devoted to the programme of special projects, which were in the national interest and included the completion and rationalization of the inter-regional water supply schemes and the provision of infrastructure in strategically important areas like Naples and the industrial zone of Calabria (Spooner 1984).

One area to receive increasing attention from the Cassa was tourism. Southern Italy, despite its relative inaccessibility, has much tourist and recreational potential, including such attributes as the sun, beaches, scenery, and a landscape heaped in history. According to Pacione (1976, p. 43), 'the increasing importance of tourism is one of the most salient features of the Italian economy in the post-war period'. The intention was to concentrate tourism into three types of selected tourist areas (Fig. 10.9):

Fig. 10.9 Tourist development zones in Italy. (*Source:* Pacione 1976)

1. Tourist development districts, where development was yet to begin.
2. Tourist development districts, where development had already started.
3. Mature tourist economy districts, where development was reaching saturation point.

Generous financial incentives were given by the Cassa after 1965, including loans of up to 70 per cent on admissible expenditure, paid back at 3 per cent interest over 20 years, and grants of 15 per cent on capital outlay.

The Cassa was in operation for 30 years and the Mezzogiorno has witnessed some notable achievements, although it is doubtful whether the area has reached 'take-off' point into sustained economic growth. To conclude this chapter, the effectiveness and limitations of regional policy in the south will be briefly examined. The term 'economic miracle' has been applied to the Mezzogiorno, but this needs to be put into perspective as the post-war record is not too encouraging. It is noticeable that the gap between the north and south has not really changed; however, this is hardly surprising when it is remembered that northern Italy experienced one of the highest rates of economic growth in the world during the 1950s and 1960s.

On the positive side, numerous advances have been made:

1. Per capita incomes increased significantly, by 67 per cent between 1951 and 1964, and 37 per cent between 1965 and the oil crisis of 1973.
2. Rates of economic growth between 1950 and 1971 stood at 5.8 per cent per year, compared to 5.4 per cent in the north, making the Mezzogiorno one of the fastest growing areas in western Europe.
3. The occupational structure has undergone a dramatic transformation, as shown in Table 10.2, and is more balanced. Employment in agriculture has been reduced substantially, with over 2.5 million people leaving the land, and there has been a large increase in the number of industrial and service jobs.

Table 10.2 *Occupational structure in the Mezzogiorno*

	% employment in:		
	Agriculture	*Industry*	*Services*
1950	57	20	23
1970	33	32	35
1973	28	32	40
1980	19	37	44

4. Agricultural output has more than doubled and productivity has been especially increased in the land-reform areas. Irrigation projects and other rural infrastructure developments have aided the rate of progress.
5. Industrial output has increased five times.
6. The economic structure of the Mezzogiorno is on a sounder basis. Infrastructure improvements have been made, unemployment has fallen, infant mortality has been more than halved, illiteracy rates have fallen by 40 per cent, overcrowding has declined by 20 per cent, and there has been an increase in the number of hospital beds, schools, and telephones.
7. The situation of the peasant has improved remarkably.

Despite these improvements, a number of problems remain and King (1981) suggests that the Mezzogiorno has modernized rather than developed. On the negative side, there are a number of limitations:

1. There was a lack of early policy for industrial development and later policy has had the effect of attracting capital-intensive industries in an area with a large labour surplus. By the early 1970s, only 30 per cent of Italian industry was in the south, and the figure for large firms (over 500 employees) was less than 10 per cent. Little private sector industry has moved to the Mezzogiorno. As Spooner (1984) remarked, policy has been 'from above', driven by external demand and dominated by capital-intensive and high technology industries and by urban–industrial growth centre strategies. What was required was development 'from below', based on the satisfaction of needs and the use of appropriate technology in labour-intensive, small-scale activity.
2. The origin of the gross domestic product contrasts with that in the north. In the Mezzogiorno, 17 per cent of the gross domestic product still comes from the primary sector, compared to 8 per cent in the north. Industry accounts for 40 per cent of the gross domestic product in the south, employing 37 per cent of the labour force, and 56 per cent in the north, employing 41 per cent of the labour force.
3. Land-reform allocations were too small and agricultural policy has not been very successful, in so far as there is still a surplus agricultural working population, a lack of adequate equipment and low farm incomes.
4. Income differentials are still large and the gap between north and south remains. In 1950, incomes in the south were 40 per cent of those in the north, and by 1978 they were still only 68.5 per cent of the national average.
5. Unemployment is still amongst the highest in western Europe and rates of migration to the north and other EC countries remained high until the early 1970s. In the 1950s, 1.8 million people migrated and in the 1960s this figure had risen to 2.3 million.
6. There has been a lack of negative controls on industry in the north, and little has been done to aid employment in the rural areas of the south. The established growth-centres contrast with the lack of development in the rural areas.
7. There is a lack of urban concentrations and

society. The growth-pole idea has not been effective in building up urban areas (Spooner 1984).

8. There has been a lack of co-ordinated planning and the Mezzogiorno remains the poorest region in the Ten. The overall situation has not improved dramatically, despite attracting huge sums of money from the EIB and ERDF, and the area has been closely watched by the European Commission on regional development.

Although the Mezzogiorno is only one of a number of problem regions, it is a microcosm of the wide range of policy measures which have been adopted in western Europe in an attempt to reduce regional disparities. It is beyond the scope of this chapter to analyse the various problem regions classified by Holland (1976a) and it was decided to concentrate upon and cover in greater depth one problem region (see Oxford's 'Problem Regions of Europe' series).

References

Alexander, I. (1979). *Office location and public policy.* Macmillan, London.

Allen, K. (1974). 'European regional policies'. In Sant, M. E. C. (Ed.) *Regional policy and planning in Europe.* Saxon House, Farnborough.

Allen, K. and MacLennan, M. C. (1970). *Regional problems and policies in Italy and France.* George Allen and Unwin, London.

Armstrong, H. W. (1978). 'Community regional policy: a survey and critique'. *Regional Studies,* **12,** 511–18.

Bannon, M. J. (1973). *Office location and regional development.* An Foras Forbartha, Dublin.

Blacksell, M. (1977). *Post-war Europe: A political geography.* Dawson, Folkestone.

Bleitrach, D. and Chenu, A. (1982). 'Regional planning-regulation or deepening of social contradictions? The example of Fos-sur-Mer and the Marseilles Metropolitan Area'. In Hudson, R. and Lewis, J. R. (Eds.) *Regional planning in Europe.* Pion, London.

Brown, A. J. (1972). *The framework of regional economics in the United Kingdom.* Cambridge University Press, Cambridge.

Brown, A. J. and Burrows, E. M. (1977). *Regional economic problems.* George Allen and Unwin. London.

Burtenshaw, D. (1974). *Economic geography of West Germany.* Macmillan, London.

Clout, H. D. (1976). *The regional problem in western Europe.* Cambridge University Press, Cambridge.

Clout, H. D. (Ed.) (1981). *Regional development in western Europe.* Wiley, London. Second edition.

Clout, H. D., Blacksell, M., King, R., and Pinder, D. (1985). *Western Europe: geographical perspectives.* Longman, London.

Commission of European Communities. (1973). *Report on the regional problem in the enlarged Community.* Brussels.

Commission of European Communities. (1977a). *Guidelines for Community regional policy.* Bulletin of the European Commission, Supplement 2/77. Brussels.

Commission of European Communities. (1977b). *Second report of the European Regional Development Fund.* Brussels.

Commission of European Communities. (1981). *The regions of Europe: first periodic report on the social and economic situation in the regions of the Community.* Brussels.

Dickinson, R. E. (1955). 'Geographic aspects of unemployment in southern Italy'. *Tijdschrift voor Economische en Sociale Geografie,* **46,** 86–97.

European Investment Bank. (1983). *25 years: 1958–83.* EIB, Luxembourg.

ERDF. (1984). *Annual report.* Luxembourg.

Flockton, C. H. (1982). 'Strategic planning in the Paris region and French urban policy'. *Geoforum,* **13,** 193–208.

Gillingwater, D. and Hart, D. A. (1978). *The regional planning process.* Saxon House, Farnborough.

Hall, P. (1976). *Urban and regional planning.* Penguin, Harmondsworth.

Hansen, J. (1983). 'Regional policy in an oil economy: the case of Norway'. *Geoforum,* **14,** 353–61.

Hayward, J. and Watson, M. (1975). *Planning, politics and public policy: the British, French and Italian experiences.* Cambridge University Press, Cambridge.

Hebbert, M. (1982). 'Regional policy in Spain'. *Geoforum,* **13,** 107–20.

Hendriks, A. J. (1974). Regional policy in the Netherlands; in Hansen, N. M. (Ed.) *Public policy and regional economic development: the experience of nine west European countries.* Cambridge, Massachusetts.

Holland, S. (1976a). *The regional problem.* Macmillan, London.

Holland, S. (1976b). *Capital versus the regions.* Macmillan, London.

Hudson, R. and Lewis, J. R. (Eds.) (1982). *Regional planning in Europe.* Pion, London.

Jarrett, R. J. (1975). 'Disincentives: the other side of regional development policy'. *Journal of Common Market Studies,* **13,** 379–90.

Keeble, D. (1976). *Industrial location and planning in the United Kingdom.* Methuen, London.

King, R. (1970). 'Structural and geographical problems of south Italian agriculture'. *Norsk Geografisk Tijdskrift,* **24,** 83–95.

King, R. (1973). *Land reform: the Italian experience.* Butterworth, London.

King, R. (1981). 'Italy'. In Clout, H. D. (Ed.) *Regional development in western Europe.* Wiley, London.

Kinsey, J. (1978). 'The application of growth-pole theory in the Aire Metropolitaine Marseillaise'. *Geoforum*, **9,** 245–67.

Kinsey, J. (1979). 'Industrial location and industrial development in the Aire Metropolitaine Marseillaise'. *Tijdschrift voor Economische en Sociale Geografie*, **70,** 272–85.

Klein, L. (1981). 'The European Community's regional policy'. *Built Environment*, **7,** 182–9.

Lever, W. F. (1974). Manufacturing linkages and the search for suppliers and markets. In Hamilton, F. E. I. (Ed.) *Spatial perspectives on industrial organisation and decision-making.* Wiley, London.

Minshull, G. N. (1978). *The new Europe: An economic geography of the EEC.* Hodder and Stoughton, London.

Moore, B. and Rhodes, J. (1973). 'Evaluating the effects of British regional economic policy'. *Economic Journal*, **83,** 87–110.

Moore, B. and Rhodes, J. (1976). 'Regional economic policy and the movement of manufacturing firms to development areas'. *Economica*, **43,** 17–31.

Moseley, M. J. (1980). 'Strategic planning and the Paris agglomeration in the 1960s and 1970s: the quest for balance and structure'. *Geoforum*, **11,** 179–223.

Mountjoy, A. B. (1973). *The Mezzogiorno.* Oxford University Press, Oxford.

Pacione, M. (1976). 'Development policy in southern Italy: panacea or polemic?' *Tijdschrift voor Economische en Sociale Geografie*, **67,** 38–47.

Pacione, M. (1982). 'Economic development in the Mezzogiorno'. *Geography*, **67,** 340–3.

Pinder, D. A. (1978). 'Guiding economic development in the EEC; the approach of the European Investment Bank. *Geography*', **63,** 88–98.

Pinder, D. A. (1983). *Regional economic development and policy: theory and practice in the European Community.* George Allen and Unwin, London.

Rodgers, A. B. (1970). 'Migration and industrial development: the south Italian experience'. *Economic Geography*, **46,** 111–35.

Sant, M. E. C. (Ed.) (1974). *Regional policy and planning for Europe.* Saxon House, Farnborough.

Spooner, D. J. (1984). 'The southern problem, the neapolitan problem and Italian regional policy'. *Geographical Journal*, **150,** 11–26.

Stenstadvold, K. (1975). 'Northern Europe'. In Clout, H. D. (Ed.) *Regional development in western Europe.* Wiley, London.

Thirlwall, A. P. (1974). 'Regional economic disparities and regional policy in the Common Market'. *Urban Studies*, **11,** 1–12.

Ulrichjung, H. (1982). 'Regional policy in the Federal Republic of Germany'. *Geoforum*, **13,** 83–96.

Yuill, D., Allen, K., and Hull, C. (1980). *Regional policy in the European Community.* Croom Helm, London.

11

Eastern Europe: Systematic Contrasts

This text has deliberately concentrated on the human geography of western Europe, but as a consequence it has omitted the remarkable developments which have been taking place in eastern Europe (Turnock 1984). The objective of this final chapter is to examine eastern Europe, in terms of the major systematic themes developed in this book, and to emphasize those characteristics which distinguish it from western Europe.

Eastern Europe is essentially a new region, which comprises eight contiguous socialist states and excludes the Soviet Union. As Turnock (1978, p. 9) states, 'the economic argument is inadequate as rationale for the post-war pattern of eastern Europe and the political aspect is fundamental, dictated by Soviet demands for a buffer zone to consolidate the regime within Russia'. Thus the apparent unity has been created by the acceptance of Soviet-style Marxist-Leninism and its imposition on different ethnic and economic backgrounds (Mellor 1975).

Comecon

The co-ordination of plans in eastern Europe and the Soviet Union is the primary function of COMECON (Council for Mutual Economic Assistance), a mechanism for consultation created in 1949 under the instigation of the Soviet Union. COMECON had six founder members—Soviet Union, Bulgaria, Czechoslovakia, Hungary, Poland, and Romania—and these were soon joined by East Germany and Albania, although the latter has not exercised her membership since 1961. Yugoslavia became an associate member in 1964. Member countries were further linked under the Warsaw Pact, signed in 1955 for political and military reasons. Although resembling trade organizations in western Europe, COMECON is quite different in that it does not provide for the easing of trade, simply because in a communist bloc all trade is state-controlled. The aims of COMECON are essentially threefold:

1. To develop inter-member trade.
2. To accelerate the development of national economies.
3. To widen co-operation, through the exchange of experience and technical aid.

Plans are submitted to COMECON by the various governments and certain schemes are given priority. These have included the development of an electric grid network to supply all member countries, the laying of an oil pipeline from the Soviet Union across Poland and into East Germany and Czechoslovakia, and the creation of canals within the bloc.

The geographic rationale of COMECON is quite clear. Member countries form a contiguous block in the European periphery and share problems in terms of population structure, agriculture, industry, and transport. These problems have helped to unite COMECON, despite tremendous diversity in human patterns. Standards of living are lower than further west and exhibit wide disparities between members. The disparities are little different from those in EFTA and the European Community, but COMECON, which suffers from an inflexible system, has recorded less success in achieving integration. This may be partly due to the dominance and diffusion of Soviet ideas and practices, which have led to certain suspicion and a lack of incentive among planners in individual countries to strive for integration. Each state is fully autonomous as regards its economic planning (Turnock 1984), and Romania, for example, has consistently objected to supranational planning, believing that it would never leave the 'bottom of the pile' if there was a single COMECON economy. Despite internal problems, rapid post-war economic growth and modernization have taken place in member countries.

Population

With a population of approximately 136 million (1983), the eight socialist states contain 28 per cent of the European population, living at a density of 105 inhabitants per km^2. Poland is the only country with more than 30 million people and the density of population in the individual countries ranges from 80 per km^2 in Bulgaria to 155 in East Germany, and falls a long way short of that in the 'core' areas of western Europe. The distribution of population reflects the physical environment, socio-economic conditions, and variations in national development policies (Mellor, 1975). There is a clear relationship between population density and agricultural potential and before the twin forces of industrialization and urbanization became increasingly prevalent in the 1960s and 1970s, the majority of urban centres were merely market towns serving the surrounding rural areas. As in the west, this relationship has become less obvious with continued industrialization and more intensive urbanization. Indeed, centrally planned industrialization, which is based on developing the most advantageous site, is causing the urban population to become increasingly divorced from the older agricultural pattern of population distribution.

Eastern Europe has followed a similar pattern of demographic transition to that experienced in western Europe, although the countries have passed through the various 'stages' at a later date. Population growth remains above-average for the continent and Albania, with an average annual growth rate of 2.5 per cent between 1970 and 1982, is in a class of its own and has yet to complete the late expanding phase of its demographic development. Death rates have been reduced to around 10 per 1000 inhabitants, except in East Germany (13), Hungary (14), and Czechoslovakia (12) where the proportion of 'old' people is similar to that in most West European countries. Indeed, rates are generally below those in western Europe and would be even lower if it were not for the comparatively high rates of infant mortality, which stand at over 25 per 1000 live births in Romania, Yugoslavia, and Albania.

In contrast to western Europe, birth rates are, on average, much higher in the socialist states, especially in Albania (28 per 1000 in 1982). With lower death rates and higher birth rates, it is not surprising to find much higher rates of natural increase in eastern Europe. In 1982, the figures for Romania and Poland were 7 and 10 per 1000 inhabitants respectively, but these were completely overshadowed by Albania's 22. The one exception is Hungary, which joined West Germany and Denmark in recording an excess of deaths over births in 1982. A detailed analysis of population structure in eastern Europe would reveal more triangular-shaped population pyramids and indicate that the area is not approaching a situation of zero population growth. Life expectancy in Europe varies by approximately 7 years, with the countries of eastern Europe grouped in the lower end of the range.

Migration and urbanization

Eastern Europe has experienced similar migration trends to western Europe, although the circumstances are naturally different. Emigration was quite high before the First World War, with America the main destination, only to be replaced by more localized movements after 1945. As a result of central planning, it has been unnecessary to operate explicit controls over migration to achieve the necessary balance between population, resources, and employment opportunities (Compton 1976).

Individuals are free to move in eastern Europe, within the framework provided by economic and social policy, but unlike the situation in western Europe, this framework is more tightly controlled by the state. Emphasis has been placed on internal rather than international migrations and rates of immigration and emigration are relatively low, except possibly in Poland where emigration takes place on a sizeable scale. International migration has also been significant in East Germany, to counteract the negative rates of natural increase in the late 1970s and early 1980s, but generally international movements have been restricted to specialists and workers sent to COMECON members to erect plants and develop new technologies.

Unlike the situation in western Europe, labour is in plentiful supply in eastern Europe. However, it is not evenly distributed and bilateral agreements exist within COMECON to regulate the movement of workers among the socialist states. These movements are short-term and East Germany is the main labour importer, having offered jobs to over 20,000 Hungarians and 2000 Czech workers in 1974-75 (Compton 1976). Hungary and Czechoslovakia have a growing labour deficit themselves, and their future demands could be met by surplus labour from Poland and Romania. The increasing concentration of production in eastern Europe has also led to a noticeable separation of work and place of residence, and distinct patterns of commuting have emerged (Fuchs and Demko 1978).

Rural–urban migration is of continued importance in COMECON and people are being 'pushed' from the countryside by agricultural modernization and rationalization, and 'pulled' to the cities by industrialization, job opportunities, better pay, and superior education. The result of this movement is urban growth and urbanization. Until twentieth century industrialization, the town was less significant than in western Europe and the main influence on urban development in eastern Europe was external. However, there was in the 1970s a growing dominance of urban centres (Kosinski 1974) and all countries experienced urban growth. Rural populations have declined and levels of urbanization have expanded rapidly in the same direction as western Europe (Elkins 1973). Comparisons are difficult, because of the varying definitions of urban, but over 70 per cent of the East German population, for example, live in urban-defined settlements. Indeed, the proportion of the total population living in such settlements increased in all eight countries between 1960 and 1982, especially in Yugoslavia (by 57 per cent), Romania (59), and Bulgaria (69). Increasing urbanization in eastern Europe is primarily the result of three factors: first, rural–urban migration; secondly, the incorporation of villages into urban settlements; and thirdly, changes in the designation of urban areas.

One of the outcomes of the urbanization process is the dominance in the urban scene of the large capital city. These cities, including Prague, Bucharest, Sofia, and Budapest, exhibit the characteristics of primate cities, but are smaller in size than the larger west European cities like Paris and London. Indeed, only five of Europe's 'millionaire' cities are found in eastern Europe and in 1982 there were only 18 cities with a population over 500 000, 14 of which were concentrated in Poland, East Germany, and Yugoslavia. Therefore, urban growth is disproportionately concentrated in the large towns and the hierarchy within the urban sector is becoming more pronounced (Turnock 1984). A second outcome of the trend towards an urbanized society is the growth of new towns, or neighbourhoods, in association with the dispersal of industrial development to green-field sites. Nowa Huta in Poland is one good example. The policy of dispersal led to the growth of medium-sized towns and helped to avoid the massive conurbations so characteristic of western Europe (Mellor 1975).

In terms of urban morphology, commercial functions are not concentrated in the CBD, as in western Europe, but are decentralized to the neighbourhoods. Lichtenberger (1976) has noted the decline in size of the apartment as one moves east, from an average of four bedrooms in Paris to two bedrooms in the socialist countries. However, the same author states that the differences in housing are much smaller than the differences in living standards and gross national products. Whilst there is some discussion on the differences between 'socialist' and 'western' cities, due to an assumed homogeneous bid rent surface and socially mixed housing policies in the former, Turnock (1984, p. 327) has argued that it is very difficult to isolate characteristics that are intrinsically 'socialist' from manifestations of national character and relatively low income levels.

Agriculture

In terms of employment and contribution to gross domestic product, agriculture plays an important but declining role in the economy of eastern Europe. Nevertheless, it is a problematical sector of the economy and one which has experienced levels of government involvement that would be unacceptable in western Europe.

The early part of the twentieth century was characterized by peasant farming and the production of grains under primitive conditions. A number of interrelated problems existed (Clout 1971), including:

1. A high level of agricultural overpopulation, due partly to a lack of alternative jobs. Agriculture accounted for up to 80 per cent of the working population in the 1930s.
2. A predominance of small farms. For example, units of less than five hectares accounted for 62 per cent of all farms in Bulgaria and 85 per cent in Hungary.
3. A lack of capital to encourage agricultural development.
4. A lack of diversification. The system of grain monoculture ignored the need for livestock and fruit.

Thus, similar agricultural problems faced western and eastern Europe, but the methods of dealing with them varied tremendously. Large-scale government intervention took place in eastern Europe after 1945, in two successive stages (Saunders 1958; Enyedi 1967; Clout 1971). First, a programme of land reform and resettlement redistributed the land, giving the peasants small units to farm. This policy of transforming the peasants into family farmers was not very successful, as they lacked the necessary experience, capital, and equipment to expand agricultural output. Warriner (1969) argued that this was simply a step to gain peasant support and an intermediate stage to more radical reform. Agricultural production fell substantially and for economic as well as political reasons a radical re-organization of agriculture began in 1947–48. Therefore, the second stage was one of collectivization, whereby farmers were encouraged to co-operate and form larger enterprises. Two types of enterprise were established—state farms and collective farms—and both were controlled by central authorities. Levels of production were fixed by the authorities, partly in the case of the collectives and wholly in the case of state farms, and the peasants, who had become family farmers under the initial land reform policy, were now reduced to the role of wage labourers.

Collectivization met with varied success and appeared to be most suited to the less-developed economies of Bulgaria (Cousens 1967) and Romania (Turnock 1970). Czechoslovakia, East Germany, and Hungary partly resisted reform and settled for a system of modified collectives, where the farmer had greater freedom of crop and livestock choice, and Yugoslavia and Poland totally opposed collectivization and maintained an important private sector. Overall, the results of collectivization were not encouraging and agricultural efficiency appeared to be sacrificed for the sake of applying state socialism. This is well demonstrated by Turnock (1984, p. 322) who noted that in Poland there is a 'dilemma between encouraging the peasant in the face of strong disapproval from the country's Warsaw Pact allies and extending the socialist sector in the teeth of bitter opposition in the countryside'.

Therefore, agriculture witnessed similar changes to those experienced in western Europe. In both areas, agriculture became more capital-intensive and there was a movement of people away from rural areas, an increase in both part-time and industrialized farming (Enyedi 1982) and much government involvement. However, it is with regard to this last point that the major difference between east and west lies.

Energy

The energy resource base in eastern Europe lacks diversity and depth (Polach 1970), and the area remained a one-fuel economy for a considerably longer period of time than western Europe. Although coal remains an important energy source, production has declined in the face of competition from alternative energy supplies. The trend is towards a multi-fuel economy, with oil and natural gas increasing in importance. Indigenous supplies of oil and natural gas are available and Romania is the largest producer, although Yugoslavia and Hungary have the greatest future potential. Similarly, large hydro-electric schemes were implemented in the 1960s and 1970s, typified by the joint Romanian and Yugoslav Iron Gates project on the Danube (Hall 1972).

However, demands for energy exceed supply and as the consumption of oil and natural gas continued to increase, eastern Europe became more dependent on imported energy, especially from the Soviet Union. The reliance on imported energy will increase throughout the 1980s and thus the socialist states will depend on the Soviet Union's ability and willingness to supply them with energy.

Although not at a similar stage of development as in western Europe, nuclear power represents an

Table 11.1 *Nuclear power in eastern Europe*

	Number of units	MWe in operation	Units under construction	MWe under construction	% electricity from nuclear 1983
Bulgaria	4	1632	2	1906	32.3
Czechoslovakia	2	762	9	4840	8.0
East Germany	5	1694	—	—	12.0
Hungary	1	408	3	1224	10.0
Poland	—	—	2	880	—
Romania	—	—	2	1320	—
Yugoslavia	1	632	—	—	6.0

Source: Atom (1984, p. 25)

important component in the energy plans of eastern Europe. A large increase in nuclear power production is planned (Table 11.1), based upon the Soviet reactor systems. East Germany and Bulgaria have the largest capacity in operation, but Czechoslovakia, with 4840 MWe under construction, is destined to become a major nuclear supplier to the Soviet Union and her east European neighbours. Major developments are also planned in Poland and Romania, which as yet have no operating nuclear stations. Three units are to be installed at Zarnowiec, near Gdansk, in Poland and five are planned in Romania, four at Cernovada and one at Olt.

Transport

Railways remain the prime haulier in eastern Europe, following Soviet planning in the 1960s which favoured public transport and railways in particular. The railway network is more widely meshed than further west and has a greater length of single track. It is still being extended in the south-eastern parts of eastern Europe; for example, all Albania's railways have been built since 1945 and a link with the Yugoslav network at Titograd in Montenegro has now been agreed (Turnock 1984). Emphasis is placed on the movement of low-value bulky goods over considerable distances (Mellor 1975), although the system has been modernized. Steam locomotives have been removed from passenger services and signalling, rolling stock, and container services have been improved.

Despite these developments, the superiority of motor transport for short-distance hauls of passengers and goods has been recognized (Turnock 1978), and the COMECON Transportation Commission (1975) recommended a system of motorways, including links from Rostock to Berlin, Prague, and Budapest, and from Italy and Austria to the Soviet Union and Turkey. The creation of a basic motorway system is still a long way behind that in western Europe and the emphasis remains on public road transport and not on the private motor car.

Waterways, pipelines, and air transport have been further developed and the socialist bloc is becoming important in world shipping, with Rostock, Gdynia and Gdansk the principal ports. The Soviet objective is to develop a unified transport system (Madeyski and Lissowska 1975), with the various modes integrated, and to persist with railways as the principal haulier. However, there has been this shift towards motorways since the late 1960s, a trend which is likely to continue into the late 1980s and beyond.

Industry

Eastern Europe lacked real industrial development until after the Second World War and emphasis was placed upon the exporting rather than the processing of raw materials and the importing of manufactures from countries which were investing their capital in the area. There was some industrialization, based on coal and iron-ore, and this was essentially concentrated in a triangular area which encompassed the Thüringen, Saxon, and Silesian regions in the northern part of eastern Europe.

After 1945, there was a planned and deliberate attempt to change the balance of economic activity, with the privileged position given to manufacturing industry (Turnock 1978). There followed a period of rapid industrialization (Hamilton 1970), based initially on the principles of autarky and dispersed heavy industry. This was reflected in the significant increase in the labour force employed in the secondary sector in the 1960s and 1970s, especially in Romania (140 per cent), Yugoslavia (94), and Bulgaria (56). Two forces were important in the pattern of industrial development:

1. Centrifugal forces, which followed the Marxist-Leninist ideas of industrial dispersal in order to eliminate the differences in standards between town and country. Dispersal was greatest in the underdeveloped regions, reducing the need for special development policies for backward areas.
2. Centripetal forces, which led to a concentration of new capacity in existing industrial areas. Development of large plants in previously established areas took place because of industrial inertia and the possibility of economies of large-scale production.

The consequence of these forces was the establishment of a concentrated pattern of industrial development in the more advanced areas, such as Upper Silesia in Poland, the Bohemian Basin in Czechoslovakia, and the Saxon triangle in East Germany, and a more widespread pattern in the underdeveloped regions.

The Soviet principles of industrial dispersal and concentration are well demonstrated in the iron and steel industry. Following the principle of autarky, there was heavy investment in iron and steel, even if natural resources were lacking. The result was wasteful investment, over-diversification and non-viable plants. After Stalin's death, the principle of trade was adopted and expansion of production was concentrated in those countries with a reasonable resource endowment and well-equipped iron and steel industries, namely Poland and Czechoslovakia. Therefore, iron and steel production in the 1970s had a clustered pattern in the developed countries of eastern Europe and a more dispersed pattern in the developing countries. Yugoslavia is a good example of the latter, where expansion at Jesenice (Slovenia), Sisak (Croatia), Smederovo (Serbia), and Zenica (Bosnia) was complemented by new units installed at Nicksic and Skopje (Hamilton 1964).

Tourism

Personal movements between COMECON countries are more restricted than in western Europe and travel to the non-socialist world is made difficult by financial restrictions and other formalities (Mellow 1975). However, considerable movement has developed since the late 1960s and an increasing proportion of people take their holidays outside the socialist bloc. Yugoslavia has for a long time been an exception and allowed free movement of its nationals into western Europe. In 1981, over 20 per cent of the Yugoslavian people taking a foreign holiday went to Greece and Italy.

Western tourists are welcome, on the hosts' terms, in most COMECON countries and tourism is very important in development plans and as a source of hard currency. Yugoslavia is the most popular destination, with over 20 million foreign visitors in 1982, followed by Czechoslovakia (12.8), Poland (7.8), and Hungary (5.2). A number of recreational facilities exist in eastern Europe, including lakes, mountains for winter sports, and spa and health centres, but it is the few coastal areas which have come under increasing pressure from national and international tourists (Turnock 1978). Tourists are particularly attracted to the Adriatic coastal resorts, which stretch from Dubrovnik to Albania and offer sandy beaches, plenty of sunshine, and low-cost hotels, and to the Black Sea resorts in Bulgaria and Romania. Priority is given to seaside resorts by the planning authorities and a new major highway, for example, has been built along the Adriatic coast partly to encourage the growth of tourism. In addition, several conservation areas, of world significance, have been recognized by UNESCO, notably the cities of Dubrovnik, in Yugoslavia, and Krakow, in Poland (Turnock 1984).

Social well-being

'The principles of Marxism-Leninism that guide development in socialist countries place great stress on the reduction of inequalities among regions and especially between town and country' (Smith 1979, p. 148–9). However, all societies, whether capitalist, state socialist, or anything else have inequalities in social well-being (Pahl 1979) the removal of which is still a future aspiration in both eastern and western Europe. Levels of social well-being vary in COMECON, at the urban, regional, and national scales.

Little attention has been given to macro-variations in social well-being in Europe, although an index of well-being, based on 27 variables, has been produced for each European country (Ilbery 1984). The results are presented in Chapter 9 (Table 9.3), and in relation to eastern Europe two important points can be made: first, levels of social well-being are, on average, much lower than further west; and secondly, considerable spatial variations exist among the eight countries, from poor index values of over 70, in Yugoslavia, Romania, and Albania to a respectable 'score' of 35 in East Germany (Fig. 9.2).

Regional variations in social well-being are also quite substantial in the eight socialist states of eastern Europe. Published work has been synthesized by Smith (1979) who concentrated on studies in Poland and Hungary. Using 22 variables, related to seven general conditions of well-being in the early 1970s, Smith derived a general level-of-living index for the five metropolitan areas and 17 voivodships of Poland. Rural-urban inequalities were one of the most noticeable features of the study, with Warsaw a long way ahead of the four remaining metropolitan areas, which in turn were well in front of the highest voivodship. In addition, levels of living were shown to be highest in the industrial areas of the south-west, suggesting that differences in well-being are related to rates of industrial development. A later study by Gruchman and Krasinski (1978), based on the new division of 49 voivodships established in 1975, highlighted similar rural–urban contrasts and marked regional variations in well-being. Comparable results were also obtained by Andorka (1976) in a study of Hungary, where living standards were shown to be highest in Budapest and the industrial areas of the west and north. Such inequalities can be partially explained first, by the fact that the needs for an urban proletariat were considered to exceed those of a peasant society; and secondly, by the adoption of a development strategy which favoured heavy industry at the expense of the countryside.

The degree of inequality in socialist cities is difficult to identify because of a lack of relevant data, but in a study of Prague's ecological structure Musil (1968) observed that the large differences in social structure which occurred between the centre and outskirts of Prague before the Second World War were being evened out. Musil pointed to two possible explanations: first, there was equal access to social services throughout Prague; and secondly, equal land values and rents militated against a spatial patterning of particular socio-economic groups. Contrary to Musil's ideas, Smith (1979) stated that one could expect inequality under socialism, both within and between cities. This can be demonstrated in relation to wage differentials, contrasting housing conditions and the hierarchy of urban centres which exists in the countries of eastern Europe. Although studies of intra-urban variations in social well-being are lacking, Smith did show how the population of Moscow displayed uneven access to services and an element of spatial socio-economic differentiation, with people in the centre 'better off' than those in the outer ring.

Regional policy

A major contrast between planning in eastern and western Europe has been shown to exist by Lichtenberger (1976). In western Europe, regional and urban planning agencies reflect the traditional division between town and country, with, little real co-operation between the two. In COMECON, state planning, based on Soviet policies, covers both urban and rural communes. During the 1950s, a wave of reform swept eastern Europe and the territorial–administrative pattern was extensively modified. Each new unit was to be planned along the following guidelines:

1. The unit should be capable of development as a complex economy, as well as fitting into the overall national plan.
2. Specialization which could be encouraged had to be developed, but not at the exclusion of other sectors.
3. Particular attention was to be given to industry and no unit should be solely, or even predominantly, agricultural.
4. Each unit was to aim towards a measure of self-sufficiency.
5. The administrative centre of each unit was to be the proletarian centre.
6. Consideration was to be given to demographic problems when planning the units, especially labour supply.

7. A balance was to be created between the units, so that none was totally dominant.

The number of units created varied from country to country, from 15 *bezirke* in East Germany to 39 *judets* (plus Bucharest) in Romania. Once reorganized, a policy of industrial dispersal was initiated in an attempt to equalize regional development and eliminate differences in the standards of living between town and country. For example, Wrobel (1980) was able to identify definite gains in the underdeveloped eastern regions of Poland, compared with the highly developed southern voivodship of Katowice. There were no special policies for backward areas and industry was developed on green-field sites as industrial inertia was rarely a restricting factor. Regional growth-points were also developed in certain areas, as in the under-industrialized regions of southern and central Poland, and a policy of 'dispersed concentration' was often pursued. In the case of Poland, Hamilton (1982) noted that between 1961 and 1975 one-third of investment was concentrated in Upper Silesia, Warsaw, and the two Baltic port areas (Gdansk and Szczecin), with a further 26 to 29 per cent dispersed, in a concentrated fashion, to just nine voivodships, mainly of an urban nature. Indeed, Fuchs and Demko (1979) have presented evidence to show that in eastern Europe generally priority is attached to growth and productivity over spatial and structural equity.

Therefore, it is not surprising that the Soviet policies of equalization met with opposition from certain COMECON members. There was a reluctance on the part of the more-developed countries to achieve a balance between all members, as this would mean the loss of trading advantages with the less-developed countries. Also, the more-developed regions within a nation preferred not to sacrifice their advantages for the benefit of the less-developed regions.

In this final chapter, the forces creating the human landscape of eastern Europe have been briefly examined. Government policy is an important controlling factor in both eastern and western Europe, but the kind of massive central intervention associated with COMECON has not taken place in western Europe because it is politically unacceptable. Despite the acceptance of Soviet-style Marxist-Leninism, eastern Europe is still characterized by wide regional disparities and core-periphery contrasts, two of the major themes developed in this book on western Europe. The systematic topics discussed above would seem to suggest that western and eastern Europe are likely to remain distinct for the foreseeable future.

References

Andorka, R. (1976). *Tendencies of regional development and differentiation in Hungary, measured by social indicators.* Central Statistical Office, Budapest.

Clout, H. D. (1971). *Agriculture.* Studies in contemporary Europe. Macmillan, London.

Compton, P. (1976). 'Migration in eastern Europe'. In Salt, J. and Clout, H. D. (Eds.) *Migration in post-war Europe.* Oxford University Press, Oxford.

Cousens, S. H. (1967). 'Changes in Bulgarian agriculture'. *Geography,* **52,** 11–22.

Elkins, T. H. (1973). *The urban explosion.* Studies in contemporary Europe. Macmillan, London.

Enyedi, G. (1967). 'The changing face of agriculture in eastern Europe'. *Geographical Review,* **57,** 358–72.

Enyedi, G. (1982). 'Part-time farming in Hungary'. *GeoJournal,* **6,** 323–6.

Fuchs, R. J. and Demko, G. J. (1978). 'The postwar mobility transition in eastern Europe'. *Geographical Review,* **68,** 171–82.

Fuchs, R. J. and Demko, G. J. (1979). 'Geographic inequality under socialism'. *Annals of the Association of American Geographers,* **69,** 301–18.

Gruchman, B. and Krasinski, Z. (1978). 'Measurement of real progress at the local level: report on the country case study in Poland'. *Measurement and analysis of progress at the local level.* UN Research Institute for Social Development, Geneva.

Hall, D. R. (1972). 'The Iron Gates scheme and its significance'. *Geography,* **57,** 51–5.

Hamilton, F. E. I. (1964). 'Location factors in the Yugoslav iron and steel industry'. *Economic Geography,* **40,** 46–64.

Hamilton, F. E. I. (1970). 'Changes in the industrial geography of eastern Europe since 1940'. *Tijdschrift voor Economische en Sociale Geografie,* **61,** 300–5.

Hamilton, F. E. I. (1982). 'Regional policy in Poland: a search for equity'. *Geoforum,* **13,** 121–32.

Ilbery, B. W. (1984). 'Core–periphery contrasts in European social well-being'. *Geography,* **69,** 289–302.

Kosinski, L. (1974). 'Urbanization in east-central Europe after World War Two'. *Eastern Europe Quarterly,* **8,** 130–53.

Lichtenberger, E. (1976). 'The changing nature of European urbanization'. In Berry, B. J. L. (Ed.)

Urbanization and counter-urbanization. Sage, London.

Madeyski, M. and Lissowska, E. (1975). 'Problems of the integrated organization of transport'. *Geographia Polonica*, **32,** 43–52.

Mellor, R. E. H. (1975). *Eastern Europe: a geography of Comecon countries*. Macmillan, London.

Musil, J. (1968). 'The development of Prague's ecological structure'. In Pahl, R. E. (Ed.) *Readings in urban sociology*. Pergamon, Oxford.

Pahl, R. E. (1979). 'Socio-political factors in resource allocation'. In: Herbert, D. T. and Smith, D. M. (Eds.) *Social problems and the city*. Oxford University Press, Oxford.

Polach, J. G. (1970). *The development of energy in eastern Europe*. Resources for the Future, Washington.

Saunders, I. T. (Ed.) (1958). *Collectivization of agriculture in eastern Europe*. Lexington, Kentucky.

Smith, D. M. (1979). *Where the grass is greener*. Penguin, Harmondsworth.

Turnock, D. (1970). 'Geographical aspects of Romanian agriculture'. *Geography*, **55,** 169–86.

Turnock, D. (1978). *Eastern Europe*. Dawson, Folkestone.

Turnock, D. (1984). 'Postwar studies on the human geography of eastern Europe'. *Progress in Human Geography*, **8,** 315–46.

Warriner, D. (1969). *Land reform in principle and practice*. Oxford University Press, Oxford.

Wrobel, A. (1980). 'Industrialization as a factor of regional development in Poland'. *Geographia Polonica*, **43,** 187–97.

Index